MITOCHONDRIAL BIOGENESIS AND BREAKDOWN

MITOCHONDRIAL BIOGENESIS AND BREAKDOWN

Valentin N. Luzikov
Moscow State University
Moscow, USSR

Translated by
Alexander V. Galkin

Translation edited by
Donald B. Roodyn
University College London
London, England

CONSULTANTS BUREAU • NEW YORK AND LONDON

Library of Congress Cataloging in Publication Data

Luzikov, Valentin N.
Mitochondrial biogenesis and breakdown.

Translation of: Regulatsiya formirovaniya mitokhondrii.
Bibliography: p.
Includes index.
1. Mitochondria. I. Roodyn, Donald B. II. Title.
QH603.M5L8913 1984 574.87′342 84-12157
ISBN 0-306-10979-4

Regulyatsiya formirovaniya mitokhondrii

The original Russian text, published by Nauka Press in Moscow in 1980, has been extensively revised by the author for the present edition. This Translation is published under an agreement with the Copyright Agency of the USSR (VAAP).

Printed in the United States of America

Foreword

It is a strange fact that many modern cell biochemists have a keen interest in biosynthetic processes, such as protein and nucleic biosynthesis or organelle biogenesis, but tend to regard degradative processes merely as irritating reactions that disrupt the flow of synthetic reactions. Historically, the elucidation of catabolic pathways preceded that of anabolic pathways, so that there is also a tendency to regard work on proteases, phospholipases, nucleases, etc., as somewhat "old-fashioned." It is the great contribution of Professor Luzikov's book to show that, at least in the case of mitochondrial research, the separation of studies on anabolic and catabolic processes has been very harmful. In an extremely erudite and measured way, the author carefully develops the argument that we can only understand mitochondrial biogenesis fully if we take into account the role of degradative processes. The action of lytic enzymes is shown not to be a random affair, but rather a process that is fully integrated into the process of mitochondrial assembly.

A second important contribution of this book is the fact that it contains a masterly review of the fundamental literature on mitochondrial structure, function, breakdown and synthesis presented in an integrated and logical manner. The author shows due regard for "seminal" papers in the field, even if they are more than two years old! "Mitochondrial Biogenesis and Breakdown" thus commendably parts from the current fashion of being only interested in the latest cloning experiment or massive DNA sequence; it places much of the subject in correct perspective and allows the reader to view the rapid advances in our understanding of the molecular genetics of mitochondrial DNA in relation to the previous more conventional developments.

Thus the reader will find in this book a rich treasure house of information on virtually every aspect of mitochondrial structure, function and development. The book should act as a catalyst to research on the role of catabolic processes in organelle biogenesis and should also help to stimulate studies on such important questions as the mechanism of assembly of multienzyme complexes; how they are integrated into membranes and how the membranes themselves are assembled or degraded under different physiological conditions.

Although the author deals almost exclusively with the mitochondrion, many of the arguments that he puts forward could apply equally to the biogenesis of other organelles and specialized membrane systems. It is to be hoped, therefore, that this work will have a general influence in the rapidly developing field of organelle bioassembly.

Donald B. Roodyn

Preface

The biogenesis of mitochondria has for the last two decades been one of the topical areas in biochemistry and molecular biology. On acquaintance with the host of material on this problem, it cannot be overlooked that one of its aspects, namely the degradation and renewal of mitochondria in the cell, has always been virtually neglected. Yet, without some notion of these questions it is impossible to understand such important aspects of mitochondrial biogenesis as transformation of the organelles during cell adaptation to changing growth conditions, adjustment of the content of mitochondrial material in the cell, control over mitochondrial assembly, and the like.

The aim of this book is to cast some light on these questions. The last one is especially appealing to me. The essence of matter is that according to the dogmas of orthodox molecular biology the fate of any protein in the cell is rigorously determined by its amino acid sequence, which sets all the levels of structural organization of protein molecules as well as their ability to interact with certain low-molecular-weight compounds and other biological macromolecules. Reasoning from this, it could be thought that the sole principle governing the assembly of various intracellular structures is the principle of "self-assembly", or more exactly, of "mutual recognition." While this principle appears to be the only one operative in the assembly of multisubunit enzymes, large and small ribosomal subunits, simplest viral particles, and some other supramolecular structures, no one has yet proved that the same is the case with the formation of subcellular organelles. This book pursues the idea that in organelle biogenesis the "mutual recognition" principle is supplemented and enforced by the principle of "selection by a performance criterion."

As to the form of presentation, it was predetermined by my wish to see among the readers not only mitochondriologists but also specialists in allied fields. For this reason the book includes several chapters of an introductory character that allow the reader not to turn to specialized sources. It seems to me that the approaches developed in this book could be applied to studying the biogenesis of other cell organelles.

Finally, I wish to express my deep gratitude to Professor I.V. Berezin for his support during the initial stage of the work on the elucidation of the molecular mechanisms of mitochondrial stabilization, to Professor V.P. Skulachev for his valuable comments on the Russian version of the book, and to Dr. A.V. Galkin for fruitful discussions, constructive criticism, and translating the book into English.

Valentin N. Luzikov
Moscow

Contents

1
Structural Organization of Mitochondria

The structural organization of mitochondria is in the focus of attention of most researchers working in mitochondriology. This problem has been treated in detail in quite a number of reviews (see, e.g., Ernster and Kuylenstierna, 1969; Racker, 1970; Harmon et al., 1974; Ernster, 1975a,b; De Pierre and Ernster, 1977) as well as in monographs by Kotel'nikova and Zvyagil'skaya (1973), Yasaitis (1973), Lloyd (1974), Munn (1974), Skulachev and Kozlov (1977), and Tzagoloff (1982). This chapter aims to point out some structural aspects, awareness of which is essential to easy reading of the book. Also, it presents some recent experimental data that may have not yet been widely recognized.

1.1. PARTITIONING OF MITOCHONDRIA

1.1.1. Resolution of Mitochondrial Compartments

A general model of the mitochondrion, which is consistent with most electron-microscopic observations, is depicted in Figure 1. According to this model (Whittaker, 1966), the mitochondrion has an outer membrane, an inner boundary membrane, and cristae. The membrane forming cristae has been reported (Hackenbrock, 1975; Hackenbrock and Miller, 1975) to differ from the inner one as such in its affinity for the polycationic reagent ferritin and hence in the surface density of anionic groups; however, such differences are at present disregarded in studies of the principles underlying the structural arrangement of mitochondrial membranes and mitochondrion as a whole.

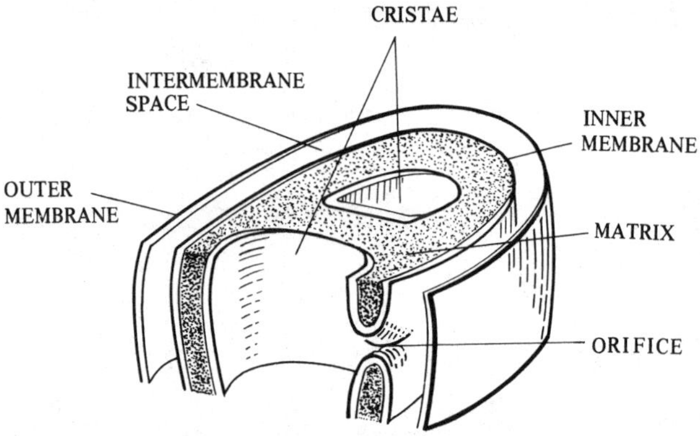

Fig. 1. A schematic cutaway representation of the mitochondrion (redrawn after Whittaker, 1966).

Several methods have been proposed for separating outer and inner mitochondrial membranes. According to one of them (Parsons and Williams, 1967; Parsons et al., 1967), mitochondria are swelled in phosphate buffer containing serum albumin, after which outer membranes are separated from the fraction of inner membranes with matrix, first by differential centrifugation and then in a sucrose density gradient. A similar technique has been described by Caplan and Greenawalt (1966) who subjected mitochondria to osmotic lysis in distilled water.

A somewhat different procedure was used by Sottocasa et al. (1976b) who first made mitochondria swell, then added ATP and Mg^{2+} ions to induce organelle contraction, and finally stripped them of outer membranes by ultrasonic treatment.

Basically new means of stepwise mitochondrial fragmentation were designed by Lévy et al. (1967, 1969); Schnaitman et al. (1967), Schnaitman and Greenawalt (1968); and Hoppel and Cooper (1968). These developments were prompted by the observation that the detergent digitonin added in small amounts selectively destroys the outer mitochondrial membrane. Excellent electron-microscopic demonstrations of the digitonin effect are given in several works (Lévy et al., 1967; Schnaitman et al., 1967; Morton et al., 1968). The inner membranes with matrix could be freed of outer membranes by differential centrifugation (Lévy et al., 1967, 1969; Schnaitman et al., 1967; Schnaitman and Greenawalt, 1968; Hoppel and Cooper, 1968) and further purified in a sucrose density gradient (Lévy et al., 1967).

Finally, Brdiczka et al. (1968) have reported that upon freezing of liver mitochondria to the temperature of liquid nitrogen in a mildly hypertonic medium the outer membranes are ruptured and partly "peeled off" whereas the inner membranes remain intact and the mitochondrial matrix retains its normal density. Centrifugation of the suspension treated in this way through a sucrose density gradient yields intact inner membranes with well-preserved matrix.

In some cases complete separation of outer and inner membranes requires use of proteinases, indicating the existence of peptide linkages between the membranes (Scholte, 1973).

As just mentioned, the outer mitochondrial membrane can be ruptured and removed with caution in such a way as to leave the inner one undamaged. The result is vesicles containing an electron-dense matrix and displaying high activities of the tricarboxylic-acid-cycle enzymes: malate dehydrogenase, (NADP) isocitrate dehydrogenase, α-ketoglutarate dehydrogenase, etc. (Sottocasa et al., 1967b; Brdiczka et al., 1968; Morton et al., 1968; Schnaitman and Greenawalt, 1968). Upon exposure of the vesicles to ultrasound (Lévy et al., 1967; Brdiczka et al., 1968) or detergents such as digitonin or Lubrol (Hoppel and Cooper, 1968; Schnaitman and Greenawalt, 1968) the above enzymes are readily released from the membranes.

Such investigations made it possible to separate the outer and inner membranes as well as the contents of the intermembrane space and matrix. Briefly summarized below are the data obtained on analysis of these fractions.

1.1.2. Localization of the More Important Mitochondrial Enzymic Systems

Studying the distribution of various enzymes in rat liver mitochondria, Schnaitman and Greenawalt (1968) estimated the portions of total protein contained in the various fractions to be as follows: outer membrane (about 4%), intermembrane space (6%), inner membrane (21%), and matrix (69%). Similar results were obtained by Brunner and Bücher (1970) who used an entirely different method of calculation. For rabbit heart mitochondria it was found (Whereat et al., 1969) that outer membranes isolated according to Sottocasa et al. (1967a,b) contribute 7.4% and the inner membrane-matrix fraction about 72% of the total mitochondrial protein. Finally, Bandlow (1972) reported the following percentages of protein in the compartments of *Saccharomyces cerevisiae* mitochondria: 6-8% in the outer membrane, about 15% in the intermembrane space, 30% in the inner membrane, and 50% in the matrix.

By now more than a hundred enzymes have been found in mitochondria from various sources (Munn, 1974). However, the exact localization of many of them is unknown. Besides, it is not quite clear in what way some of these enzymes are related to the main metabolic processes taking place in mitochondria. It would therefore

Outer membrane	Intermembrane space	Inner membrane	Matrix
Monoamine oxidase	ATP:AMP	Respiratory-chain	Krebs cycle enzymes:
Kynurenine	phospho-	components:	(NAD) L-Malate dehydrogenase
hydroxylase	transferase	NADH dehydrogenase	(NADP) L-Isocitrate
NADH-cytochrome *c*	ATP:nucleo-	Succinate dehydrogenase	dehydrogenase
reductase (rotenone-	sidediphos-	Cytochromes *b*, c_1, *c*,	Citrate synthetase
insensitive)	phate	and aa_3	Fumarase
Acyl:Co A ligase	phospho-	Iron-sulfur proteins	Aconitase
(ATP-dependent)	transferase	NADP:NAD reductase	Krebs-cycle-linked enzymes:
Fatty acid elonga-	ATP:creatine	(pyridine nucleotide	(NADP) l-Malate dehydrogenase
tion system (C_{14}, C_{16})	phospho-	transhydrogenase)	(NAD) L-isocitrate
	transferase	Energy-transducing	dehydrogenase
Acetyl-Co A:		machinery:	(NADP) L-Glutamate
glycero-3-phosphate		ATPase complex	dehydrogenase
O-acetyltransferase		Pyrophosphate phospho-	Carbamyl phosphate-ornithine
Acetyl-Co A: lyso-		hydrolase	carbamoyltransferase
lecithin O-acetyl-		Other redox enzymes:	GTP:AMP phosphotransferase
transferase		(NAD) D-β-Hydroxybutyrate	Malonyl-Co A decarboxylase
Acetyl-Co A: lyso-		dehydrogenase	(ATP) Acetyl-Co A synthetase
phosphatidic acid		Glycero-3-phosphate	(acetate:Co A ligase)

O-acetyltransferase	dehydrogenase	(GTP) Succinyl-CoA synthetase
Choline phosphate	Choline dehydrogenase	(succinate:Co A ligase)
transferase	L-Proline-NADP reductase	Pyruvate carboxylase
L-α-Phosphatidate	Steroid 11β hydroxylase	Fatty acid oxidation system:
phosphohydrolase	L-3-Hydroxyacyl-Co A	(ATP) Acyl-Co A synthetase
	hydrolyase	(fatty acid:Co A ligase)
Phospholipase A_2	Miscellaneous:	
Phosphoacyl hydrolase	Acetyl-Co A: carnitine	(NAD) L-3-Hydroxyacyl-Co A
ATP:D-hexose-6-phospho-	O-acetyltransferase	dehydrogenase
transferase	Fatty acid elongation	3-Ketoacyl-Co A thiolase
	system (C_{10})	Miscellaneous:
	Ferrochelatase	δ-Aminolevulinic acid
		synthetase(?)
	δ-Aminolevulinic acid	Electron-transfer flavo-
	synthetase(?)	proteins
		(Fatty acyl-Co A dehydro-
		genases)(?)

be reasonable to restrict ourselves to the basic features of the distribution of the key enzymic systems among the mitochondrial compartments.

1.1.2.1. *Enzymes of the Outer Mitochondrial Membrane*

These enzymes can be arbitrarily divided into two groups. The first comprises specific oxidative enzymes: monoamine oxidase (Okamoto et al., 1967; Sottocasa et al., 1967b; Schnaitman et al., 1967; Schnaitman and Greenawalt, 1968; Brdiczka et al., 1968; Parkes and Thompson, 1970), kynurenine hydroxylase (Okamoto et al., 1967; Mayer et al., 1968; Schnaitman and Greenawalt, 1968), and rotenone-insensitive NADH-cytochrome *c* reductase (Parsons et al., 1967; Sottocasa et al., 1967a,b; Schnaitman and Greenawalt, 1968). The second group includes numerous enzymes of lipid metabolism: acyl-coenzyme A synthetase (Norum et al., 1966; Yates and Garland, 1967; Skrede and Bremer, 1970) specific for long-chain fatty acids (Aas and Bremer, 1968; Haddock et al., 1970); enzymes involved in fatty-acid elongation (Hinkle and Pullman, 1968; Whereat et al., 1968, 1969; Colli et al., 1969; Howard, 1970) and synthesis de novo (Howard, 1970); enzymes performing acylation of glycerophosphate, lysophosphatidic acid, and lysophosphatidylcholine, and addition of choline phosphate to diglycerides (Stoffel and Schiefer, 1968); phospholipase A_2 (see Chapter Two); and hexokinase, or ATP: D-hexose phosphotransferase (Kropp and Wilson, 1970; Mayer and Hübscher, 1971). A characteristic feature of many of the enzymes listed (for instance, acyl-coenzyme A synthetase and fatty-acid elongation enzymes) is that they require ATP for their functioning.

1.1.2.2. *Enzymes of the Intermembrane Space*

Enzymes held to belong in the space between the outer and inner mitochondrial membranes are those readily released into solution upon selective disruption of the outer membrane. It cannot though be excluded that in intact mitochondria some of them are to a certain extent membrane-bound. Such enzymes are adenylate kinase, or ATP: AMP phosphotransferase (Sottocasa et al., 1967b; Brdiczka et al., 1968; Lima et al., 1968; Schnaitman and Greenawalt, 1968; Schnaitman and Pedersen, 1968) and ATP : nucleoside diphosphate phosphotransferase (Pfaff and Schwalbach, 1967; Lima et al., 1968; Schnaitman and Greenawalt, 1968; Schnaitman and Pedersen, 1968), the latter being partly recovered in the outer membrane fraction (Schnaitman and Greenawalt, 1968). Creatine phosphokinase (ATP: creatine phosphotransferase) has been reported to be easily extractable from cardiac mitochondria with phosphate buffer, without any ultrasonic or detergent treatment (Klingenberg and Pfaff, 1966), and therefore to float, like adenylate kinase, in the intermembrane space. However, other authors (Farrell et al., 1972; Jacobus and Lehninger, 1973; Scholte,

1973; Scholte et al., 1973; Lyengar and Lyengar, 1980) are of the opinion that creatine phosphokinase is rather bound to the exterior surface of the inner mitochondrial membrane.

Enzymic reactions taking place in the outer mitochondrial membrane and intermembrane space with the participation of exogenous ATP (or ADP) are unaffected by atractyloside (Chappell and Crofts, 1965; Garland and Yates, 1967) which inhibits the adenine nucleotide carrier.

1.1.2.3. *Enzymes of the Inner Mitochondrial Membrane*

The main function of the inner mitochondrial membrane is to generate the electrochemical potential of H^+ ions and to perform at its expense phosphorylation of ADP. The inner membrane contains (Parsons et al., 1967; Sottocasa et al., 1967a,b; Lévy et al., 1967, 1969; Schnaitman et al., 1967; Schnaitman and Greenawalt, 1968; Schnaitman and Pedersen, 1968; Hoppel and Cooper, 1968) the components of the respiratory chain - NADH dehydrogenase, succinate dehydrogenase, iron-sulfur proteins, cytochromes b, c_1, c, a and a_3, etc. - and of the energy transformation system: soluble ATPase, oligomycin-sensitivity-conferring protein, other constituents of the ATPase complex and coupling factors (see further the section devoted to the ATPase complex, 1.3.1.8.). Also associated with the inner membrane are the electron-transfer flavoproteins participating in fatty-acid oxidation (Garland et al., 1967), and dehydrogenases of choline (Frisell et al., 1965), glycerophosphate (Klingenberg, 1970), and β-hydroxybutyric acid (Norum et al., 1966; Hoppel and Cooper, 1968; Schnaitman and Greenawalt, 1968; McIntyre et al., 1978). Another enzyme of the inner mitochondrial membrane is pyridine nucleotide transhydrogenase (Danielson and Ernster, 1963).

A most important function of the inner membrane is to maintain the integrity of the matrix compartment and to transport various metabolites. As mentioned above, careful disruption of outer membranes does not appreciably release matrix enzymes (Sottocasa et al., 1967b; Brdiczka et al., 1968; Hoppel and Cooper, 1968; Schnaitman and Greenawalt, 1968) and cofactors, in particular adenine and pyridine nucleotides (Pfaff and Schwalbach, 1967; Meisner, 1970). There are grounds for believing that the inner membrane has specific carriers for phosphate, calcium ions, adenine nucleotides, ornithine, α-ketoglutarate, L-glutamate, L-aspartate, di- and tricarboxylic acids, and other metabolites (see reviews by Lehninger et al., 1967; Chappell, 1968; Quagliariello et al., 1969; Klingenberg, 1970; Passarella and Quagliariello, 1976). Such carriers, or translocases, share a number of features with enzymes: (1) substrate apecificity, (2) saturability with substrate, (3) sensitivity to specific inhibitors, and (4) a rather high activation energy (compared to diffusion processes). An example of a translocase is acyl-coenzyme A : carnitine transferase responsible for fatty acid transport across the inner membrane (Norum et al., 1966; Haddock et al., 1970). Notably, the transport of metal ions, fatty acids, and intermediates of the tricarboxylic-acid cycle is energy-dependent and requires ATP or coupled electron transfer (Chappell, 1968; Pressman, 1970).

Finally, the inner membrane performs a number of essential biosynthetic functions. It carries the bulk of mitochondrial DNA (Kazakova et al., 1969, 1977; Nass M., 1969; Van Tuyle and Kalf, 1972) and is the place of the template synthesis of mitochondrial RNA (Kalf and Faust, 1969; Kazakova et al., 1972; Van Tuyle and Kalf, 1972). Mitochondrial protein synthesis most probably also takes place on the inner membrane (see Chapter Six). The inner-membrane machinery is capable of carrying out synthesis de novo (Whereat et al., 1968; Wit-Peeters, 1969) and elongation (Quagliariello et al., 1968; Colli et al., 1969) of some fatty acids, phospholipid hydrolysis, and lysophosphatide acylation (see Chapter Two). The membrane contains ferrochelatase (McKay et al., 1969; Jones and Jones, 1969) and, probably, δ-aminolevulinic acid synthetase (McKay et al., 1969), which means that the inner membrane, together with the contents of the matrix, is actively involved in heme synthesis (Meyer and Schmid, 1973) and thereby in the regulation of cytochrome formation.

1.1.2.4. *Enzymes of the Matrix*

The first to be referred to as matrix components are the enzymes of the tricarboxylic-acid cycle: citrate synthetase, or citrate-oxaloacetate lyase; aconitase, or citrate (isocitrate) hydrolyase; isocitrate dehydrogenase, or L-isocitrate:NADP reductase; fumarase, or L-malate hydrolyase; and malate dehydrogenase, or L-malate:NAD reductase (Parsons and Williams, 1967; Parsons et al., 1967; Sottocasa et al., 1967b; Schnaitman et al., 1967; Schnaitman and Greenawalt, 1968; Brdiczka et al., 1968; Haddock et al., 1970; Landriscina et al., 1970). The matrix contains a number of "ancillary" enzymes in one way or another connected with the tricarboxylic-acid cycle and controlling its work: glutamate dehydrogenase, or L-glutamate:NADP reductase; aspartate aminotransferase, or L-aspartate : 2-ketoglutarate aminotransferase; pyruvate carboxylase, or pyruvate-CO_2 ligase; phosphopyruvate carboxylase, etc. (Norum et al., 1966; Sottocasa et al., 1967b; Aas and Bremer, 1968; Hoppel and Cooper, 1968; Morton et al., 1968; Schnaitman and Greenawalt, 1968; Landriscina et al., 1970).

A very important role in mitochondrial metabolism is played by fatty acids, oxidation of which is the most effective means of feeding substrates into the respiratory chain. The matrix is the place of fatty-acid activation involving a GTP- and ATP-dependent acyl-coenzyme A synthetase specific towards acids with short and medium-length chains (Norum et al., 1966; Yates and Garland, 1967,

Aas and Bremer, 1968; Garland et al., 1969; Haddock et al., 1970), and some steps in fatty-acid β-oxidation. In particular, the matrix of rat liver mitochondria was found to contain L-3-hydroxyacetyl-coenzyme A : NAD reductase and β-ketothiolases (acyl-coenzyme A : acetyl-coenzyme A acetyltransferases) of various substrate specificities (Beattie, 1968; Brdiczka et al., 1968; Garland et al., 1969; Haddock et al., 1970).

In characterizing the enzymes of mitochondrial matrix, it must not be ignored that they differ from one another in the ease with which they are removed from the inner membrane-matrix fraction, or mitoplasts (Brdiczka et al., 1968; Jones and Jones, 1969; Haddock et al., 1970). For instance, during sonication of rat liver mitoplasts malate dehydrogenase is more readily released into solution than glutamate dehydrogenase. On the other hand, it was shown (Matlib and O'Brien, 1975) that solubilization of aconitase, (NAD) isocitrate dehydrogenase, α-ketoglutarate dehydrogenase, fumarate dehydrogenase, palmitoyl-coenzyme A dehydrogenase, and ornithine transcarbamylase requires much lower concentrations of digitonin than solubilization of malate dehydrogenase, citrate synthetase, isocitrate dehydrogenase, aspartate-glutamate transaminase, and other enzymes. This may mean that some matrix enzymes are more or less loosely bound to the inner mitochondrial membrane. In other words, the conventional grouping of enzymes into "membrane" and "soluble" ones is rather arbitrary.

SUMMARY

Thus, the mitochondrion is a system highly organized in the functional sense and marked by a very definite arrangement of enzymes and other components in its particular regions. It is easy to see that such an arrangement makes for the highest efficiency of the organelle. Enzymes that are closely coupled functionally usually share the same mitochondrial compartment. Indeed, enzymes of the tricarboxylic-acid cycle and linked systems are localized in the matrix; enzymes carrying out fatty-acid β-oxidation are also concentrated in the matrix and partly in the inner membrane; enzymes utilizing ATP and responsible for the metabolic coupling of mitochondria and cytoplasm (e.g., creatine phosphokinase, nucleoside diphosphokinase, hexokinase) occupy the outer mitochondrial compartment. A simplified scheme summarizing these data is given in Figure 2.

A topic calling for attention is the structure-function interdependency of the mitochondrial main enzymic systems. Indeed, the respiratory chain and the energy-transducing machinery cannot function without the matrix which harbors substantial amounts of substrates, cofactors, and "soluble" enzymes. On the other hand, preservation of the matrix, and in particular maintenance in it of suf-

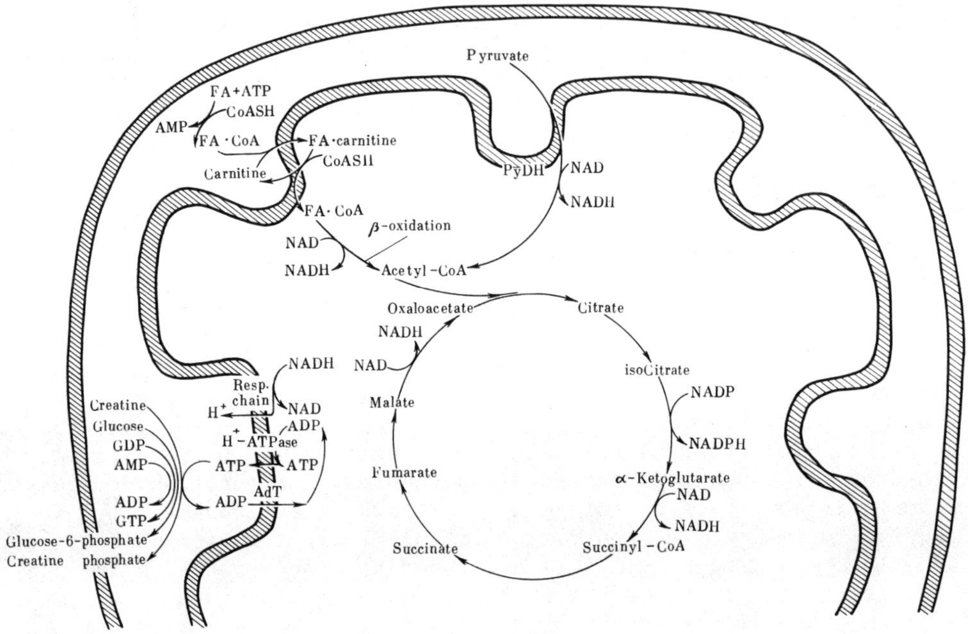

Fig. 2. Compartmentation of some metabolic processes taking place in mitochondria. (AdT) Adenine nucleotide translocator; (PyDH) pyruvate dehydrogenase; (FA) fatty acid.

ficient concentrations of tricarboxylic-acid-cycle intermediates, cofactors, and many other components, requires energy expenditure and thereby depends on the respiratory chain and the energy-transducing machinery. Functioning of the latter is in no small measure conditioned by the enzymes of the intermembrane space and outer membrane such as hexokinase, creatine phosphokinase, adenylate kinase, etc., which by consuming ATP and controlling its level in mitochondria can affect the overall oxidative phosphorylation rate. At the same time, the very existence of these enzymes in the intermembrane space is indirectly dependent on the energy transformation system since shutoff of this system leads to swelling of the matrix (see Chapter Two), rupture of the outer membrane, and dissipation of the intermembrane-space contents in the surrounding medium. Thus, the integrity of the outer membrane is provided by the joint operation of the components of the matrix, inner membrane, and intermembrane space. Again, it is the outer membrane that confines in the intermembrane space the enzymes regulating the energetic functions of the inner membrane (see above) and that is therefore indirectly responsible for preservation of the matrix.

1.1.3. Lipid Composition of Mitochondrial Membranes

The mitochondrial outer and inner membranes significantly differ in specific gravity: values reported were 1.094 to 1.122 for the outer and 1.192 to 1.230 for the inner membrane (Parsons and Yano, 1967; Parsons et al., 1967). The difference becomes understandable on considering that the phospholipid-to-protein ratio [phospholipids amount to no less than 75% of the total lipids of mitochondria from various sources (Munn, 1974)] is 0.829 for the outer membrane and only 0.275 for the inner (Parsons and Williams, 1967; Parsons et al., 1967). In other words, the portion of phospholipids in the outer membrane is 2-3 times that in the inner one. It is therefore not surprising that upon removal of 80-95% of lipids mitochondria lose their outer membranes while the inner membranes retain their macrostructure (Fleischer et al., 1967).

Substantial differences are also found in the phospholipid composition: a characteristic feature of the inner membrane is a high content of cardiolipin, whereas the outer membrane is enriched in phosphatidylinositol (Parsons et al., 1967; Newman et al., 1968; Stoffel and Schiefer, 1968; Lévy et al., 1969). The outer membrane also contains a considerable amount of cholesterol (Parsons and Yano, 1967; Lévy et al., 1969) which is the probable cause of its enhanced susceptibility to digitonin (Schnaitman et al., 1967). It should be noted that a small portion of lysophosphatides - lysophosphatidylcholine, lysophosphatidylethanolamine, and lysophosphatidylserine - have been detected among mitochondrial phospholipids (Lévy et al., 1969).

Analysis of the fatty acids in phosphatidylcholine, phosphatidylethanolamine, cardiolipin, and phosphatidylinositol showed that the relation of different saturated and unsaturated fatty acids substantially varies for these phospholipids. However, no appreciable differences in the fatty acid composition could be found between phospholipids of the same class contained in the outer and inner mitochondrial membranes (Munn, 1974).

1.2. STRUCTURAL ORGANIZATION OF THE OUTER MITOCHONDRIAL MEMBRANE

As shown above, the outer mitochondrial membrane displays about ten enzymic activities. Schnaitman (1969) described a simple method of solubilizing membrane proteins for polyacrylamide gel electrophoretic separation, and found that the outer membrane of rat liver mitochondria comprises 12 protein components. The comparison of the polypeptide sets in the outer and inner membranes suggested that they may have one common constituent. The bulk of the membrane protein is represented by polypeptides with apparent molecular weights of 37,000 to 68,000, the portion of each fraction not exceeding 18%. Similar results were obtained by Hayashi and Capaldi (1972) for beef heart mitochondria. The outer membrane fraction from this source was found to contain 14 polypeptides, three of which (apparent molecular weights of 27,000, 34,000, and 46,000) constituted more than half of the total outer-membrane protein.

According to Freitag et al. (1982a), a major protein of the outer mitochondrial membrane is the so-called pore protein. This polypeptide of molecular weight 31,000 can form channels 2 nm in diameter in artificial lipid bilayers. Such channels allow free diffusion of, e.g., polyethylene glycol molecules of molecular weight up to 3400. Purified 31,000-molecular-weight pore protein accounts for about 15% of the total outer-mitochondrial-membrane protein in *Neurospora crassa*.

Intense bands at the same molecular weight were observed upon sodium dodecyl sulfate-polyacrylamide gel electrophoresis of mitochondrial outer membrane preparations from mung bean and potato (Manella and Bonner, 1975), rat liver (Zalman et al., 1980), and beef heart (Hayashi and Capaldi, 1972). The membrane fraction containing the 31,000-molecular-weight protein, being fused with liposomes of soybean phospholipids, formed vesicles permeable for saccharides with molecular weights of 2000-8000 (Zalman et al., 1980). Recently the pore protein was reported to be probably identical with the hexokinase-binding protein of the rat liver mitochondrial outer membrane (Fiek et al., 1982; Lindén et al., 1982).

Huber and Morrison (1973) were the first to make an attempt at elucidating the arrangement of particular proteins with respect to the outer and inner sides of the membrane. These authors subjected the polypeptides on the exterior surface of the membrane to selective labeling with ^{125}I by the lactoperoxidase technique. Studies were performed on rat liver mitochondria whose integrity was assessed at different stages of modification by retention of adenylate kinase, a soluble enzyme of the intermembrane space. Such a control gave grounds for believing that lactoperoxidase did not penetrate the outer membrane and modification therefore took place only on the exterior surface. The bulk of the label (about 70%) in the outer-membrane fraction isolated from the iodinated mitochondria turned out to be associated with polypeptide(s) of apparent molecular weight 14,000. This result is likely to mean that not all types of polypeptides are exposed on the outer side of the membrane; in other words, the membrane is asymmetric in the transverse distribution of its protein components. More detailed data on the arrangement of individual proteins in the outer mitochondrial membrane are lacking at present.

1.3. STRUCTURAL ORGANIZATION OF THE INNER MITOCHONDRIAL MEMBRANE

The inner mitochondrial membrane carries quite a number of enzymic systems responsible for energy generation and transformation, metabolite transport, various biosynthetic processes, etc., (see 1.1.2.3. above). Polyacrylamide gel electrophoresis of mitochondrial inner membrane preparations solubilized with sodium dodecyl sulfate and mercaptoethanol reveals 21-25 polypeptides with molecular weights ranging from 10,000 to 80,000 (Schnaitman, 1969; Capaldi, 1974; Hare and Crane, 1974; Harmon et al., 1974). This number increases to 33 if gel electrophoresis is performed separately for the fraction of protein easily extractable from the membrane (with urea, urea plus trypsin, or acetic acid) and for the fraction remaining after such extraction. From this it follows that preliminary membrane fractionation enhances the resolving power of the method, and that in reality the set of polypeptides is still more numerous. Naturally, the elucidation of the structural design of such an intricate system, including at that the phospholipid components (about 25% of the overall weight), is an extremely complicated task. The experimental data available thus far permit us to discuss mainly the arrangement within the membrane of the constituents of the respiratory chain and energy-transducing machinery.

1.3.1. Mitochondrial Inner-Membrane Components

This section will briefly describe the components of the respiratory chain and energy transformation system that have by now been obtained in a more or less purified state. As to the respiratory chain, it is represented in sufficient detail, albeit not exhaustively, by the scheme of Ohnishi (1979) which may serve as a guide to a major portion of the section.

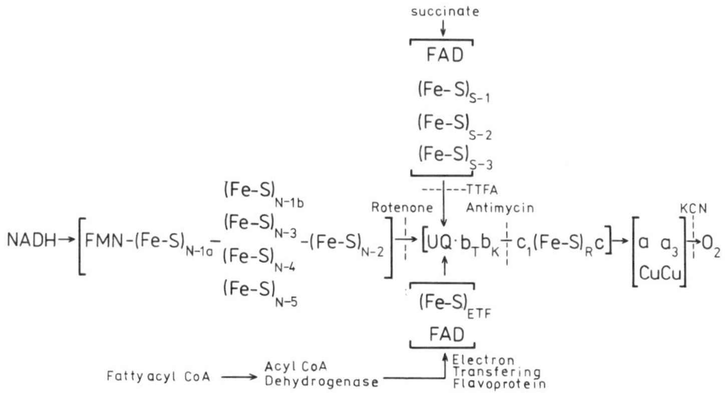

Respiratory-chain components present in the inner mitochondrial membrane. (Fe-S) Iron-sulfur centers; (UQ) ubiquinone; (b_T, b_K) cytochromes b_T and b_K; (c_1, c) cytochromes c_1 and c; (aa_3) cytochrome aa_3; (TTFA) tenoyl trifluoroacetone. Reaction sites of the respiratory-chain inhibitors are also indicated. After Ohnishi (1979).

1.3.1.1. *NADH-Ubiquinone Reductase*

Initially a low-molecular-weight NADH dehydrogenase (type II) was isolated by extracting mitochondria and submitochondrial particles with 10% ethanol in an acidic medium (Edelhoch et al., 1952; Mahler et al., 1952; De Bernard, 1957; Pharo et al., 1966, 1968; Kumar et al., 1968). The molecular weight of different enzyme preparations varied from 70,000 to 98,000 (Mahler et al., 1952; Pharo et al., 1966; Kumar et al., 1968). A molecule of such mass carried one flavin mononucleotide group (Cremona and Kearney, 1963; Rao et al., 1963) and small amounts of nonheme iron (2-4 atoms) and labile sulfides (1-5 residues). The type II NADH dehydrogenase differed in catalytic properties from the larger fragments of the inner mitochondrial membraning have NADH dehydrogenase activity. First, its preparations did not show high selectivity towards acceptors, reducing at comparable rates either ferricyanide, menadione, 2,6-dichlorophenol indophenol, ubiquinone (coenzyme Q), or cytochrome *c* (Kumar et al., 1968; Pharo et al., 1968). Secondly, the NADH- ubiquinone reductase and NADH-cytochrome *c* reductase activities of type II NADH dehydrogenase were but slightly sensitive to amytal or rotenone (Pharo et al., 1968), whereas NADH oxidation in the respiratory chain is completely blocked by these compounds.

The type II enzyme is most probably a constituent of the high-molecular-weight NADH dehydrogenase (type I) isolated in Singer's laboratory (Kearney et al., 1962; Watari et al., 1962; Minakami et al., 1963; Ringler et al., 1963; Cremona and Kearney, 1964; Singer, 1966) and of NADH-ubiquinone reductase (complex I) obtained according to Hatefi et al. (1962a). Type I dehydrogenase is marked by an exceptionally high specificity: it displays a high reduction rate only with ferricyanide and no appreciable activity with menadione, 2,6-dichlorophenol indophenol et seq. (Minakami et al., 1963). Complex I is specific towards ferricyanide and coenzyme Q (ubiquinone), its NADH-ubiquinone reductase activity being completely inhibited by amytal (Hatefi et al., 1962a).

Besides the noncovalently bound flavin (FMN), both enzymes contain considerable amounts of nonheme iron: complex I has 17 g-atoms of iron per mole FMN (Hatefi et al., 1962a) and an equivalent amount of sulfur, and in the high-molecular-weight NADH dehydrogenase FMN, nonheme iron, and labile sulfides are related as 1:17:28 (Lusty et al., 1965; Singer, 1966). The bulk of the nonheme iron is associated with special proteins which can be obtained as a separate fraction after treating complex I with urea (Hatefi and Stempel, 1967, 1969). The other product of such fractionation is an iron-sulfur flavoprotein containing one flavin mononucleotide group, four atoms of iron, and four labile sulfide residues per a molecular mass of 70,000 daltons. This component of complex I resembles in its composition, substrate specificity, and inhibitor sensitivity the above-described type II NADH dehydrogenase (Hatefi and Stempel, 1967, 1969;

Hatefi 1968). According to Ragan (1976a), the flavoprotein fraction of complex I upon polyacrylamide gel electrophoresis yields two protein bands corresponding to molecular weights of 53,000 and 26,000. Analogous results were later obtained in Slater's laboratory (Dooijewaard et al., 1978) upon fragmentation of complex I with sodium perchlorate. Galante and Hatefi (1979) found in the low-molecular-weight NADH dehydrogenase three polypeptides of molecular weights 51,000, 24,000, and 10,000. Recently it was shown (Ragan et al., 1982a) that the fraction comprising the second and third of these polypeptides contains iron and acid-labile sulfides which probably form binuclear clusters; the largest polypeptide was also found to carry an iron-sulfur center of unknown structure.

Since FMN is lost during separation of NADH dehydrogenase subunits, its localization in the intact enzyme was not elucidated. Chen and Guillory (1981) showed, by using an NAD^+ analog covalently attachable to NADH dehydrogenase, that the pyridine nucleotide-binding site is on the 57,000-molecular-weight subunit which appears to be identical to the above-mentioned 51,000-molecular-weight polypeptide.

The iron-protein fragment of the complex accounting for about 20% of its protein includes six polypeptides with molecular weights of 75,000, 49,000, 30,000, 18,000, 15,000, and 13,000 (Ragan et al., 1982b). Three fractions were derived from this fragment, the first containing mostly one polypeptide of molecular weight 75,000, the second a polypeptide of 49,000, and the third two polypeptides with molecular weights of 30,000 and 13,000 (Ragan et al., 1982b). All fractions displayed visible spectra characteristic of iron-sulfur proteins. However, it is at present impossible to identify among them any centers responsible for any particular signals in the epr spectrum of complex I.

Thus, the data now available shed some light on the possible functions of six polypeptides of the NADH-ubiquinone reductase segment of the respiratory chain. Yet, according to Ragan (1976a), it comprises at least 16 polypeptides; the function of most of them is still quite obscure*.

1.3.1.2. *Succinate-Ubiquinone Reductase*

Initially the simplest fragment of the inner mitochondrial membrane capable of performing the first stages of succinate oxidation was considered to be the succinate dehydrogenase isolated by the method of Singer. The properties of this enzyme are described in detail in a number of reviews (see, e.g., Singer and Kearney, 1963; Singer, 1966). It is a homogeneous entity with a molecular weight of about 200,000 containing one flavin-adenine dinucleotide (FAD) group

*See Addendum

covalently bound to the protein, four atoms of nonheme iron, and labile sulfides. Such a fragment can catalyze electron transfer from succinate to phenazine methosulfate, or less efficiently to ferricyanide, but cannot interact with the mitochondrial electron-transfer system.

The succinate dehydrogenase of King (1966), which has 1 mole FAD, 8 g-atoms nonheme iron, and 8 moles labile sulfides per 320,000g protein, retains the ability to associate with the alkali-treated preparation of submitochondrial particles and to restore its respiratory activity.

Also known is the succinate-ubiquinone reductase complex, or complex II (Ziegler and Doeg, 1962). Assuming a molecular weight of 240,000, the unit contains one FAD residue, eight atoms of nonheme iron, and one molecule of cytochrome *b* (Ziegler and Doeg, 1962). Fractionation of complex II yields a succinate dehydrogenase of molecular weight 100,000 containing FAD, nonheme iron, and labile sulfides in a molar ratio of 1:8:8 and completely devoid of cytochrome *b* (David and Hatefi, 1969). This enzyme, like Singer's succinate dehydrogenase, is capable of reducing phenazine methosulfate but lacks succinate-ubiquinone reductase activity and is inadequate for reconstituting the succinate oxidase system (Baginski and Hatefi, 1969).

The low-molecular-weight (100,000) enzyme is still not the minimal indivisible unit. As shown by Davis and Hatefi (1971), it includes two components, one of which - a ferroflavoprotein FP_S of molecular weight 70,000 - contains FAD, nonheme iron, and labile sulfides in a ratio of 1:4:4, while the other one (IP_S), having a molecular weight of 27,000, is an iron-sulfur protein entirely devoid of flavin. Identical results were published by Righetti and Cerletti (1971).

According to Ohnishi et al. (1974a,b), the soluble dehydrogenase of King suitable for reconstituting the succinate oxidase system displays three epr signals. Two of these correspond to ferredoxin-like iron-sulfur centers with half-reduction potentials of -5 mV (S-1) and -400 mV (S-2) at pH 7.4 (Ohnishi et al., 1974a, 1976b), and the third one to the high-potential iron-sulfur center with a half-reduction potential of +60 mV (Ohnishi et al., 1974b, 1976a).

Direct analysis of redox components in isolated subunits FP_S and IP_S is impossible because their separation involves drastic procedures which render the subunits catalytically inactive. Nevertheless, analysis of interaction between the flavin and the S-1, S-2, and S-3 iron-sulfur centers of succinate dehydrogenase shows that the flavin together with S-1 and S-2 are part of FP_S whereas S-3 is localized in IP_S (Ohnishi, 1979). Ackrell et al. (1980), Vinogradov et al. (1980), and Yu and Yu (1980a,b) have described the isolation of a specific coenzyme Q-binding protein (molecular weight 13,000-

15,000), designated QP_S, addition of which to the soluble two-subunit succinate dehydrogenase endows it with succinate-ubiquinone reductase activity. The three polypeptides mentioned were suggested to constitute the succinate-ubiquinone reductase segment of the respiratory chain (Yu and Yu, 1981).

Comparing the results obtained for NADH dehydrogenase and for succinate dehydrogenase, one can easily note the similarity in the molecular organization of these enzymes. Indeed, both include a flavoprotein with a molecular weight of about 70,000 containing flavin, nonheme iron, and labile sulfides in a ratio of 1:4:4, and an iron-sulfur protein. The flavoproteins are not fragmented even under drastic conditions causing their denaturation (see, e.g., Davis and Hatefi, 1971; Hanstein et al., 1971). The composition and various properties of mitochondrial flavin dehydrogenases have been extensively treated by Ohnishi (1979).

There are a number of indirect indications that the NADH-ubiquinone reductase segment of the respiratory chain may contain a Q-binding protein analogous to QP_S (see a review by Yu and Yu, 1981).

1.3.1.3. *Cytochromes b*

A rather complicated question is the one concerning the *b*-type cytochromes: their number, differences in chemical properties, participation in electron transfer and energy transformation, and the like. The data of a host of works published in this field have been summarized by Chance et al. (1970), Slater (1971, 1972, 1973), Slater et al. (1972), Slater and Lee (1973), Erecińska et al. (1973), Wikström (1973), Wilson et al. (1973), Mitchell (1976), Rieske (1976), and Weiss (1976). Rat liver, beef heart, and pigeon heart mitochondria were generally supposed to possess three *b*-type cytochromes (see, e.g., Wikström, 1973), though an even greater number has been suggested (Davis et al., 1973; Slater, 1973; Wikström, 1973). These cytochromes have absorptions α-bands at 557-559 nm (*b*-558), 561-562 nm (*b*-562), and 565-566 nm (*b*-566) at room temperature; at 77°K the bands are shifted to 554-555, 558-559, and 562.5 nm, respectively. Band intensities of *b*-558, *b*-562, and *b*-566 relate as 1:3:2 (Wikström, 1973). Cytochromes *b*-562 and *b*-566 are distinguishable by their epr spectra (Orme-Jonson et al., 1971, 1974; Siedow et al., 1978) as well as by thermodynamic (Wilson et al., 1973) and kinetic (Erecińska et al., 1973) parameters. It must be noted that the actual number of *b*-type cytochromes is debatable. Thus, Sato et al. (1971) believed that absorption bands at 558 and 566 nm belonged to one and the same cytochrome *b*-566. A similar conclusion has been later reached by Arutjunjan et al. (1978).

Cytochromes *b*-562 and *b*-566 are found not only in intact mitochondria but also in an oligoenzyme complex - ubiquinol-cytochrome *c* reductase (complex III) - which can be obtained by fractionating the inner mitochondrial membrane with bile acid salts (Davis et al., 1973; Slater, 1973; Nelson and Gellerfors, 1974), or in the bc_1 complex (Riccio et al., 1977). Davis et al. (1972, 1973) also indicated that there is a special cytochrome *b*-557 (77°K) belonging to complex II. This point of view was criticized by Slater (1973) who argued that cytochrome *b* found in complex II is a denatured form of cytochrome *b* from complex III and can be regarded as an impurity.

The most obscure issue is the protein characteristics of the above cytochromes. Attempts to separate them and to isolate them in native form have so far failed. Goldberger et al. (1961) proposed a method of cytochrome *b* isolation from beef heart mitochondria. The cytochrome *b* described by them was homogeneous by the data of analytical ultracentrifugation and had a molecular weight of 28,000 calculated from the iron or heme content. It showed an absorption maximum at 562 nm (α-band) at room temperature. Davis et al. (1973) have tried to use this technique, as well as that of Yamashita and Racker (1969), to separate cytochromes *b*-562 and *b*-566 of complex III. Products obtained in both cases displayed absorption spectra different from that of initial cytochromes. This means that either cytochromes are substantially modified during fragmentation of the complex, or that their spectral features depend on their environment in the complex. To pursue this idea, one can refer to Wikström (1971,1973) who granted that the three *b*-type cytochromes discerned spectrally may actually be one and the same protein situated in three essentially different membrane milieus. For instance, it is known from the work of Goldberger et al. (1962) that binding of purified cytochrome *b*-562 with mitochondrial structural protein markedly changes its redox potential.

Considering the question of the existence of chemically different forms of cytochrome *b*, mention must be made of the data of Yu et al., (1975), according to which two forms of cytochrome *b* can be isolated from succinate-cytochrome *c* reductase preparations, having molecular weights of 37,000 and 17,000. The first polypeptide displayed a distinct α-band of absorption at 560 nm (β- and γ-bands at 530 and 427 nm); as to the second one, its α- and β-bands were poorly discernible, and the γ-band was at 411 and 420 nm in oxidized and reduced states, respectively.

Indirect evidence for the existence of the high (28,500-34,000) and low (9000-14,500)-molecular-weight cytochromes *b* is provided by Gellerfors and Nelson (1975) and by Bell and Capaldi (1976).

According to Weiss and Ziganke (1974a) and to Lin and Beattie (1978), polyacrylamide gel electrophoresis of highly purified cytochrome *b* preparations from mitochondria of Neurospora crassa and Saccharomyces cerevisiae reveals a single major polypeptide having a molecular weight of 30,000 and 28,000, respectively. A preparation of cytochrome *b* with a molecular weight of 31,000 was isolated by Von Jagow et al. (1978) from beef-heart mitochondrial complex III. All these preparations can be identified as cytochrome *b*-562 by their spectral characteristics.

Finally, one should mention cytochrome *b*-560 isolated by Hatefi and Galante (1980) from beef heart succinate-ubiquinone reductase (complex II). Sodium dodecyl sulfate-polyacrylamide gel electrophoresis of this preparation revealed two polypeptides (molecular weights 13,500 and 15,500) in roughly equimolar amounts. Unfortunately, the authors did not properly determine the heme content in the protein molecules, despite the fact that this point is of principal importance for elucidating the nature of these polypeptides. The function of such cytochrome *b*-560 is still obscure.

Summing up the material of this subsection, it can be said that the apoprotein of *b*-562 has a molecular weight about 30,000. All other issues await further clarification (see, e.g., the discussion in Section IIIC-1 in Yu and Yu, 1981). The major complication in interpreting the experimental data is that heme *b* is noncovalently bound to protein and can therefore be somehow redistributed among particular polypeptides during solubilization and fractionation of the mitochondrial inner membrane.

In the presence of mild detergents cytochrome *b* from Neurospora crassa behaves as a protein of molecular weight 55,000. This may mean that normally cytochrome *b* is a dimer. Indeed, if a preparation of cytochrome *b* prior to dissociation in sodium dodecyl sulfate is treated with a bifunctional reagent (e.g., dimethyl suberimidate) forming links between subunits, subsequent polyacrylamide gel electrophoresis shows a protein band corresponding to a molecular weight of 55,000 (Weiss, 1976). Von Jagow et al. (1978) have reported that cytochrome *b* from beef heart mitochondria also exists as a dimer. The cytochrome *b* subunits are supposed (Weiss, 1976) to be nonidentical. For instance, they are separable by hydroxylapatite chromatography in the presence of sodium dodecyl sulfate. At the same time, no distinction could be found between peptide fragments obtained from the subunits, which testifies to their very similar primary structure.

1.3.1.4. *Cytochrome* c_1

Isolation of cytochrome c_1 with an appreciably high heme content (about 27 nmoles per mg protein) was described by Bomstein et al. (1961). Calculation of molecular weight on the basis of iron content gave 37,000. Later Yu et al. (1972) obtained a preparation of cytochrome c_1 which had a high heme content and could readily serve as electron donor or acceptor in interaction with the respiratory chain. Gel electrophoresis in the presence of sodium dodecyl sulfate revealed two bands, one of which corresponded to a hemoprotein with a molecular weight of 29,000. The identity of the second band (molecular weight 15,000) was not clear; it could be an impurity, but the possibility could not be excluded that this glycoprotein was a subunit of cytochrome c_1.

Similar results were published later by Bell and Capaldi (1976) who studied the composition of ubiquinol-cytochrome *c* reductase from beef heart mitochondria and found that heme c_1 is bound to a polypeptide having a molecular weight of 27,000-29,000. The latter was usually accompanied through all purification stages by a polypeptide of 14,000-15,000.

Ross and Schatz (1976a) isolated a highly purified cytochrome c_1 from yeast mitochondria. According to these authors, treatment of the preparation with sodium dodecyl sulfate and polyacrylamide gel electrophoresis yields a single polypeptide of molecular weight 31,000 carrying the covalently attached heme. Some preparations also contained a polypeptide of 18,500; however, as it was readily removed upon even mild treatment, it was not considered a cytochrome c_1 subunit. It is important to note that a homogeneous preparation of cytochrome c_1 containing a single polypeptides species could only be obtained if all isolation and purification steps were performed in the presence of an inhibitor of neutral proteinases (phenylmethylsulfonyl fluoride). In the absence of the inhibitor a set of polypeptides was obtained with molecular weights of 29,000, 27,000, and 25,000, indirectly evidencing an extremely high sensitivity of cytochrome c_1 to proteinases.

Finally, one should cite the work of Trumpower and Katki 1975) who isolated a preparation of cytochrome c_1 containing one hemoprotein with a molecular weight of 30,600. It was easily cleaved by trypsin, its molecular weight decreasing to 29,000.

1.3.1.5. *Rieske Iron-Sulfur Protein*

As indicated in the scheme of the respiratory chain (see the beginning of Section 1.3.1.), its bc_1 segment includes a specific iron-sulfur protein, which gives an epr signal with a g-factor of 1.90 at 77°K. Data are available that give grounds for regarding the component producing this signal as a constituent part of the respiratory chain functionally linked with cytochrome c_1 (Rieske et al., 1964a). Rieske et al. (1964b) have shown that complex III can be resolved with bile acid salts in the presence of antimycin A into two fractions, one of which contains cytochromes *b* and c_1, and the other contains nonheme iron and shows an epr signal with a g-factor of 1.90. Further development of this method made it possible to isolate an iron-sulfur protein homogeneous in analytical ultracentrifugation (Rieske et al., 1964c). The protein was devoid of heme or flavin and had a minimal molecular weight of 12,000-16,000 as judged by the content on nonheme iron. A twice larger value (26,000) was obtained from the results of analytical ultracentrifugation.

Later these studies were taken on by Trumpower and coworkers who elaborated techniques for isolating the iron-sulfur protein of the respiratory-chain bc_1 segment in a reconstitutionally active form (Trumpower and Edwards, 1979) and investigated in detail its physicochemical properties as well as its role in electron transfer through the respiratory chain and in energy transduction in the mitochondrial inner membrane (see a review by Trumpower, 1981). According to these authors, the purified protein is a homogeneous polypeptide showing a molecular weight of 24,000 in sodium dodecyl sulfate-polyacrylamide gel electrophoresis, displaying an epr signal with a g-factor of 1.90, and containing 50 nmoles of nonheme iron and 36 nmoles of labile sulfides per mg protein.

Treatment of isolated succinate-cytochrome *c* reductase with guanidine chloride plus cholate entails loss of activity together with the disappearance of the g=1.90 epr signal and the 24,000-molecular-weight polypeptide. Addition of the purified protein to such depleted complex restores its succinate-cytochrome *c* reductase activity. The reconstituted complex again contains the 24,000-molecular-weight polypeptide and shows the g=1.90 epr signal. Trumpower (1981) believes the iron-sulfur protein to be a component of the protonmotive Q cycle where it plays the role of ubisemiquinone-cytochrome *b* / ubiquinol-cytochrome c_1 oxidoreductase:

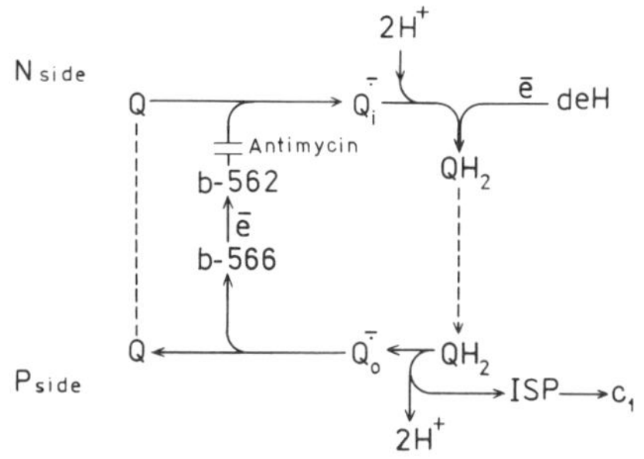

where deH is dehydrogenase, ISP is the iron-sulfur protein, $Q_o^{\bar{\cdot}}$ and $Q_i^{\bar{\cdot}}$ are ubisemiquinones on the outer and inner (electropositive, P) (electronegative, N) sides of the membrane, respectively.

1.3.1.6. *Other Components of the bc_1 Segment*

Besides cytochromes *b* and c_1 and a specific iron-sulfur protein, a "core" protein apparently performing no catalytic function has been isolated from complex III. Silman et al. (1967) found the protein to be homogeneous in polyacrylamide gel electrophoresis and to have a molecular weight of 40,000-50,000. However, Das Gupta and Rieske (1973) and Gellerfors and Nelson (1975) proposed that the core protein fraction contains two polypeptides with molecular weights of 50,000 (46,000) and 47,000 (43,000). The core protein carries no prosthetic group. Nevertheless, there are grounds for believing that it is an indispensable constituent of complex III necessary for its catalytic functions. Indeed, Gellerfors et al. (1976) have shown that the only component acylated upon treatment of complex III with iodo[^{14}C]acetamide is a polypeptide of molecular weight 50,000 identified as one of the two types of core proteins. Modification of this polypeptide is accompanied by the loss of complex III enzymic activity and a decrease in the rates of cytochrome *b* and c_1 reduction by ubiquinol.

Of other components of complex III, one should mention a protein (11,000 daltons), reported by Das Gupta and Rieske (1973), which specifically binds antimycin A (see the scheme in the beginning of Section 1.3.1).

1.3.1.7. *Cytochrome c Oxidase*

The terminal segment of the respiratory chain catalyzes the transfer of four electrons from ferrocytochrome *c* to an oxygen molecule and simultaneously "pumps" protons across the inner mitochrondrial membrane, generating a transmembrane proton gradient. These reactions can be described by the equation (see references in Azzi, 1980):

$$4 \text{ cyt } \underline{c} \ (Fe^{2+}) + n\, H_i^+ + O_2 \rightarrow$$

$$4 \text{ cyt } \underline{c} \ (Fe^{3+}) + (n - 4)H_o^+ + 2\, H_2O$$

where H_i^+ and H_o^+ are protons in the matrix (inside) and in the sytosol (outside), respectively, and n is the number of protons translocated per molecule of oxygen reduced.

Cytochrome *c* oxidase isolated from beef heart mitochondria has a molecular weight of 200,000-240,000 as determined by physical measurements (Tzagoloff et al., 1965; Love et al., 1970; Vail and Riley, 1974; Phan and Mahler, 1976a; Saraste et al., 1981). For yeast cytochrome *c* oxidase, molecular weights of 226,000 (Rubin and Tzagoloff, 1973a) and 260,000 (Phan and Mahler, 1976a) were reported. Such preparations usually contain about 10 nmoles of heme *a* per mg protein and a roughly equimolecular amount of copper. From this it follows that one enzyme molecule carries two hemes and two atoms of copper (Hartzell et al., 1978). However, cytochrome *c* oxidase preparations have been described in which the content of these one-electron centers is twice as high (Rubin and Tzagoloff, 1973a; Saraste et al., 1981).

More than twoscore years ago Keilin and Hartree (1939) suggested that the terminal segment of the respiratory chain includes two components. According to the classical concept, one of these (cytochrome *a*) gives in the reduced state the major contribution to the absorption band at 605 nm of various mitochondrial preparations. The other one (cytochrome a_3) is responsible for about half the absorbance at 415 nm and interacts with oxygen as well as with carbon monoxide, cyanide, azide, etc. The differences between cytochromes *a* and a_3 have been extensively reviewed (see, e.g., Lemberg, 1969; Malmström, 1974; Wikström et al., 1976). Nevertheless, despite repeated attempts cytochromes *a* and a_3 could not be separated into individual proteins. In particular, Horie and Morrison (1964) proposed a means of obtaining cytochrome *a* free of cytochrome a_3; however, these results have been criticized (see Lemberg, 1969).

As shown by Love et al. (1970) and Chan et al. (1970), in alkaline solutions containing nonionic detergents (e.g., Emasol) cytochrome *c* oxidase dissociates to give a product with a molecular weight of 100,000 which retains 83-97% of the activity of the dimer. As this takes place, the amount of carbon monoxide bound by the oxidase increases twice, and the spectral properties of the dissociated form begin to resemble to a certain extent that of cytochrome a_3 in the dimer. On the basis of these and other data (Love and Auer, 1970; Chan et al., 1970) it was concluded that the features usually ascribed to cytochrome aa_3 result from association of two identical monomeric units. From this point of view the incomplete binding of carbon monoxide by the dimer (one CO molecule per two heme residues) is due to the asymmetric aggregation of two identical monomeric units owing to which one of them loses the capacity of additional coordination. This may happen if the heme iron is bonded at the fifth and/or sixth coordination site with a ligand (see also Wikström et al., 1976). The idea of the existence of two identical units in cytochrome *c* oxidase is not consistent, however, with the data on the stoichiometry of polypeptides constituting the complex (see further).

According to present-day notions, electrons of the substrate (ferrocytochrome *c*) are accepted by heme *a*, passed to cytochrome a_3 with the participation of copper, and finally accepted by oxygen which binds to the heme iron of cytochrome a_3 (see references in Azzi, 1980). These interactions can be depicted by the scheme:

$$\begin{array}{ccc} \text{cyt } \underline{c} \longrightarrow \text{cyt}\,\underline{a} & & \text{cyt } \underline{a}_3 \longrightarrow O_2 \\ \rightleftharpoons & & \rightleftharpoons \\ \text{Cu A} & \longrightarrow & \text{Cu B} \end{array}$$

Research conducted in the last decade has shown cytochrome *c* oxidase to comprise a complex set of subunits (see for review Azzi, 1980; Kadenbach and Merle, 1981). The enzyme isolated from yeast mitochondria contains seven types of polypeptide with molecular weights of 40,000-42,000 (I), 27,000-34,000 (II), 22,000-27,400 (III), 13,800-14,600 (IV), 12,700-14,600 (V), 10,200-12,300 (VI), and 4600-10,600 (VII) (Rubin and Tzagoloff, 1973a; Poyton and Schatz, 1975a; Phan and Mahler, 1976a). Similar data were obtained for cytochrome *c* oxidase from Neurospora crassa mitochondria (Sebald et al., 1973). Seven types of polypeptides were also found in rat liver cytochrome *c* oxidase (Höchli and Hackenbrock, 1978). For cytochrome *c* oxidase from beef heart mitochondria, six (Rubin and Tzagoloff, 1973a; Yamamoto and Orii, 1974; Briggs et al., 1975), then seven (Downer et al., 1976; Yu and Yu, 1977) and eight subunits (Bucher and Pennial, 1975; Steffens and Buse, 1976) were reported. Use of more sophisticated polyacrylamide gel electrophoretic techniques has in recent years led to discovery of 10 (Verheul et al., 1981) and 12 (Merle and Kadenbach, 1980) types of polypeptides in mammalian cytochrome *c* oxidase. As evident from Table 1 summarizing these data, the increase in the number of polypeptides is due to the enhanced resolving power of the gel systems for the low-molecular-weight polypeptides.

Methods have been developed for isolating individual subunits (Sebald et al., 1973; Briggs et al., 1975; Poyton and Schatz, 1975a; Downer et al., 1976; Steffens and Buse, 1976; Werner, 1977). The most efficient techniques proved to be column chromatography in dissociating media using molecular sieves such as Bio-Gel P60 or Sephadex G-100, and preparative sodium dodecyl sulfate-polyacrylamide gel electrophoresis. Amino acid analysis of subunits I, II, and III shows a relatively low content of amino acid residues with hydrophilic side chains. For cytochrome *c* oxidase from beef heart mitochondria, the respective values obtained were 32-35%, 41-44%, and 37-40% (Briggs et al., 1975; Downer et al., 1976; Steffens and Buse, 1976). Subunits IV to VII have higher polarities: 48%, 45-48%, 45-50%, and 44-54%, respectively (Briggs et al., 1975; Steffens and Buse, 1976). Rather similar data were obtained for cytochrome *c* oxidase from yeast mitochondria (Poyton and Schatz, 1975a; George-Nascimento and Poyton, 1981). Information on the amino acid composition of cytochrome *c* oxidase subunits is of principal importance for elucidating their stoichiometry in the enzymic complex (see below).

To date, complete amino acid sequences have been determined for subunits II (Steffens and Buse, 1979), IV (Sacher et al., 1979), V (Tanaka et al., 1979), VI (Biewald and Buse, 1982), VII_{ala} (Steffens et al., 1979), and VII_{ser} (Buse and Steffens, 1978). The nucleotide sequences have been elucidated for mitochondrial DNA coding for subunits I, II, and III in yeast (Coruzzi and Tzagoloff, 1979; Bonitz et al., 1980; Thalenfeld and Tzagoloff, 1980) and human placenta (Anderson et al., 1981). The gene for subunit III of the Neurospora crassa enzyme has recently been sequenced (Browning and RajBhandary, 1982). In some cases the knowledge of the amino acid sequences helps to understand the functions of subunits and their topography in the inner mitochondrial membrane (see further).

Table 1. Subunit Composition of Cytochrome *c* Oxidase from Different Sources. Molecular weight values are given in thousands. Asterisked numbers have been determined from amino acid sequences; other molecular weights have been calculated from gel electrophoretic mobilities. Components a, b, and c reported by Downer et al. (1976) are regarded as impurities by Ludwig et al. (1979).

Sub-units	S. cerevisiae			N. crassa	Beef heart		Rat liver	
	Rubin and Tzagoloff (1973a)	Poyton and Schatz (1975a)	Phan and Mahler (1976a)	Werner (1977)	Downer et al. (1976)	Steffens and Buse (1976)	Merle and Kadenbach (1980)	Höchli and Hackenbrock (1978)
I	40	40	42.4	41	35.3	36	41.7	34
II	27.3	33	34.1	28.5	24.2	26*	25	26.8
III	25	22	27.4	21	21	22	24.5	23.7
IV	13.8	14.5	14.6	16	16.2	17.1*	15.8	17
a					13			
V	13	12.7	14.6	14	12.1	12.4*	11.5	12.5
b					10.4			
c					8.5			
VI	10.2	12.7	12.3	11.5	6.7	14	10.3	9.5
						11	9.5	
						11	8.9	
						10*		
VII	9.5	4.6	10.6	10	4.4	5.5	5.9	3.6
				10	3.4	5.5	5.9	
					3.0		5.9	
VIII	-	-	-	-	-	-	3.9	-

Knowing the molecular weights of all subunits and the portion of protein each of them accounts for in the cytochrome *c* oxidase preparation, one can determine their stoichiometry in the enzyme. For the enzyme from beef heart mitochondria, 12 polypeptide chains of 10 (Verheul et al., 1981) or 12 types (Merle and Kadenbach, 1890) were found. The total molecular weight of the enzyme calculated from these data (Verheul et al., 1981) amounts to 172,000 (or 198,000 if molecular weights for subunits I and III are taken to be 57,000 and 30,000 on the basis of the nucleotide sequences of corresponding genes). The molecular weight calculated from the heme *a* content (10-11 nmoles/protein) constitutes 180,000-200,000 assuming two hemes per enzyme molecule. This means that all hemes and copper atoms are distributed among a single set of subunits, and that the model of cytochrome *c* oxidase implying two identical associated units each carrying one heme (see Love and Auer, 1970) does not hold true. This also applies to yeast cytochrome *c* oxidase, since simple summation of subunit molecular weights (Poyton and Schatz, 1975a) gives 140,000, which agrees nicely with the value of 143,000 obtained from the heme content (14 nmoles/protein) in this cytochrome *c* oxidase preparation.

The question arising is: with which subunits are associated the redox components of cytochrome *c* oxidase? Earlier experiments on the covalent linking of the heme with subsequent identification of its polypeptide carriers (Schatz et al., 1972), on isolation of active cytochrome *c* oxidase with a reduced number of subunits (Komai and Capaldi, 1973; Phan and Mahler, 1976b), and on separation of subunits in dissociating media (Werner, 1977; Yu et al., 1977b) showed heme *a* and copper to be bound to subunits IV-VII; subunits V (Tanaka et al., 1977) and VII_{ala} (Steffens et al., 1979) were considered the most likely candidates for the part of the heme carrier, as their primary structures resembled in some features those of the hemoglobin β-chain and of the proteins of the cytochrome *c* family. According to the later data (Boonman et al., 1979; Winter et al., 1980), heme and copper are bound to subunits I and II. The copper is most probably carried by subunit II which has amino-acid-sequence homology with blue-copper proteins including azurins, plastocyanins, and stellacyanins (Tanaka et al., 1975; Steffens and Buse, 1979). It has also been reported that upon cytochrome *c* oxidase fragmentation heme *a* is found in the fraction of subunits I-III while copper goes with subunits IV-VII (Fry, 1979). It is even possible to obtain fractions of subunits I-III and subunits IV-VII which both contain heme *a* and copper in comparable amounts (Tada et al., 1981).

These discrepancies are in no small measure due to the lack of control of the redistribution (i.e., removal from the natural binding sites and attachment to new reaction sites) of the noncovalently bound heme *a* and copper among the subunits during the experimental procedures. As shown by Winter et al. (1980), the pattern of heme *a* distribution among the subunits in sodium dodecyl sulfate-polyacrylamide gel electrophoresis depends on buffer concentration, composition of the detergent system used to dissociate the enzyme, temperature, conditions of preelectrophoresis run to remove free radicals and peroxides from the gel, and possibly other factors. Upon electrophoresis in mild conditions specially selected to minimize redistribution of heme *a* and copper, about 90% of copper was found on subunit II while heme *a* was shared equally by subunits I and II. Both subunits displayed properties (absorption spectra and CO-binding ability) resembling those of cytochrome a_3. It must be added that, according to a number of studies on the covalent linking of cytochrome *c* to cytochrome oxidase (Bisson et al., 1978a,b, 1980; Briggs and Capaldi, 1978), it is bound to subunit II which thereby plays the part of cytochrome *a*. Not going into the details of the structural and functional arrangement of the two-subunit cytochrome *c* oxidase (for this see Winter et al., 1980), one can mention that such a composition has been found for some microbial cytochrome *c* oxidases (Yamanaka et al., 1979; Fee et al., 1980; Ludwig and Schatz, 1980). The final proof of the validity of the two-subunit concept requires reconstitution of cytochrome *c* oxidase from this minimal set of components.

As indicated in the beginning of this section, cytochrome *c* oxidase performs not only ferrocytochrome *c* oxidation with molecular oxygen (dioxygen), but also proton transport across the inner mitochondrial membrane. Data are available that this function is carried out by subunit III (for review see Azzi, 1980; Kadenbach and Merle, 1981). In particular, Saraste et al. (1980) showed that selective extraction of subunit III from beef heart cytochrome *c* oxidase does not impair the redox activity of the enzyme but abolishes its ability to pump protons across the membrane. In accordance with this, N,N'-dicyclohexyl carbodiimide, which is known as an inhibitor of proton transport by mitochondrial ATPase or cytochrome *c* oxidase, binds with subunit III in a 1:1 molar ratio (Casey et al., 1980) just as it interacts with the proton-translocating proteolipid in H^+-ATPase. The proton-translocating function of subunit III is compatible with its probable transmembrane position in the inner mitochondrial membrane (see further).

Thus, only three polypeptides (subunits I, II, and III) appear to be required for the now-known functions of cytochrome *c* oxidase. Still, as pointed out by Poyton and Schatz (1975b) and by Merle and Kadenbach (1980), the other subunits are not casual impurities. The seven polypeptides of yeast cytochrome *c* oxidase and the 12 polypeptides of the beef-heart mitochondrial enzyme behave as a unity. Kadenbach and Merle (1981) ascribe regulatory functions to the "noncatalytic" subunits. Notwithstanding that this suggestion is too vague at present, one should not forget that the model of a two-subunit catalytic ensemble, albeit most popular today, has not been conclusively proved. Among other things, no cytochrome *c* oxidase preparation is yet available that contains only subunits I and II and has high catalytic activity.

1.3.1.8. *ATPase Complex*

Isolation of oligomycin-insensitive ATPase (factor F_1) from mitochondria or submitochondrial particles has been described by quite a number of authors (Pullman et al., 1960; Schatz et al., 1967; MacLennan et al., 1968; Sone et al., 1969; Datta and Penefsky, 1870; Horstman and Racker, 1970; Senior and Brooks, 1970; Brooks and Senior, 1971; Catterall and Pedersen, 1971; Lambeth and Lardy, 1971; Tzagoloff and Meagher, 1971; Knowles and Penefsky, 1972a). Enzyme preparations obtained from various sources (beef heart, rat liver, yeast) were homogeneous by gel filtration, sucrose density gratient centrifugation, or analytical ultracentrifugation, and had molecular weights of 340,000-360,000 (Catterall and Pedersen, 1971; Lambeth and Lardy, 1971; Lambeth et al., 1971, Tzagoloff and Meagher, 1971; Knowles and Penefsky, 1972b). Polyacrylamide gel electrophoresis in the presence of sodium dodecyl sulfate revealed five types of polypeptides with molecular weights of 53,000-62,500, 49,000-57,000, 25,000-39,000, 12,500-18,000, and 3500-12,000 (Senior and Brooks, 1970, 1971; Brooks and Senior, 1971; Knowles and Penefsky, 1972b; Van de Stadt et al., 1972; Capaldi 1973; Catterall et al., 1973; Jackl and Sebald, 1975; Stutterheim et al., 1981). These polypeptides were termed subunits A, B, C, D, and E or subunits α, β, γ, δ, and ε (or, still otherwise, subunits 1, 2, 3, 4, and 5). All five subunits of factor F_1 from beef heart (Brooks and Senior, 1972; Knowles and Penefsky, 1972b) and the three larger subunits of the rat liver enzyme (Catterall et al., 1973) were obtained as individual proteins. For all subunits of yeast F_1, amino acid compositions have been determined (see, e.g., Stutterheim et al., 1981). According to some data (Knowles and Penefsky, 1972b; Catterall et al., 1973), subunits A, B, and C account for 93-97% of the total F_1 protein. The stoichiometry of polypeptides in F_1 corresponds to the formula A_3B_3CDE for enzymes from rat-liver (Catterall et al., 1973), beef-heart (Senior and Brooks, 1971), and yeast mitochondria (Stutterheim et al., 1981). This formula is consistent not only with the data on the content of individual polypeptides but also with the results of electron-microscopic observations (Senior, 1973; Pedersen, 1975). Other stoichiometric models have been considered; Kozlov and Mikelsaar (1974) suggested that the actual formula of F_1-ATPase is A_3B_3C', and during enzyme isolation subunit C' is broken down into three polypeptides (C, D, and E) usually observed on electrophoresis with sodium dodecyl sulfate. Mention should also be made of the formula $\alpha_2\beta_2\gamma_2$ based on a more exact molecular weight estimate for F_1 (see the discussion in Penefsky, 1979).

The functions of particular F_1-ATPase subunits have been comprehensively treated by Skulachev and Kozlov (1977) and Kozlov and Skulachev (1982).

Besides the five polypeptide types indicated, there is another low-molecular-weight protein capable of suppressing the ATPase activity of isolated factor F_1 or submitochondrial particles (Pullman and Monroy, 1963; Horstman and Racker, 1970; Brooks and Senior, 1971; Nelson et al., 1972; Satre et al., 1975; Chan and Barbour, 1976). The ATPase protein inhibitor has a molecular weight of about 10,000 (Brooks and Senior, 1971; Nelson et al., 1972; Cintron and Pedersen, 1979), being thus close in dimensions to the small ATPase subunit; however, it differs from the latter in electrophoretic mobility and amino acid composition (Brooks and Senior, 1971, 1972; Knowles and Penefsky, 1972b). The inhibitor is readily removed from factor F_1 during purification of the latter, and therefore some preparations of F_1 (as well as of the ATPase complex) do not contain it at all. Frangione et al. (1981) elucidated the complete amino-acid sequence of the beef heart ATPase inhibitor. It proved to be built of 84 amino acid residues and to have a molecular weight of 9578.

Factor F_1 is a constituent of the ATPase complex - a fragment of the inner mitochondrial membrane also called CF_0-F_1 (Kagawa and Racker, 1966a), oligomycin-sensitive ATPase (Kopaczyk et al., 1968a; Tzagoloff et al., 1968a; Tzagoloff and Meagher, 1971), complex V (Stiggall et al., 1978), or H^+-ATPase (Skulachev and Kozlov, 1977). The ATPase complex catalyzes ADP phosphorylation at the expense of the transmembrane electrochemical potential of H^+ ions, or hydrolyzes ATP to generate such a potential on the inner mitochondrial membrane. The isolated complex is characterized by an ability to carry out oligomycin-sensitive ATP hydrolysis, to catalyze the ATP-P_i exchange reaction and several other processes.

By the data of sedimentation analysis and electron microscopy, the complex from *Saccharomyces cerevisiae* has a molecular weight of about 468,000 when corrected for the lipid content (Tzagoloff and Meagher, 1971; Ryrie, 1975).

Polyacrylamide gel electrophoresis in the presence of sodium dodecyl sulfate reveals in different preparations of the ATPase complex from 10 to 16 types of polypeptides (Tzagoloff and Meagher, 1971; Capaldi, 1973; Senior, 1973; Swanlijung et al., 1973; Jackl and Sebald, 1975; Koch, 1976; Serrano et al., 1976; Galante et al., 1977; Marahiel et al., 1977; Sebald, 1977; Berden and Voorn-Brouwer, 1978; Stiggall et al., 1978; Galante et al., 1979; Ryrie and Gallagher, 1979; Ludwig et al., 1980; Todd et al., 1980). Five of these belong to factor F_1, the polypeptide with a molecular weight of approximately 10,000 is the ATPase inhibitor (see above) having regulatory functions, and the others constitute the so-called membrane sector, or F_0 sector, or the complex (Kagawa and Racker, 1966a; Bulos and Racker, 1968; Tzagoloff et al., 1968b; Glaser et al., 1980). One component of this sector is easily removed with e.g., ammonium hydroxide. This water-soluble protein of molecular weight 18,000-22,000 (MacLennan and Tzagoloff, 1968; Russell et al., 1976) confers sensitivity to oligomycin upon ATPase and is accordingly termed "oligomycin-sensitivity-conferring protein", abbreviated OSCP.

Table 2. Molecular Weights and Identity of Polypeptides in Oligomycin-Sensitive ATPase from Beef Heart Mitochondria (after Ludwig et al., 1980).

Band No.	Molecular weight from:			Identity of polypeptide
	Weber-Osborn gels	Swank-Munkres gels	Amino acid composition	
1	54,000	48,000	60,200	α subunit of F_1
2	48,500	44,000	53,300	β subunit of F_1
3	31,500	33,500	33,160	γ subunit of F_1
4	29,000	30,500		uncoupler-binding protein
5	24,500	25,000		
6	21,500	20,800		OSCP
7	22,000	20,500		
8	29,000	20,000		
9	15,000	19,500	16,100	δ subunit of F_1
10	13,000	14,700		
11	10,000	12,300		inhibitor of F_1
12	15,500	12,300		
13	13,000	11,800		
14	7,000	10,700		} F_6 ?
15	7,000	10,200		} P_i carrier ?
16	10,000	8,500	7,627	DCCD-binding protein
17	7,000	7,700	5,850	ε subunit of F_1
18	6,500	6,500		

Isolation of particular "intrinsic membrane" components of F_0 is described in a number of works (see, e.g., Bronnikov and Evtodienko, 1977; Sebald, 1977).

The only component of the F_0 sector that has been studied well is the polypeptide covalently binding N,N'-dicyclohexyl carbodiimide (DCCD) and extractable from mitochondria or isolated ATPase complex with a chloroform-methanol mixture. The molecular weight of this proteolipid (the so-called ninth subunit according to the notation of Tzagoloff et al., 1973) ranges by the data of different authors for different cell types from 6500 to 14,00 (Cattell et al., 1971; Stekhoven et al., 1972; Sierra and Tzagoloff, 1973; Sebald and Jackl, 1975; Dianoux et al., 1978; Graf and Sebald, 1978; Aroskar and Avadhani, 1979; Sebald et al., 1979a; Glaser et al., 1980, 1981). Amino acid composition and sequence have been elucidated for the DCCD-binding protein (Sebald et al., 1979b). The proteins from yeast and *Neurospora crassa* mitochondria comprise 76 and 81 amino acid residues, respectively, of which only 25% have hydrophilic side chains. It is supposed that two hydrophobic segments traverse the lipid bilayer of the inner mitochondrial membrane while the polar regions are either exposed to the aqueous phase or are in contact with other H^+-ATPase subunits.

According to present-day concepts (see reviews by Fillingame, 1980; Houstek et al., 1982), the DCCD-binding polypeptide is the ionophoric component of H^+-ATPase and performs proton transport across the inner mitochondrial membrane. In particular, it was shown by Criddle et al. (1977) to collapse the transmembrane potential and to increase the electric conductivity of a model membrane system built of a Millipore filter impregnated with phospholipids and containing bacteriorhodopsin as a generator. Having been incorporated into phospholipid vesicles, the proteolipid can function as an oligomycin-sensitive protonophore (Criddle et al., 1979).

Of other H^+-ATPase membrane-sector components, one should mention the 8500-molecular-weight polypeptide providing for the binding between F_1 and F_0 (Russell et al., 1976). This polypeptide may be identical with the coupling factor F_6 (Kanner et al., 1976). It was demonstrated (Guerrin and Napias, 1978; Kuzela et al., 1980) that, besides the DCCD-binding protein, chloroform-methanol extracts from the ATPase complex contain a polypeptide (molecular weight 10,000 in yeast and 9000 in rat liver) which has a high affinity for phosphate and is likely to act as a phosphate carrier (Guerrin and Napias, 1978; Velours et al., 1980). Alfonso et al. (1981) suppose that a polypeptide of molecular weight 28,000, marked by high sensitivity to trypsin, serves as a structure linking F_1 and OSCP with the DCCD-binding proteolipid. Finally, among the components of the F_0 sector there is a polypeptide with a molecular weight of some 30,000 (Serrano et al., 1976; Berden and Voorn-Brouwer, 1978; Galante et al., 1979) identified as the component binding the uncoupler 2-azido-4-nitrophenol (Galante et al., 1979).

Thus far, there is more or less definite information on the structural or functional purposes of 12 out of 18 polypeptides seen upon maximal resolution of H^+-ATPase components by two-dimensional electrophoresis in sodium dodecyl sulfate-polyacrylamide gels (Ludwig et al., 1980). It should also be borne in mind that the unidentified polypeptides may include aggregation products of the identified ones. Specifically, this applies to the oligomeric form of the DCCD-binding protein (see, e.g., Graf and Sebald, 1978; Glaser et al., 1981).

The details of the composition and structural arrangement of the mitochondrial F_1-ATPase and ATPase complex have been extensively reviewed (Senior, 1973; Beechey, 1974; Catterall and Pedersen, 1974, Pedersen, 1975; Penefsky, 1979; Fillingame, 1980; Houštek et al., 1982).

1.3.1.9. *Adenine Nucleotide Carrier*

Tightly coupled respiring mitochondria perform an asymmetric exchange of endogenous ATP for exogenous ADP, which ensures a certain steady-state level of oxidative phosphorylation.

The transport of adenine nucleotides across the inner mitochondrial membrane is known (see a review by Vignais, 1976) to be specifically inhibited by atractyloside (AT), carboxyatractyloside (CAT), or bongkrekic acid (BA). Therefore the search for the components of the translocating system amounted to isolation from mitochondria of protein(s) specifically binding these inhibitors. There are two basically different methods of solving this task. One of these makes use of the exceptionally low affinity of the AT(or CAT)-binding protein for hydroxylapatite (Riccio et al., 1975a,b; Aquila et al., 1978; Klingenberg et al., 1978). The key step of the other technique is the selective adsorption of the protein from detergent-solubilized submitochondrial particles on carboxyatractylate-Sepharose 4B (Brandolin et al., 1974; Bojanovski et al., 1975, 1976) followed by elution with 2% sodium dodecyl sulfate.

The AT(CAT, BA)-binding protein in polyacrylamide gel-sodium dodecyl sulfate electrophoresis displays a molecular weight of 26,000 in the case of rat liver mitochondria (Bojanovski et al., 1976), 30,000 with beef-heart (Klingenberg et al., 1978) and 37,000 with yeast mitochondria (Boulau et al., 1979). Since ADP and ATP (but not ITP) displace CAT from the complex with the protein (see, e.g., Klingenberg et al., 1978), the latter can be regarded as the adenine nucleotide carrier or at least as the key component of the translocase. The protein is hydrophobic (39% polarity) and shows a tendency to dimerization (Bojanovski et al., 1975, 1976; Klingenberg et al., 1978). Its content in the inner mitochondrial membrane reaches 10-12% of the total protein (Bojanovski et al., 1976; Klingenberg et al., 1978), making it the major protein of the outer side of this membrane (Boxer et al., 1977).

The adenine nucleotide carrier has been supposed to include two protein subunits having molecular weights about 30,000 (see a review by Vignais, 1976). It is hard to be certain yet that the carrier does not contain any other protein components. Krämer and Klingenberg (1977) showed that the activity of the purified CAT-binding protein incorporated into phospholipid vesicles constitutes only 8% of the initial mitochondrial activity. Still, Serrano et al. (1976) obtained a fraction from beef heart mitochondria that was highly active in adenine nucleotide transport into liposomes. The main protein component of this fraction was a polypeptide with a molecular weight of some 30,000.

Recently the complete amino acid sequence has been determined for the translocase (Aquila et al., 1982). It turned out that among the 297 residues there are at least two hydrophobic tracts more than 20 residues long, which in principle can traverse the lipid bilayer of the inner mitochondrial membrane.

1.3.1.10. *Inorganic Phosphate Carrier*

Considerable advances have been made in recent years in isolating the protein which is likely to perform the transport of inorganic phosphate across the inner mitochondrial membrane. This process, which is extremely important for the maintenance of oxidative phosphorylation in mitochondria, takes place as a phosphate-hydroxyl exchange or as phosphate-H^+ symport (Chappell and Crofts, 1965). It is suppressed by sulfhydryl reagents such a N-ethyl maleimide, mersalyl, or p-hydroxymercuribenzoate (Coty and Pedersen, 1975). Wohlrab (1980) has found that the protein responsible for phosphate transport, like the CAT-binding protein, has an exceptionally low affinity for hydroxylapatite, which permits these two proteins to be separated in a single step from other mitochondrial proteins after thorough solubilization of mitochondria.

The protein fraction containing only these two proteins (by the data of polyacrylamide get electrophoresis) can be incorporated into phospholipid vesicles; such a reconstituted system is capable of carrying out the phosphate exchange between the vesicle contents and the surrounding medium. The exchange is inhibited by N-ethyl maleimide, mersalyl, p-hydroxymercuribenzoate, or $HgCl_2$, but is insensitive to carboxyatractyloside. The molecular weight of the carrier was estimated at 34,000 (Wohlrab, 1980). Later Touraille et al. (1981) developed a method allowing the phosphate carrier and the CAT-binding protein to be separated. According to these authors, the proteins from porcine heart mitochondria have molecular weights of 32,000 and 30,000, respectively.

1.3.1.11. *Pyridine Nucleotide Transhydrogenase*

This enzyme, which is one of the generators of the transmembrane electrochemical potential on the inner mitochondrial membrane, catalyzes in submitochondrial particles the transhydrogenation between NADPH and NAD^+ coupled with proton transport according to the equation:

$$n\,H^+_{in} + NADPH + NAD^+ \rightleftharpoons n\,H^+_{out} + NADP^+ + NADH$$

Rydström et al. (1975) described conditions for extracting from submitochondrial particles a fraction accounting for 90% of their total transhydrogenase activity. Then Höjeberg and Rydström (1977) proposed a method for obtaining a preparation which had transhydrogenase activity 50 times that of the initial submitochondrial vesicles. It was a homogeneous protein forming on polyacrylamide gel electrophoresis in the presence of sodium dodecyl sulfate a single band corresponding to a molecular weight of 97,000. Upon incorporation into lecithin liposomes this protein could generate a transmembrane potential, being provided with substrates (NAD^+ and NADPH). The transhydrogenase is supposed to account for about 2% of the total mitochondrial inner membrane protein.

Anderson and Fisher (1978) have obtained a preparation of beef heart mitochondrial transhydrogenase that had a molecular weight of 120,000 and upon incorporation into liposomes proved capable of H^+ transport across their membranes (Earle et al., 1978).

As shown later by the same authors (Anderson and Fisher, 1981), the isolated transhydrogenase, as well as the liposome-incorporated enzyme, exists as a dimer which can be demonstrated with bifunctional linking reagents. Interestingly, larger oligomers were not detected.

1.3.1.12 *Inorganic Pyrophosphatase*

According to Shakhov et al. (1974), preparations of ultrasonic submitochondrial particles can catalyze the synthesis of inorganic pyrophosphate coupled with electron transfer through the respiratory chain, and pyrophosphate hydrolysis. This pyrophosphatase has been isolated and partially purified by Mansurova et al. (1977). It proved to be so weakly bound to the membrane that it could be extracted with 0.25 M sucrose without any additional treatment. Two isoenzymes were revealed during purification, having the same molecular weight (75,000 by gel filtration) but resolvable on DEAE cellulose. One of the enzymes displayed only pyrophosphatase activity, while the other, upon being inserted into the membrane of submitochondrial vesicles, catalyzed pyrophosphate synthesis coupled with electron transfer.

1.3.1.13. *Carriers of Ca^{2+} Ions*

Lehninger (1971) demonstrated that rat liver mitochondria subjected to osmotic shock in distilled water lose the ability to efficiently bind Ca^{2+} ions. Components responsible for this ability were recovered in the supernatant obtained on pelleting the membranes. Fractionation of the aqueous extract yielded a protein factor having high affinity for calcium (Gomez-Puyou et al., 1972). The preparation contained up to 27% phospholipids, more than 8% hexosamines, free fatty acids, and some other components. The main protein constituents showed molecular weights of 67,000 ± 7000 in sodium dodecyl sulfate-polyacrylamide gel electrophoresis. Since Ca^{2+} binding by the isolated glycoprotein (or, more exactly, glycolipoprotein) was inhibited by La^{3+} ions or ruthenium red, this factor could be thought to be responsible for Ca^{2+} transport in mitochondria.

Isolation of a glycoprotein with high avidity for Ca^{2+} was described by Sottocasa et al. (1972). The source was bovine liver mitochondria. The glycoprotein contained up to 30% phospholipids and about 5% carbohydrates. As judged by polyacrylamide gel electrophoresis in the presence of sodium dodecyl sulfate, the molecular weight of the protein moiety was close to 42,000, or by more recent data 30,000-33,000 (Panfili et al., 1974). Calcium binding by the glycoprotein was sensitive to the inhibitors of Ca^{2+} transport in mitochondria: La^{3+} ions, ruthenium red, or butacaine (Panfili et al., 1974). It is of interest that upon stepwise fractionation of bovine liver mitochondria the glycoproteins specifically binding Ca^{2+} were found in all fractions: outer membranes, intermembrane space, inner membranes, and matrix (Sandri et al., 1974).

According to the data of Sottocasa et al. (1972), the content of the Ca^{2+}-binding glycoprotein is not less than 0.5-1% of the total mitochondrial protein.

The role of the glycoprotein in Ca^{2+} transport was demonstrated in a number of experiments (see a review by Tedeshi, 1981). In particular, it can be mentioned that the glycoprotein fraction added to phospholipid bilayers enhances their Ca^{2+}-electric conductivity, this increase being blocked by ruthenium red (Prestipino et al., 1974; Lukyanenko et al., 1980). It has also been shown (Panfilini et al., 1976) that antibodies against the Ca^{2+}- binding glycoprotein suppress metabolically dependent Ca^{2+} transport into mitochondria.

Nevertheless, on the basis of the good solubility of the glycoprotein in water, Jeng and Shamoo (1980) argue that it cannot be an intrinsic membrane protein and therefore is not fit for the part of calcium carrier. These authors have isolated from calf-heart mitochondrial inner membrane a low-molecular-weight (assessed at 3000 by sodium dodecyl sulfate-polyacrylamide get electrophoresis and amino acid analysis) hydrophobic polypeptide soluble in diethyl ether and displaying high affinity and selectivity for Ca^{2+} ions. This protein, named calciphorin, was regarded as the Ca^{2+} carrier across the inner mitochondrial membrane. Unfortunately, data on reconstitution of Ca^{2+} channels from calciphorin are still lacking, and the situation with the mitochondrial inner-membrane carriers for Ca^{2+} is still obscure.

1.3.1.14. *Other Mitochondrial Inner-Membrane Components*

The mitochondrial inner membrane contains quite a number of components whose properties have not been sufficiently studied. This applies in particular to the systems performing the transport of amino acids, dicarboxylic and tricarboxylic acids, etc. Here mention can be made of a report by Palmieri et al. (1974) on the isolation from submitochondrial particles of some proteins that could specifically bind inorganic phosphate, aminophosphate, and glutamate. Later a proteolipid showing high affinity for glutamate was obtained from porcine heart mitochondria (Julliard and Gautheron, 1978).

Finally, also associated with the inner mitochondrial membrane are some dehydrogenases (choline dehydrogenase, β-hydroxybutyrate dehydrogenase, sarcosine dehydrogenase, etc.), carnitine acyltransferases, and other enzymes. However, most of them are not regarded as integral membrane proteins which determine the basic structural features of the membrane. Besides, the molecular design of many of the enzymes mentioned is not yet clearly understood.

1.3.2. Principles of the Mitochondrial Inner-Membrane Structural Organization

There are grounds for believing that the protein components listed in the preceding section are arranged in the mitochondrial inner membrane in a strictly defined way. The experimental data available permit three main problems to be discussed: (1) how these components are positioned with respect to the outer and inner faces of the membrane, (2) how individual electron carriers and coupling factors are subintegrated in complexes, and (3) how such complexes are arrayed with respect to each other within the membrane.

1.3.2.1. *Mitochondrial Inner-Membrane Asymmetry*

The nonuniform distribution of protein particles between the inner and outer sides of the inner mitochondrial membrane is dramatically demonstrable by electron microscopy of membrane preparations subjected to the "freeze-fracture" procedure. Such data obtained in Packer's laboratory (Wrigglesworth et al., 1970; Melnick and Packer, 1971) are presented in Figure 3. Obviously the convex fracture surface of the membrane (the side facing the matrix) contains approximately twice more particles than the concave one (the side facing the intermembrane space). Incidentally, the inner side of the outer membrane is also poor in protein particles as compared to its outer side.

Mitochondrial inner-membrane asymmetry is also revealed by the experiments of Clarke (1976) who subjected rat liver mitoplasts (inner membranes with matrix) to iodination via the lactoperoxidase system. Since only seven out of more than 30 polypeptides were iodinated, the protein components could be thought to be arranged in such a way that many of them have no access to the exterior surface of the membrane which is in contact with the surrounding medium.

Purposeful studies of the position of particular proteins within the inner mitochondrial membrane became possible after methods had

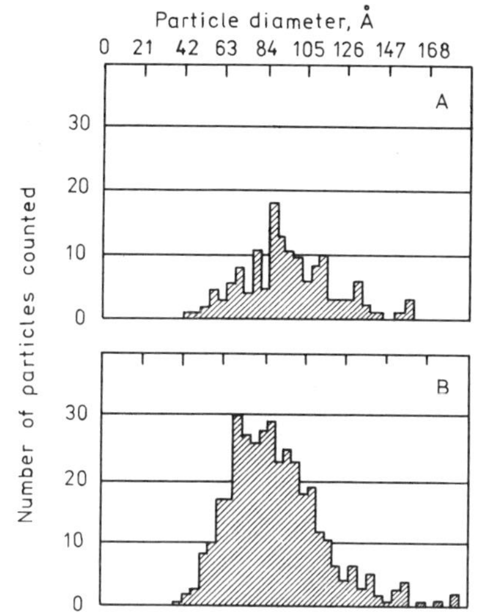

Fig. 3. Distribution of protein particles in the mitochondrial inner membrane (redrawn after Packer, 1972). (A) Concave fracture surface: particles on the outer side of the inner membrane (particle density 920 μm^{-2}); (B) convex fracture surface: particles facing the matrix (particle density 2360 μm^{-2}). Areas examined: (A) 0.162 μm^2; (B) 0.157 μm^2.

been devised for obtaining submitochondrial preparations represented by vesicles in which the membrane had different orientation with respect to the surrounding medium, i.e. vesicles with the membrane orientated as in intact mitochondria ("right-side-out") or in quite the opposite way ("inside-out"). Before giving an account of the experimental data, a few words must be said about the criteria for this or that membrane orientation in submitochondrial vesicles. The most graphic test appears to be the presence or absence of "mushroom" formations covering the vesicle exterior surface. These formations are observed electron-microscopically in submitochondrial preparations obtained by sonication or mild alkalization of mitochondria and negatively contrasted with phosphotungstate (Parsons, 1963; Smith, 1963; Stoeckenius, 1963; Fernàndez- Moràn et al., 1964). The exterior surface of intact inner mitochondrial membranes and of vesicles obtained by mitochondrial fragmentation with digitonin lack the "mushroom" formations (Malviya et al., 1968). However, the morphological criterion is not quite reliable. Specifically, it cannot distinguish between vesicles with normal membrane orientation and inside-out vesicles devoid of "mushrooms" owing to exposure to ultrasound, urea, some salts (KI or KSCN), etc. (Chance et al., 1963; Stasny and Crane, 1964; Racker et al., 1965; Kagawa and Racker, 1966b).

Harmon et al. (1974) have proposed to assess membrane orientation in given vesicles by the occurrence or lack of stimulation of mitochondrial respiration (i.e. NADH oxidase or succinate oxidase activity) by exogenous cytochrome *c*. Let us introduce the terms generally accepted in the literature: "C-side" for the outer side of the inner membrane, facing in mitochondria the intermembrane space; and "M-side" for its inner side facing the matrix. It is believed that the decrease in the respiration rate upon washing with dilute salt solutions and its increase upon subsequent addition of cytochrome *c* to the medium are characteristic of vesicles with normal membrane orientation (i.e. the same as in mitochondria). With inside-out submitochondrial vesicles, neither treatment should induce any changes in their NADH oxidase or succinate oxidase activities. Intermediate effects are held to indicate that the preparation studied is a mixture of two types of vesicles with opposite membrane orientation. This would be true assuming that cytochrome *c* is localized on the C-side of the inner mitochondrial membrane (see below) and that it does not penetrate the membrane.

Thus, having at one's disposal vesicles with membranes oriented in two opposite ways towards the surrounding medium, and a collection of reagents not penetrating (or rather slowly penetrating) the membrane and interacting with definite components of the respiratory chain, energy transformation system and the like, one can assess the arrangement of these components with respect to the inner and outer sides of the membrane. Such data are summarized, for instance, in reviews by Racker (1970), Harmon et al. (1974), and De Pierre and Ernster (1977). Considered below are the most important of these data together with the results of recent studies encompassing a broader set of membrane components.

1.3.2.1a. NADH Dehydrogenase. According to present-day notions, NADH dehydrogenase is localized (at least in heart and liver mitochondria) on the M-side of the inner mitochondrial membrane. Already the early work of Lehninger demonstrated that oxidation of only endogenous NADH in animal cell mitochondria is accompanied by ADP phosphorylation (see, e.g., Lehninger, 1955). The exogenous substrate is oxidized by intact mitochondria at a low rate (Ernster and Nazavio, 1956; Van den Bergh and Slater, 1962) which increases on addition of cytochrome *c*. Such oxidation is insensitive to antimycin A and is probably caused by the enzymic systems of the outer mitochondrial membrane or contaminating microsomes (Lehninger, 1955). The intact inner mitochondrial membrane is held to be impermeable for NADH (Von Jagow and Klingenberg, 1970), and the antimycin A-sensitive oxidation of exogenous NADH by oxygen or cytochrome *c* is considered an indication of a certain impairment of the membrane integrity.

Thus, the component of NADH dehydrogenase responsible for NADH binding is localized on the M-side of the inner mitochondrial membrane. An exception to this rule is yeast mitochondria which, as shown by Von Jagow and Klingenberg (1970), can oxidize exogenous NADH through a special NADH dehydrogenase situated on the C-side of the membrane. This 'exterior' dehydrogenase, like the usual 'interior' one, is linked to the cytochrome chain via the ubiquinone pool (Figure 4).

The question arises of the intramembrane arrangement of other NADH dehydrogenase components. The definite location is at present known only for the component responsible for the enzyme interaction with ferricyanide. As shown by Mitchell and Moyle (1969) and by Klingenberg and Buchholz (1970), ferricyanide or ferrocyanide cannot penetrate the mitochondrial membrane and therefore can only interact with its components exposed on the outer surface. Experiments on $Fe(CN)_6^{3-}$ reduction by intact mitochondria and inside-out submitochondrial vesicles led Grinius et al. (1971) to the conclusion that the NADH dehydrogenase site interacting with ferricyanide is on the M-side of the membrane. In these experiments the submitochondrial vesicles performed ferricyanide reduction by NADH even in the presence of rotenone. In the case of mitochondria, addition of a glutamate-malate mixture also led to ferricyanide reduction, but it did not take place immediately on NADH dehydrogenase but through a longer respiratory-chain segment including cytochrome *c*, since the reaction was sensitive to antimycin (Figure 4).

These data to an extent allow one to imagine how the flavoprotein moiety of NADH dehydrogenase is positioned in the inner membrane. Still, they provide no information on the arrangement in the membrane of other components of the complete NADH-ubiquinone

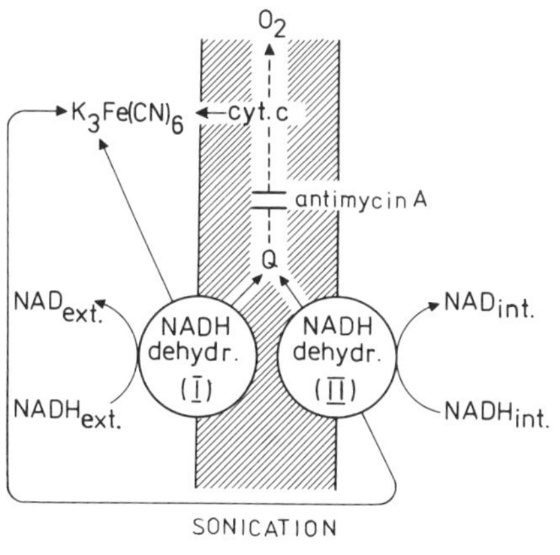

Fig. 4. Types of NADH dehydrogenase in yeast mitochondria (adapted from Von Jagow and Klingenberg, 1970). NADH dehydrogenase I directly interacts with NADH and ferricyanide in the surrounding medium. NADH dehydrogenase II interacts with ferricyanide via a pathway that can be blocked with antimycin A, and becomes accessible to ferricyanide upon sonication of mitochondria.

reductase complex which includes, besides the flavoprotein, several iron-sulfur proteins. An attempt to locate these proteins in the membrane was made by Harmon and Crane (1976) who studied NADH oxidation by various acceptors in mitochondria and submitochondrial vesicles and the sensitivity of these redox processes to nonpenetrating hydrophilic chelators inactivating the membrane iron-sulfur components. From the results of these experiments it was concluded that iron-sulfur proteins are found both on M- and C-sides of the membrane, i.e. that NADH-ubiquinone reductase has a transmembrane arrangement. However, De Pierre and Ernster (1977) doubt the validity of this conclusion. Unlike Harmon and Crane, Case et al. (1976) failed to find any iron-sulfur centers belonging to the NADH-ubiquinone reductase segment of the respiratory chain on the membrane surface in either mitochondria or submitochondrial vesicles.

1.3.2.1b. Succinate Dehydrogenase. Succinate dehydrogenase was initially supposed to be entirely localized on the M-side of the inner membrane. This point of view was based in particular on the experimental data concerning the enzyme extraction from inside-out submitochondrial vesicles with its subsequent reincorporation into the vesicles which was accompanied by restoration of succinate oxidase activity, respiratory control, and ability to catalyze the energy-dependent transhydrogenase reaction coupled with succinate oxidation (Lee et al., 1969, 1971). Such an extraction performed in a mildly alkaline medium (pH 9.5) does not impair vesicular membrane integrity or catalytic properties. Indeed, after the removal of succinate dehydrogenase the vesicles retain respiratory control in NADH oxidation and the system necessary for the transport of K^+ ions across the membrane.

Independent data on the arrangement of succinate dehydrogenase in the inner mitochondrial membrane were obtained while elucidating the localization of the enzyme sites responsible for succinate oxidation and ferricyanide reduction. Since the rate-limiting step in succinate oxidation in mitochondria is its transport across the membrane (Quagliariello and Palmieri, 1968), it can be thought that the succinate-binding site is on the M-side of the membrane. It is also there that the site performing ferricyanide reduction is found. Indeed, this process is almost completely inhibited by antimycin A in mitochondria and only partially in submitochondrial vesicles (Estabrook, 1962; Walter and Lardy, 1964; Lee et al., 1967; Klingenberg and Buchholz, 1970; Grinius et al., 1971).

More detailed information in the succinate dehydrogenase sites interacting with ferricyanide is presented by Vinogradov et al. (1976). The authors believe that there are two such sites, one of them being marked by enhanced affinity for ferricyanide. This site is readily demonstrable in isolated succinate dehydrogenase but becomes inaccessible for ferricyanide on enzyme incorporation into the membrane. Special experiments have shown that the site in question is most probably the iron-sulfur center S-3 (Vinogradov et al., 1975). From this it follows that the S-3 center appears to be exposed neither on the M-side nor on the C-side of the membrane. This suggestion is consistent with the more recent data of Case et al. (1976).

On the basis of the results obtained in studies of magnetic interaction (cross relaxation or dipolar splitting phenomena of epr signals) between succinate-dehydrogenase nearest components, Ohnishi (1979) has proposed a spatial scheme of the enzyme (Figure 5) in which center S-1 is placed so as to interact with the flavin (the distance between them is estimated at 10 Å) and with center S-2. As to center S-3, it is not likely to be positioned near enough to S-1 to permit magnetic interaction between them. At the same time S-3 is in proximity of 10 Å to the spin-coupled QH˙QH˙pair.

As pointed out in Section 1.3.1.2, succinate-coenzyme Q reductase, besides the flavoprotein and the iron-sulfur protein, includes a coenzyme Q-binding protein QP_S having a molecular weight of 13,000-15,000. The work of Merli et al. (1979) shows that of all components of the complex only one polypeptide (molecular weight 13,500) interacts with a nonpenetrating label diazobenzene [^{35}S]-sulfonate on the C-side of the inner mitochondrial membrane. These data together with those on the ability of individual polypeptides of succinate-ubiquinone reductase to interact with arylazidophospholipids (Girdlestone et al., 1981) served as a basis for the model of succinate-ubiquinone reductase in which the complex spans the membrane, the FP_S flavoprotein is exposed on its M-side while the 13,500-molecular-weight polypeptide resembling QP_S comes out on the C-side.

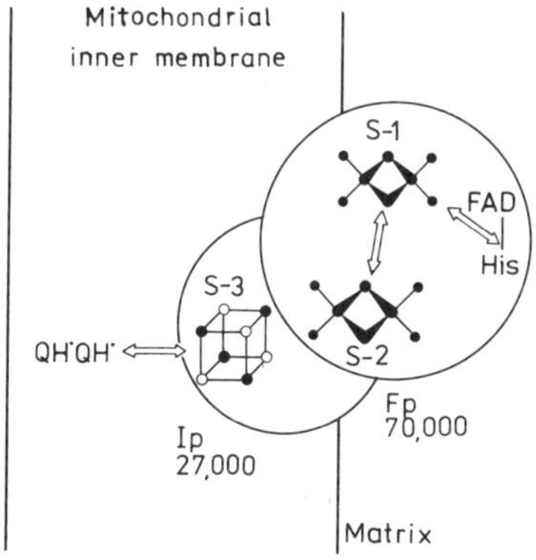

Fig. 5. Schematic disposition of succinate dehydrogenase in the inner mitochondrial membrane (after Ohnishi, 1979).

This model contradicts the data of Vignazia et al. (1981) who have shown that highly purified antibodies against the FP_s and IP_s subunits of succinate dehydrogenase recognize their antigenic counterparts exposed on both sides of the inner mitochondrial membrane. This implies that both subunits span the membrane.

1.3.2.1c. Cytochromes *b* and *c*$_1$, Rieske Iron-Sulfur Protein. Harmon et al. (1974) presented some indirect evidence for the localization of cytochrome c_1 on the C-side of the membrane. Schneider and Racker (1971) arrived at this conclusion on the basis of experiments with antibodies to cytochrome c_1. Later the same authors (Schneider and Racker, 1973) published results indicating that cytochrome c_1 is found both on C- and M-sides of the membrane. Such dual localization of cytochrome c_1 was considered an artifact resulting from its redistribution in the membrane during sonication of mitochondria to produce submitochondrial particles.

More recent data of Kiselev and Konstantinov (1976) rule out the possibility of cytochrome c_1 being localized on the M-side of rat-liver mitochondrial inner membrane. It was shown that extraction of cytochrome *c* with a hypotonic potassium chloride solution does not appreciably decrease the antimycin A-sensitive succinate-ferricyanide reductase activity of mitochondria, while their succinate oxidase activity falls almost 30-fold.

The C-surface position of cytochrome c_1 is consistent with and hence probably determined by its structure. By subjecting Neurospora crassa cytochrome c_1 to limited proteolysis, Li et al. (1981a) found that it consists of a large hydrophilic moiety (molecular weight 24,000) carrying the heme and interacting with cytochrome *c*, and a small hydrophobic moiety serving to anchor cytochrome c_1 to the lipid bilayer.

A rather promising technique for locating cytochromes, iron-sulfur proteins, and coenzyme Q in the mitochondrial inner membrane was proposed by Case and Leigh (1976) and Case et al. (1976). The method is based on the fact that paramagnetic ions (e.g., Ni^{2+}), being adsorbed by mitochrondria or inside-out submitochondrial vesicles, can change (owing to dipole-dipole interactions) the epr spectra of the electron carriers provided that they are at a distance of not more than 1 nm from the paramagnetic centers of the carriers. As follows from the experiments of Case and Leigh (1976), the hemin groups of cytochromes *b*-562, *b*-566, and c_1 are accessible for the aqueous medium only from the C-side of the membrane. As to the iron-sulfur centers of the bc_1 segment, one of them (the high-potential iron-sulfur center *bc*-3) is found on the surface of the C-side whereas all others have no access to the aqueous medium either on the C-side or on the M-side of the membrane.

It must be emphasized that the lack of effect of Ni^{2+} ions on the epr spectrum of a certain cytochrome or iron-sulfur protein does not necessarily mean that this component does not reside near the membrane surface. It is quite possible that the paramagnetic center of the electron carrier is in a hydrophobic environment, and other functional groups of the carrier directly interact with reagents in the surrounding medium. It is possibly just for this reason that Case and co-workers failed to find cytochromes *c*, *a*, and a_3 on either side of the membrane; these results are in conflict with the data obtained for these cytochromes by other methods (see below).

The localization of cytochrome *b*-566 according to Case and Leigh (1976) is in agreement with the data of Grigolava and Konstantinov (1977) who found that in rat liver mitochondria cytochrome *b*-566 is susceptible to inactivation by a nonpenetrating reagent p-diazobenzene sulfonate and is consequently exposed on the C-side of the inner mitochondrial membrane. This work, however, leaves it unknown whether there are any *b*-type cytochromes on the M-side of the membrane.

Finally, in a short communication Di Jeso (1978) reports that antibodies against cytochrome *b* suppress electron transfer in intact mitochondria but are ineffective with inside-out submitochondrial vesicles. These data also testify to cytochrome *b* being localized on the C-side of the inner mitochondrial membrane. Unfortunately, this report does not answer the question of the particular type of cytochrome *b* under consideration: *b*-562 or *b*-566.

In recent years data have appeared implying a transmembrane position of cytochrome *b*. Figure 6 depicts the schematic arrangement of the components of complex III in the beef-heart mitochondrial inner membrane, which was envisaged by Capaldi's group (Bell et al., 1979) from the data on the binding of diazobenzene [^{35}S]sulfonate to individual polypeptides in mitochondria and inside-out submitochondrial vesicles, and from the results obtained with the intermolecular linking technique. As follows from their scheme, most of the components (Rieske iron-sulfur protein, core protein I, cytochrome c_1 and polypeptide VII associated with it) have access only to the outer side of the inner mitochondrial membrane. Cytochrome *b* and core protein II span the membrane; none of the subunits of the complex is situated exclusively on the M-side of the membrane.

Rather similar results have later been reported by Beattie et al. (1981) for yeast mitochondria. The two core proteins and cytochrome *b* have been shown to be labeled with [^{125}I]diazobenzene sulfonate both in mitoplasts and in inside-out submitochondrial vesicles. Consequently, these proteins must span the inner mitochondrial membrane. In contrast, cytochrome c_1 and the iron-sulfur protein are labeled in mitoplasts but not in submitochondrial vesicles; hence they are positioned on the C-side of the membrane. Going

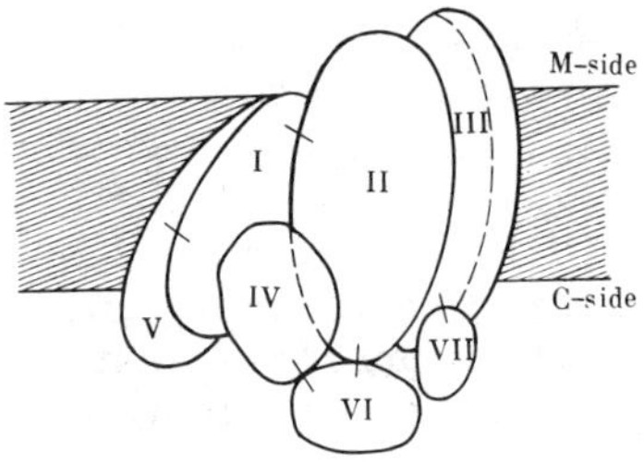

Fig. 6. Arrangement of complex III in the inner mitochondrial membrane. (I, II) Core proteins; (III) cytochrome *b*; (IV) cytochrome c_1; (V) Rieske iron-sulfur protein; (VI) polypeptide associated with cytochrome c_1; (VII) polypeptide associated with cytochrome *b*. The sites of intermolecular link formation between individual polypeptides are indicated by lines. Redrawn after Bell et al. (1979).

back to cytochrome *b*, it should be said that Chen and Beattie (1981) found two forms (molecular weights 31,000 and 29,000) for it differing by a single amino acid sequence of 2000 daltons. Experiments with a nonpenetrating label [^{125}I]diazobenzene sulfonate revealed that both forms of cytochrome *b* are exposed in mitochondria (i.e. on the C-side of the inner mitochondrial membrane) and only the 29,000-molecular-weight form is exposed in submitochondrial vesicles (i.e. on the membrane M-side).

Some other material pertaining to the arrangement of the ubiquinol-cytochrome *c* reductase segment of the respiratory chain (see the scheme in Section 1.3.1) in the inner mitochondrial membrane can be found in Mendel-Hartvig and Nelson (1978) and Li et al. (1981b).

1.3.2.1d. Cytochrome *c*. Cytochrome *c* is held to be localized on the outer side (C-side) of the inner membrane in intact mitochondria and accordingly on the interior surface of the inside-out (e.g., ultrasonic or alkaline) submitochondrial vesicles. This issue demands especially thorough argumentation since, as pointed out above, the position of cytochrome *c* is used as a biochemical criterion for assessing membrane orientation in various mitochondrial preparations. Presented below are only the most essential and representative facts underlying the generally accepted point of view.

First of all it should be noted that cytochrome *c* can be easily extracted from intact mitochondria by washing them with dilute solutions of potassium chloride. As a result of such treatment, the mitochondria almost completely lose their respiratory activity (Jacobs and Sanadi, 1960; MacLennan et al., 1966). On the other hand, cytochrome *c* is not removed from ultrasonic inside-out submitochondrial vesicles upon washing of the latter with an isotonic potassium chloride solution (Lenaz and MacLennan, 1966; Lee, 1971). In conformity with this, exogenous cytochrome *c* activated respiration in mitochondria but not in the inside-out submitochondrial vesicles (Crane et al., 1956; Lenaz and MacLennan, 1966; MacLennan et al., 1966).

As shown by Jacobs and Sanadi (1960) and by Estabrook (1961), cytochrome *c* stimulates ferricyanide reduction and ferrocyanide oxidation in mitochondria. On the basis of these and some other data it was suggested that the compounds mentioned interact with the respiratory chain at the level of cytochrome *c* (Estabrook, 1961). Taking into account that ferricyanide and ferrocyanide do not penetrate the inner mitochondrial membrane (Klingenberg and Buchholz, 1970), it can be concluded that cytochrome *c* must be situated on the C-side of the membrane.

More direct evidence for the localization of cytochrome *c* was furnished by experiments with antibodies to this protein and with a nonpenetrating radioactive label diazobenzene [^{35}S]sulfonate. Di Jeso et al. (1969) and Racker et al. (1971) ascertained that antibodies against cytochrome *c* inhibit NADH oxidase, succinate oxidase, and phenazine methosulfate oxidase activities in mitochondria but have practically no effect on the same activities in ultrasonic submitochondrial vesicles. This signifies that cytochrome *c* is on the C-side of the membrane. Schneider et al. (1972) subjected submitochondrial vesicles, mitochondria, and cytochrome *c* isolated therefrom to treatment with diazobenzene [^{35}S]sulfonate. After extracting cytochrome *c* from the labeled membranes, its radioactivity was measured in all three cases. The radioactivity levels for cytochrome *c* from mitochondria, submitochondrial vesicles, and in isolation turned out to be related as 1:0.13:1. The results presented argue in favor of cytochrome *c* being confined to the outer side (C-side) of the inner mitochondrial membrane. Incorporation of some label into cytochrome *c* of submitochondrial vesicles may mean that either the submitochondrial preparation used contained some mitochondria, or the membrane of the submitochondrial vesicles was to some extent permeable for diazobenzene sulfonate.

1.3.2.1e. Cytochromes *a* and a_3. Important information about the position of cytochromes *a* and a_3 in the inner mitochondrial membrane was gained in experiments with antibodies raised against highly purified cytochrome *c* oxidase. According to the early data of Racker et al. (1971), such an antiserum produces an equal extent of cytochrome *c* oxidase inhibition in mitochondria and inside-out submitochondrial vesicles, testifying to a roughly similar accessibility of the enzyme from the M- and C-sides of the membrane, i.e. to its transmembrane arrangement. An obscure aspect in the work mentioned was that the inhibition did not exceed 57-65%. Rather similar results were reported by Di Jeso et al. (1969). The more recent studies of Hackenbrock and Hammon (1975) are free from this shortcoming. These authors have shown that a highly purified immunoglobulin of the G class monospecific for cytochrome *c* oxidase quickly and completely suppresses succinate oxidase and (ascorbate) cytochrome *c* oxidase activities on interaction with a biochemically and morphologically intact fraction of mitoplasts (inner membranes with matrix) derived from rat liver mitochondria, and with a homogeneous preparation of inside-out vesicles obtained by ultrasonic treatment of this fraction. These data gave grounds for concluding that the entire cytochrome *c* oxidase participating in succinate and ascorbate oxidation is completely accessible for antibodies from both sides of the inner mitochondrial membrane, and that there are no cytochrome *c* oxidase molecules with antigenic sites exposed only on one membrane side. In compliance with this, the same amount of the immunoglobulin was required to fully inhibit cytochrome *c* oxidase in mitoplasts and in the inside-out submitochondrial vesicles.

Another important issue evolving from the work of Hackenbrock and Hammon is that cytochrome *c* oxidation takes place only on the outer side of the membrane. Indeed, an immunoglobulin monospecific for cytochrome *c* immediately and completely blocks succinate or ascorbate oxidation, as well as ferricyanide reduction, by mitoplasts but in no way affects the rate of succinate oxidation by the vesicular preparation with inside-out membranes. With certain allowances this may be interpreted to mean that all molecules of cytochrome *a* (which is directly involved in cytochrome *c* oxidation) come out on the C-side of the inner mitochondrial membrane.

As addition of cytochrome *c*, or antibodies against cytochrome *c*, to inside-out vesicles has no effect on their succinate oxidase activity, cytochrome *c* can be supposed to take no part in electron transfer from the M-side of the inner mitochondrial membrane. At the same time, the immunoglobulin specific for cytochrome *c* oxidase completely inhibits the succinate oxidase and (ascorbate) cytochrome *c* oxidase activities of the inside-out vesicles. Assuming that the immunoglobulin exercises its inhibitory action by binding to an active site on cytochrome oxidase, it must be concluded that this site must be cytochrome a_3 directly involved in oxygen reduction.

Another proof of the localization of cytochrome *a* on the C-side of the inner mitochondrial membrane was presented by Konstantinov et al. (1975). According to their results, the basic protein protamine efficiently suppresses the cytochrome *c* oxidase activity of mitochondria and only slightly decreases that of inside-out submitochondrial vesicles. As protamine hinders electron transfer between cytochrome *c* and cytochrome *a*, it is to be thought that the latter, like cytochrome *c*, resides on the C-side of the membrane.

As an independent line of support for the transmembrane arrangement of cytochrome *c* oxidase, one can mention the results of experiments with diazobenzene [^{35}S]sulfonate, a reagent that does not penetrate the membrane. Schneider et al. (1972) have shown that cytochrome *c* oxidase preparations isolated from mitochondria or ultrasonic submitochondrial vesicles treated with this reagent display close levels of radioactivity, which indicates an approximately equal extent of the enzyme accessibility from both sides of the membrane.

Thus, experimental data are available (see also a review by Harmon et al., 1974) which make it possible to assume that cytochrome *a* is on the C-side while cytochrome a_3 is on the M-side of the inner mitochondrial membrane. In recent years the 'classical' scheme of cytochrome *c* oxidase arrangement in the membrane (see, e.g., Racker, 1970) has been undergoing reconsideration. In this connection one should first of all cite the work of Artsatbanov et al. (1977) who have shown the normal redox potential of heme *a* to depend on the activity of H^+ ions in the mitochondrial matrix but not in the incubation medium. Taken together with the above-listed evidence that cytochrome *a* oxidizes cytochrome *c* on the outer side of the mitochondrial membrane, this implies a transmembrane arrangement no longer of the whole cytochrome oxidase but of the cytochrome *a* monomer.

Bearing in mind the complex subunit composition of cytochrome *c* oxidase, we shall have to dwell on what is known about the disposition of its individual polypeptides in the inner mitochondrial membrane. Eytan et al. (1975) took beef heart mitochondria and inside-out submitochondrial vesicles derived therefrom and specially purified, treated them with diazobenzene [^{35}S]sulfonate, then dispersed the membrane with cholate, isolated cytochrome *c* oxidase by immunoprecipitation, resolved its subunits by sodium dodecyl sulfate-polyacrylamide gel electrophoresis, and assayed the radioactivities of individual fractions. It turned out that in mitochondria the label is preferentially incorporated in subunits II, V, and VI, while in submitochondrial particles it concentrates in subunit III. In the light of the data of Downer et al. (1976) this notation should be changed as follows: I→I, II→II+III, III→IV, IV→V, V→VI, and VI→VII. With these corrections, the data of Eytan et al. (1975) mean that subunits II, III, VI, and VII are on the C-side of the inner mitochondrial membrane, subunit IV is on its M-side, and subunits I and V are buried in the lipid bilayer (Table 3).

Using a similar technique, Ludwig et al. (1979) showed later that in mitochondria diazobenzene [^{35}S]sulfonate reacts with cytochrome *c* oxidase subunits II and III, that in inside-out submitochondrial vesicles subunits II, III, IV, V, and VII are labeled, and that subunits I and VI are labeled in neither case (Table 3).

Like results were reported by Chan and Tracy (1978) who studied the inhibition of cytochrome *c* oxidase activity in mitochondria and

Table 3. Topology of Cytochrome *c* Oxidase* Subunits According to the Data of Experiments with Hydrophilic and Hydrophobic Labels (adapted from Azzi, 1980). C, Cytosolic side; M, matrix side; +, positive reaction; −, nonreactive.

Labeling technique	Subunits							Reference
	I	II	III	IV	V	VI	VII	
Hydrophilic labels:								
diazobenzenesulfonate	−	C	C	M	−	C	C	Eytan et al. (1975)
diazobenzenesulfonate	−	C	C+M	M	M	−	M	Ludwig et al. (1979)
antibodies to subunits	−	C	C	M	C+M	−	C+M	Chan and Tracy (1978)
NAP-taurine	+	++	++	++	+	++	++++	Prochaska et al. (1980)
Hydrophobic labels:								
azidophospholipids	+++	++	++++	++	−	−	++++	Bisson et al. (1979); Prochaska et al. (1980)
iodonaphthylazide	+++	++	++++	−	−	−	++++	Cerletti and Schatz (1979)

* All data are for cytochrome *c* oxidase from beef heart mitochondria.

mitoplasts by antisera against individual subunits of the enzyme, and the interaction of inside-out submitochondrial vesicles with ^{125}I-labeled immunoglobulins to individual subunits. By their data, subunits II and III are found exclusively on the C-side while subunits I and VI are buried within the membrane. As to subunits V and VII, they were detectable on both sides of the inner mitochondrial membrane (Table 3).

Additional information on the topology of cytochrome *c* oxidase subunits in the mitochondrial inner membrane was obtained in experiments with probes for the parts of integral membrane proteins intercalating the lipid bilayer. These probes are labeled derivatives of phospholipids (Bisson et al., 1979; Prochaska et al., 1980) or naphthalene (Cerletti and Schatz, 1979) which are capable of covalent binding to polypeptides showing enhanced affinity for these hydrophobic compounds. As shown by Bisson et al. (1979) and by Prochaska et al. (1980), 'deep' and 'shallow' probes (arylazidophospholipids) react with all subunits of cytochrome *c* oxidase built into lecithin vesicles except subunits V and VI. The labels are best of all incorporated into subunits I, III, and VII, to a lesser extent into subunit II, and still worse into subunit IV (Table 3). Subunit II is better labeled with a 'deep' probe whereas subunit VII with a 'shallow' one. Cerletti and Schatz (1979) obtained similar results using naphthylazide labels.

Summing up the data obtained with hydrophilic labels [especially those of Ludwig et al. (1979) and of Chan and Tracy (1978) which agree quite nicely] and with hydrophobic ones (Table 3), Bisson et al. (1979) proposed a scheme (Figure 7) in which subunit III spans the membrane, subunit II comes out on the C-side, subunit IV on the M-side, and subunit I is completely embedded in the lipid bilayer. As to the subunits of group VII (see Table 1), they are partly immersed in the surface zones of the bilayer on both sides of the membrane. The data on other subunits are still rather vague.

This scheme is consistent with what is known about the amino acid sequence of some subunits. Specifically, subunit III was found to have seven segments of 20-30 amino acids each composed exclusively of hydrophobic residues (Azzi, 1980). Such segments may traverse the bilayer, which is necessary for the hypothetical protonophoric function of subunit III. Subunit IV comprises 147 amino acid residues (Sacher et al., 1979) of mostly hydrophilic nature. However, it has a segment of 19 hydrophobic residues in positions 80-98 which may act as an anchor fastening this subunit to the surface of the lipid bilayer.

Finally, in support of the scheme of the topological arrangement of cytochrome *c* oxidase subunits in the membrane depicted in Figure 7, data can be presented on the binding of cytochrome *c*. As stated above, the latter is localized exclusively on the C-side of the

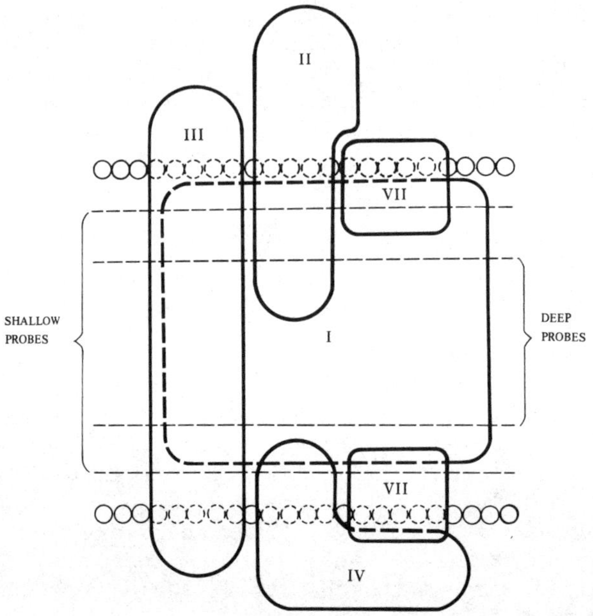

Fig. 7. Disposition of cytochrome *c* oxidase subunits in the inner mitochondrial membrane. Circles delimit the membrane lipid bilayer. Roman numbers denote individual subunits. After Bisson et al. (1979).

mitochondrial inner membrane. On the other hand, data are available that cytochrome *c* binds to cytochrome *c* oxidase subunits II and III. Indeed, in a series of works by Bisson et al. (1978a,b, 1980), a lysine-13 derivative of cytochrome *c* (arylazidocytochrome *c*) covalently binds to cytochrome oxidase subunit II, while cysteine-107 of iso-1-cytochrome *c* activated with 5,5'-dithiobis(2-nitrobenzene) can be linked to a cysteine in subunit III (Birchmeier et al., 1976). This means that both subunit II and subunit III are exposed to the aqueous phase on the C-side of the membrane.

1.3.2.1f. ATPase Complex. Let us first of all consider the question of the localization of F_1-ATPase in the inner mitochondrial membrane. The most convincing results appear to have been obtained in experiments with antibodies to F_1 and with diazobenzene [^{35}S]-sulfonate - reagents that do not penetrate the membrane and can only interact with components having surface position. As shown in Racker's laboratory (Christiansen et al., 1969; Racker et al., 1971), antibodies against F_1 suppress the phosphorylating ability of inside-out submitochondrial vesicles but do not appreciably affect phosphorylation in intact mitochondria. These data are evidence for the accessibility of the ATPase for antibodies on the M-side and its absence from the C-side of the membrane. In accordance with this, treatment of inside-out submitochondrial vesicles with diazobenzene [^{35}S]sulfonate entailed considerable labeling of F_1, which could be easily detected by measuring the radioactivity of the factor after its isolation from the particles. With mitochondria, label incorporation into F_1 was negligible whereas cytochrome *c* localized on the outer surface of the mitochondrial membrane acquired a high degree of radioactivity (Schneider and Racker, 1971; Schneider et al., 1972).

Harmon et al. (1974) attempted to get information on the position of F_1 in the inner mitochondrial membrane by using the ATPase protein inhibitor. It was supposed that since the inhibitor poorly penetrates the membrane, it would be more effective with inside-out submitochondrial vesicles than with mitochondria if F_1-ATPase is localized on the M-side of the membrane. Indeed, the ATPase of the inside-out vesicles was shown to be affected by the inhibitor to a greater extent than that of mitochondria. Chan and Barbour (1976) also treated rat liver mitochondria and inside-out submitochondrial vesicles derived therefrom with the ATPase protein inhibitor and found that the inhibitors has no effect on the mitochondrial ATPase activity in conditions when the activity of the vesicles is completely blocked.

Trypsin has also been used as a nonpenetrating reagent (see Maïrouch and Godinot, 1977). It promotes F_1-ATPase detachment from inside-out submitochondrial vesicles but not from the membrane of mitoplasts (a preparation of normally oriented inner membranes enclosing mitochondrial matrix) obtained from porcine heart mitochondria. This further confirms the suggestion that F_1 is localized on the M-side of the inner mitochondrial membrane.

The experimental data available at present allow us to discuss not only the general aspects but also some details of the arrangement of factor F_1 in the inner mitochondrial membrane. As pointed out above, the negatively contrasted inside-out submitochondrial vesicles exhibit on their surface numerous "mushroom" formations consisting of spheres about 90 Å in diameter attached to the membrane by cylindrical "stalks" some 50 Å long and 30-40 Å thick. The same formations are observable on the M-side of the inner membrane in mitochondria (see, e.g., Fernández-Morán et al., 1964). By the data of Racker's laboratory, the spheres represent F_1-ATPase. In particular, exposure of submitochondrial vesicles containing tritium-labeled F_1 to urea was shown to cause a loss of the label (Kagawa and Racker, 1966b) and ATPase activity (Racker et al., 1965; Racker and Horstman, 1967) by the vesicles concomitantly with the removal of the spheric particles from their surface. Some salts (e.g., KI or KSCN) speed up the cold inactivation of ATPase in submitochondrial vesicles which is due to enzyme dissociation into subunits; by the results of morphological examinations, this is accompanied by complete disappearance of the "mushroom" particles (Kagawa and Racker, 1966b). It should also be noted that the dimensions of isolated ATPase correspond to the dimensions of these particles (Racker et al., 1965; Kagawa and Racker, 1966b), and that addition of the enzyme to vesicle preparations treated with ultrasound, urea, or silicotungstate and thereby depleted of the "mushroom" particles, restores their initial appearance (Racker et al., 1965, 1969; Kagawa and Racker, 1966b; Racker and Horstman, 1967).

Thus, the question of the identity of "mushroom" formations observed on the M-side of the mitochondrial inner membrane or on the exterior of inside-out submitochondrial vesicles may now be considered settled. The same unfortunately cannot be said of the actual position of the 90 Å-diameter particles in intact membranes. It has long been maintained that the "mushroom" protrusions are created on the membrane surface in the course of the negative-contrasting procedure. However, according to the more recent data of several authors, such membrane structural elements are also demonstrable in preparations of mitochondria or submitochondrial vesicles fixed with glutaraldehyde and resin-embedded for subsequent making of thin sections (Ashhurst, 1965; Stiles and Crane, 1966; Teleford and Racker, 1973).

New difficulties in interpreting the results cropped up with the introduction into the electron-microscopic practice of the "freeze-fracture" and "freeze-etching" techniques. It is presumed that, since these methods employ no chemicals for specimen processing, they produce results that to the most degree approximate the real mitochondrial structure. Nevertheless, the results obtained thus far are controversial. In fact, according to Wrigglesworth et al. (1970), the inside-out submitochondrial vesicles have a smooth surface devoid of spherical particles. On the other hand, Vail et al. (1972) have found such particles on the M-side of the inner mitochondrial membrane. Fikhte and Zaichkin (1972) were also able to discern on the mitochondrial inner-membrane surface certain particles having an appearance of cylindrical bodies (diameter 100 Å, length 150 Å) with rounded free ends. These particles differed in shape from those usually observed in negative contrast. In particular, they lacked the characteristic distal "mushroom" thickenings. An analogous conclusion was arrived at by Garber and Stepokus (1974) on examining the structure of chloroplast thylakoids isolated from spinach leaves. These authors have shown by negative contrast and freeze-fracture that the thylakoid exterior surface carries the coupling factor F_1 particles which can be reversibly removed by treating thylakoids with silicotungstate. It is interesting that using the deep-etching technique Garber and Stepokus failed to detect in thylakoids a component playing the part of the "stalk"; at the same time it was distinctly seen in negatively contrasted specimens.

Hatase and Oda (1972) associate the presence or absence of "mushroom" formations on the surface of submitochondrial vesicles with the energetic state of the latter. According to these authors, vesicle energization with ATP and succinate leads to the disappearance of "mushrooms", and subsequent de-energization elicits their reappearance. These data imply conformational changes in the inner mitochondrial membrane, which are attended by sinking of the spherical particles into the membrane depth and their subsequent ejection, and as a consequence, by variations of approximately 30 Å in the membrane thickness (Harmon et al., 1974).

Advancing the idea of the fluctuating character of F_1-ATPase position in the inner mitochondrial membrane depending on the extent of membrane energization, one should also refer to Kozlov and Skulachev (1977) who are of the opinion that operation of the mitochondrial ATPase involves such rearrangements of the ATPase complex as a whole that factor F_1 moves in the direction perpendicular to the membrane-water interface. These authors believe the "mushroom" protrusions to correspond only to a certain functional state of the ATPase complex.

Though a body of evidence is available for the relatively free access to F_1-ATPase from the matrix and its relative inaccessibility for nonpenetrating reagents from the C-side of the mitochondrial membrane (see above), this issue requires additional clarification. It is quite possible that factor F_1 is to a considerable degree submerged into the membrane. Indeed, Kozlov and Chernyak (1976) showed that the catalytic site of isolated ATPase can be easily blocked with hydrophilic reagents, N-cyclohexyl-N'-β-(4-methylmorpholinium) ethylcarbodiimide (CMCD) and a mixed anhydride of ATP and mesitylenic acid (ATP-MA). Neither of these compounds inhibits ATPase in mitochondria or inside-out submitochondrial vesicles. With mitochondria, sensitivity to the inhibitors is elicited on addition of butanol to the medium by virtue of the appearance of free proton conductance between the ATPase catalytic site and the aqueous medium. In submitochondrial vesicles butanol produces no significant increase in ATPase sensitivity to CMCD or ATP-MA, indicating the existence of a strong hydrophobic barrier between the ATPase catalytic site and the aqueous medium. These data were considered a proof of the transmembrane arrangement of factor F_1. Since butanol sensitizes ATPase to the inhibitors in mitochondria but not in inside-out submitochondrial vesicles, the catalytic site of the enzyme can be supposed to be nearer to the C-side of the inner mitochondrial membrane (Figure 8).

Positions of other components of the ATPase complex in the inner mitochondrial membrane are even less studied. Some time ago there was a wide discussion on the localization of the protein endowing the soluble ATPase with sensitivity to oligomycin. One of the viewpoints is that the protein (OSCP) serves as the "stalk" in the "mushroom" protrusion. Indeed, MacLennan and Asai (1968) showed that in the absence of OSCP, F_1 is poorly bound to the membrane sector of the ATPase complex. On addition of OSCP to the two other components of

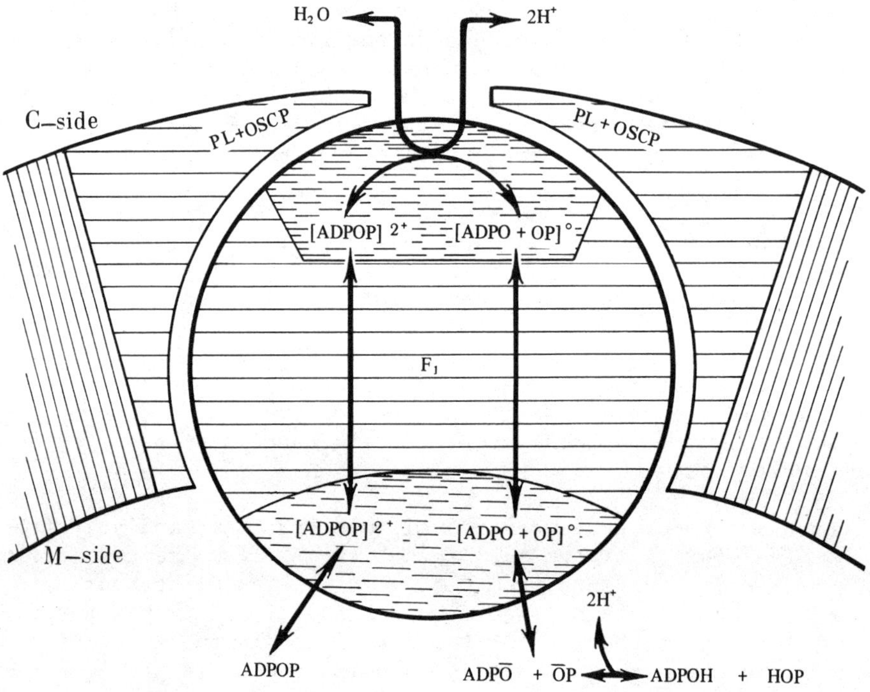

Fig. 8. A scheme of the arrangement of the ATPase complex in the inner mitochondrial membrane. (PL) Proteolipids; (F_1) ATPase (factor F_1); (OSCP) oligomycin-sensitivity-conferring protein. After Skulachev and Kozlov (1977).

the complex, "mushroom" structures appear, and the ATPase acquires oligomycin sensitivity and cold stability, i.e. the usual features characteristic of the enzyme bound to the intact mitochondrial membrane. Besides this, one may be reminded that factor F_1 forms a complex with OSCP (Van de Stadt et al., 1972) and that isolated OSCP can form aggregates (tetramers) which have dimensions corresponding to those of the "stalk" of the "mushroom" protrusions (MacLennan and Asai, 1968).

The point of view presented above was subject to criticism (Racker, 1970). Bulos and Racker (1968) argued that mitochondrial vesicles devoid of factor F_c, which performs the functions of OSCP, readily bind F_1, and the reconstituted vesicles are morphologically indistinguishable from the initial ones containing all the ATPase-complex components. Later Russell et al. (1976) isolated from the mitochondrial membrane fraction a specific protein with a molecular weight of 8500 that promoted F_1 binding to submitochondrial vesicles (the so-called TUA particles) deprived of OSCP. At the same time isolated OSCP displayed no such effect. On this basis it was concluded that OSCP can hardly serve as a linkage between F_1 and the membrane. Glazer et al. (1977a) also decided that different proteins are responsible for F_1 binding to the membrane and its susceptibility to oligomycin inhibition. In particular, they found that trypsinization of complex V (ATP synthetase complex) from which F_1-ATPase has been preliminarily removed yields a "membrane sector" which is still able to bind F_1, but the reconstituted system no longer shows oligomycin sensitivity.

It should be admitted that all the results listed provide persuasive evidence only for that the "membrane sector" of the ATPase complex includes a certain protein essential for F_1 binding to the inner mitochondrial membrane. Yet they do not rule out that OSCP is also necessary for proper binding and that it enters into the composition of the "mushroom stalk." This agrees with the model of Skulachev and Kozlov (1977) which implies that when factor F_1 is submerged into the membrane, OSCP and the proteolipids of the membrane sector form a hydrophobic sheath enclosing F_1 (see Figure 8). This model is capable of explaining the results of Maïrouch and Godinot (1977) on the decline of the ATPase sensitivity to oligomycin upon tryptic treatment of mitoplasts. These data were interpreted as indicating hydrolysis of a certain protein situated in the vicinity of the exterior surface (C-side) of the inner mitochondrial membrane and at least indirectly responsible for the oligomycin sensitivity of the ATPase complex.

Direct information on the localization of individual H^+-ATPase components in the membrane could be gained in experiments with nonpenetrating labels. As shown by Ludwig et al. (1980), treatment of inside-out submitochondrial vesicles with diazobenzene [^{35}S]sulfonate brings about labeling of F_1-ATPase β subunit, and to a lesser extent α, γ, and ε subunits. The label is also incorporated into OSCP and at least one polypeptide of of F_0 sector (molecular weight 25,000). At the same time, exposure of intact mitochondria to this reagent is accompanied by labeling of a single membrane-sector component (molecular weight 12,300) of unknown function. Thus at present it can only be said that one component of the membrane sector is positioned on the C-side of the inner mitochondrial membrane. This information does not seem to be exhaustive; among other things, it is to be expected from the concept of the protonophore function of the DCCD-binding protein that it also has access to the C-side of the membrane.

1.3.2.1g. Transhydrogenase. Summing up the data on the properties of the mitochondrial transhydrogenase, Ernster (1975a) came to the conclusion that the pyridine nucleotides NAD(H) and NADP(H) interact with the enzyme on the M-side of the inner mitochondrial membrane, and this does not have to be preceded by nucleotide transport across the membrane. This issue is supported by the more recent results of Höjeberg and Rydström (1977). On the other hand, Sweetman et al. (1974) studied the inhibition of transhydrogenase activity by 2-phenylisatogen and found it to be very weak with inside-out submitochondrial vesicles. The extent of inhibition increased 5-fold if 2-phenylisatogen was added to mitochondria during sonication and was hence trapped within the resulting submitochondrial vesicles. The results of Sweetman et al. (1974) indicate that on the C-side of the inner mitochondrial membrane there is a certain site of the transhydrogenase that is responsible for 2-phenylisatogen binding. Thus, the experimental data available at present are compatible with the transmembrane arrangement of the transhydrogenase.

1.3.2.1h. Adenine Nucleotide Carrier. Since atractyloside and carboxyatractyloside do not penetrate the membrane, it was possible to elucidate the position of the adenine nucleotide carrier with respect to the membrane outer and inner sides. According to Vignais (1976), there is the following experimental evidence for the sites responsible for inhibitor binding being situated on the C-side of the membrane. First, atractyloside impedes nucleotide transport in intact mitochondria but not in inside-out submitochondrial vesicles; at the same time atractyloside proves active when enclosed in the inside-out vesicles during their preparation. Secondly, sonication of mitochondrial in the presence of atractyloside or carboxyatractyloside does not increase the amount of bound inhibitor, which would have been the case if the binding sites occurred on both outer and inner sides of the inner mitochondrial membrane.

Later more direct proof was presented for the adenine nucleotide carrier being on the C-side of the inner mitochondrial membrane. As shown by Boxer et al. (1977), upon iodination of beef heart mitochondria via the lactoperoxidase system, which takes place mostly on the C-side of the inner membrane, up to 85% of the radioactive iodine is associated with a polypeptide having a molecular weight of 29,500. This polypeptide behaves quite similarly to the carboxyatractyloside-binding protein at different stages of purification and therefore can be thought to be identical to it.

With all this, some peculiarities of the primary sequence of the adenine nucleotide carrier (Aquila et al., 1982) suggest that this protein may traverse the inner mitochondrial membrane. Indeed, it contains at least two hydrophobic segments of more than 20 amino acid residues. Such segments can form helices of lengths exceeding the thickness of the membrane bilayer. Besides, the carboxyatractyloside-binding protein incorporated into liposomes is known to be able to carry out, albeit at a rather low rate, the transport of adenine nucleotides, which points to its transmembrane position (Boxer et al., 1977).

1.3.2.1i. Inorganic Phosphate Carrier. The data on the arrangement of the inorganic phosphate carrier in the inner mitochondrial membrane are very scanty. According to Rhodin and Racker (1974), phosphate uptake by submitochondrial vesicles is suppressed by a hydrophobic compound, N-benzyl maleimide, capable of penetrating lipid membranes, but is insensitive to its hydrophilic analog N-ethyl maleimide. This fact was considered evidence for the asymmetric orientation of the phosphate carrier in the membrane.

SUMMARY

Thus, to date the position in the inner mitochondrial membrane has been more or less reliably determined for about half of the twoscore of its polypeptide constituents. It can be stated with certainty that some proteins are exposed on the M-side of the membrane (FP_N; FP_S; α, β, γ, and ε subunits of F_1-ATPase; OSCP; subunits IV and VII of cytochrome oxidase), others on the C-side (PQ_S; cytochrome *c*; cytochrome c_1; Rieske iron-sulfur protein; subunit II of cytochrome *c* oxidase), and still others span the membrane (cytochrome *b*; subunit III of cytochrome *c* oxidase; adenine nucleotide carrier). Such an arrangement provides for the transmembrane electron transfer and coupled proton transport, giving rise to the electrochemical proton potential on the membrane (Mitchell, 1966, 1972).

1.3.2.2. *Oligoenzyme Complexes of the Inner Mitochondrial Membrane*

1.3.2.2a. Complexes Constituting the Respiratory Chain. As already outlined in this chapter, many of the respiratory-chain protein components appear to be subintegrated into complexes that can catalyze the partial redox reactions of the electron-transfer system. Such preparations of various compositions and different degrees of complexity can be obtained by mild fragmentation of the inner mitochondrial membrane followed by separation of the fragments. The task of isolating and purifying the oligoenzyme complexes that constitute the respiratory chain has been dealt with in most detail in Green's laboratory, where methods have been developed for isolating NADH-ubiquinone reductase (complex I, see Hatefi et al., 1962a), succinate-ubiquinone reductase (complex II, see Ziegler and Doeg, 1962), ubiquinol-cytochrome *c* reductase (complex III, see Hatefi et al., 1962b), cytochrome *c* oxidase (complex IV, see Griffiths and Wharton, 1961), and two larger fragments: NADH-cytochrome *c* reductase (complex I + III, see Hatefi et al., 1961) and succinate-cytochrome *c* reductase (complex II + III, see Tisdale et al., 1963).

In the years that followed these studies, the complexes became conventional objects of investigation in many laboratories. Among the publications in which isolation techniques were improved or modified to suit new sources of mitochondria, one can mention: for complex I Hatefi (1978); for complex II Hatefi and Stiggall (1978); for complex III Siedow et al. (1978), Engel et al. (1980), Bill et al. (1982); for complex IV Hartzell et al. (1978), Rubin and

Tzagoloff (1978), Weiss and Sebald (1978), Bill et al. (1982); for complex II+III Yamashita and Racker (1969).

Some features of the complexes, and in particular the data on their composition, are given in Table 4. Based on these data together with those (Vinogradov and King, 1979) concerning the contents of individual redox components in mitochondrial inner membrane preparations (submitochondrial vesicles), the following estimates are obtained for the proportion of each complex in beef heart mitochondria: complex I 7-8%, complex II 3%, complex III 7-8%, and complex IV 11-12% of the overall membrane protein.

Green and co-workers (Green et al., 1965; Green and Tzagoloff, 1966) suggested that the complexes are the actual catalytic units of the membrane. The following experimental facts argue in favor of this concept. Firstly, the stoichiometric relations between the respiratory-chain components in the complexes correspond to those in whole mitochondria. This applies in particular to complexes II and

Table 4. Composition of Complexes I, II, III, and IV of the Respiratory Chain (after Hatefi and Galante, 1978).

Complex	Mol. wt.	Number of polypeptides	Component	Concentration (per mg protein)
I (NADH-ubiquinone reductase)	700,000	16-18	FMN	1.4-1.5 nmoles
			nonheme iron	23-26 ng-atoms
			labile sulfides	23-26 nmoles
			ubiquinone	4.2-4.5 nmoles
			lipids	0.22 mg
II (succinate-ubiquinone reductase)	200,000	4-6	FAD	4.6-5.0 nmoles
			nonheme iron	36-38 ng-atoms
			labile sulfides	32-38 nmoles
			cytochrome *b*	4.5-4.8 nmoles
			lipids	0.2 mg
III (ubiquinol-cytochrome *c* reductase)	250,000	8-9	cytochrome *b*	8.0-8.5 nmoles
			cytochrome c_1	4.0-4.2 nmoles
			nonheme iron	10-12 ng-atoms
			labile sulfides	6-8 nmoles
			ubiquinone	2 nmoles
			lipids	0.4 mg
IV (cytochrome *c* oxidase)	150,000	7(?)	cytochrome aa_3	10.5-10.9 nmoles
			copper	10-12 ng-atoms
			lipids	0.3-0.4 mg

IV (Green and Wharton, 1963; De Pierre and Ernster, 1977). Secondly, the dimensions of at least complexes III and IV, which are to a first approximation spherical particles 50-80 Å in diameter (Green and Tzagoloff, 1966; Sun et al., 1968; Vail et al., 1975), match the dimensions of protein globules most frequently encountered in the membrane (Wrigglesworth et al., 1970). Thirdly, isolated complexes can be further fragmented to individual proteins only under rather drastic conditions so that the dissociation proves irreversible. At any rate, none of the complexes have so far been reconstituted from the individual polypeptides entering into its composition (see Chapter Seven). Fourthly and finally, the complexes can under certain conditions be made to form membranous multienzyme systems that are capable of performing the functions of the respiratory chain including the transport of H^+ ions across the membrane and generation of the transmembrane potential.

Some notion of the general structure of some complexes was gained in studies on the "two-dimensional membrane crystals" of the complexes using electron microscopy and image-reconstruction techniques. According to Fuller et al. (1979), the cytochrome *c*-oxidase monomer in a two-dimensional crystal obtained by treating mitochondria with deoxycholate is some 110 Å long and resembles a lopsided letter Y (Figure 9). The domains forming the hands of the Y have about 55 Å (center-to-center). Both these domains are on the M-side of the membrane and are almost completely submerged in the lipid bilayer. The "cytoplasmic side" of cytochrome *c* oxidase is represented by a single large domain forming the leg of the Y. This region of the complex protrudes up to 60 Å from the bilayer (see also Vanderkooi, 1974; Henderson et al., 1977; Frey et al., 1978).

Similar experiments were carried out by Leonard et al. (1981) on membrane crystals of ubiquinol-cytochrome *c* reductase prepared by mixing a Triton X-100 solution of the complex with phospholipid-Triton micelles and then removing Triton. In this system the enzyme appears as a dimer in which the monomers are coupled by a second-order rotation axis perpendicular to the plane of the membrane (Figure 10). The protein within the complex is distributed so that

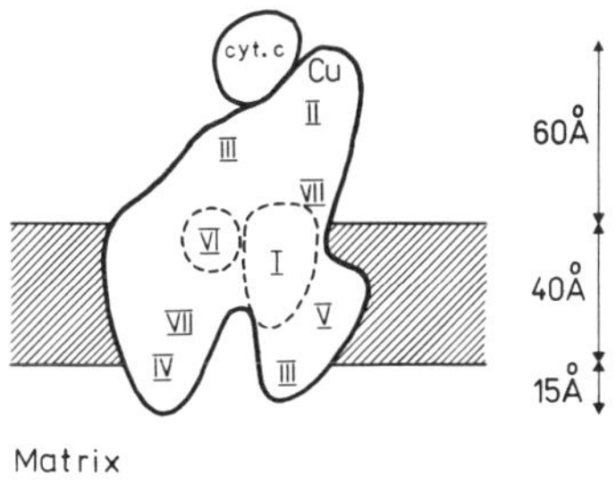

Fig. 9. Arrangement of cytochrome *c* oxidase in the inner mitochondrial membrane. After Azzi (1980).

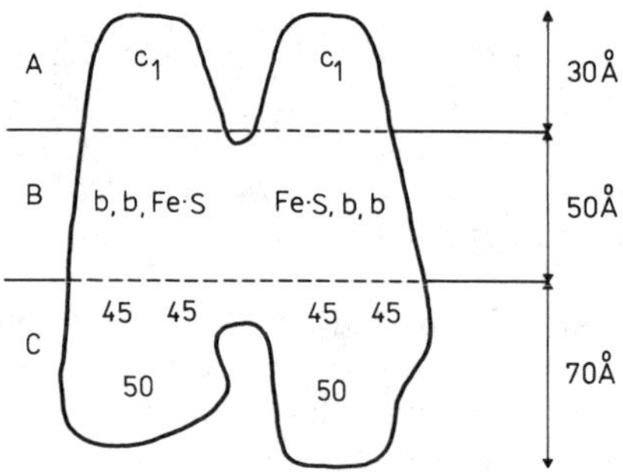

Fig. 10. Disposition of complex III in the phospholipid bilayer. (50) Core protein of molecular weight 50,000; (45) core protein of molecular weight 45,000; (b) cytochrome *b*; (c_1) cytochrome c_1; (Fe•S) Rieske iron-sulfur protein. The upper aqueous region (A) corresponds to cytochrome c_1 on the cytoplasmic side of the inner mitochondrial membrane. The central region (B) corresponds to the membrane subunits, mainly cytochrome *b* and iron-sulfur protein. The lower region (C) corresponds to the large water-soluble subunits on the matrix side of the membrane. After Leonard et al. (1981).

approximately 30% of its overall mass is embedded in the bilayer, 50% is in the part extending 70 Å beyond one side of the bilayer, and 20% in that extending 30 Å beyond the other side. Unfortunately, no such data are as yet available for complexes I or II.

Thus there are grounds for believing that the protein complexes indeed exist in the mitochondrial inner membrane and are not artifacts arising during its fragmentation. The question arises whether any individual respiratory-chain components (cytochromes, flavoproteins, etc.) may be not integrated into complexes. Regarding the complexes as assemblies with constant stoichiometric relations between components and assuming that "free" respiratory-chain components do not exist in the mitochondrial membrane, one should expect that the molar contents of flavins, cytochromes, nonheme iron, etc. in mitochondria would be related as simple whole numbers. However, according to the data of quite a few authors (see, e.g., King et al., 1964; Williams, 1968; Kröger and Klingenberg, 1970), this is not the case. Later King (1973) pointed out the pitfall of simplifying the experimental data in order to approximate simple stoichiometry, warning that this might lead to an erroneous conception of the structural and functional design of the mitochondrial membranes.

1.3.2.2b. <u>ATPase Complex</u>. Isolation of the ATPase complex has been described by quite a number of researchers (Kagawa and Racker, 1966a; Kopaczyk et al., 1968a; Tzagoloff et al., 1968a; Tzagoloff and Meagher, 1971; Swanljung et al., 1973; Hatefi et al., 1974; Sadler et al., 1974; Ryrie, 1975; Serrano et al., 1976). As has been said before, this complex (H^+-ATPase) comprises from 10 to 18 types of polypeptides and has a molecular weight of about 460,000. According to Soper et al. (1979), fresh preparations of the complex being in a highly dispersed, membrane-free form are represented by tripartite particles with discernible headpiece, basepiece, and a connecting stalk. The basepiece is thick enough (about 60 Å) to span the membrane bilayer, while the headpiece has the same dimensions at F_1-ATPase.

To elucidate the molecular arrangement of H^+-ATPase, Todd and Douglas (1981) used the technique of intermolecular linking with bifunctional reagents followed by electrophoretic resolution of the reaction products. Based on the data obtained, they proposed a scheme (Figure 11) that accounts for the mutual position of subunits (nearest neighbors) and their stoichiometry in the complex (see Todd et al., 1980). Subunits on the scheme are numbered in the order of decreasing molecular weight: 1, 56,000 (F_1 α subunit); 2, 52,000 (F_1 β subunit); 3, 32,000 (F_1 γ subunit); 4, 25,300; 5, 32,700; 6, 21,900 (OSCP); 7, 13,700 (F_1 δ subunit?); 8, 9600 (F_1 inhibitor; F_1 ε subunit?); and 9, 8000 (DCCD-binding protein). It should be particularly noted that, according to Todd and Douglas (1981), α and β subunits of F_1-ATPase are in close enough contact to form intermolecular links with all other subunits except two: subunit 7 which cannot be linked to any other one, and subunit 9 which is the DCCD-binding protein acting as the proton channel in H^+ - ATPase. Considered in the light of the tripartite structure of H^+ - ATPase, these data prompt one to conclude that the DCCD-binding protein is an obligatory basepiece constituent and that the stalk can hardly be composed only of OSCP (see above the discussion on the arrangement of H^+ - ATPase subunits in the inner mitochondrial membrane).

1.3.2.2c. Mutual Arrangement of the Complexes in the Inner Mitochondrial Membrane. A natural sequel of the topic of the oligoenzyme complexes would be to discuss their mutual arrangement in the inner mitochondrial membrane. The data along this line are quite scarce. Of direct experimental observations, mention can so far be made only of the work of Hackenbrock and Hammon (1975) who have used ferritin conjugates of antibodies to cytochrome *c* oxidase to show that the enzyme molecules form hexameric or octameric clusters unevenly distributed along the membrane place: the minimal distance between them is 100 Å and the maximal 5000 Å.

Of special interest is the question of how the respiratory chain and the energy-transducing machinery are mutually arranged in the membrane. Ernster (1975a) has summarized the evidence for the interaction between the energy-transducing components taking place not only at the level of individual proteins within the simplest enzymic complexes but at the level of whole complexes as well. Substrate oxidation in the respiratory chain is known to change some charac-

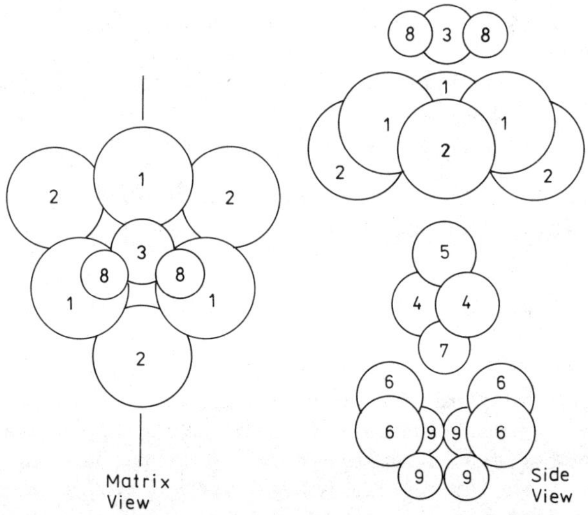

Fig. 11. A schematic representation of H^+-ATPase (after Todd et al., 1980). (1,2,3,5 and 8) F_1 subunits α,β,γ,δ and ε, respectively; (6) OSCP; (9) DCCD-binding protein.

teristics of the ATPase complex. This manifests itself as the altered ability of factor F_1 to bind aurovertin (Bertina et al., 1973; Chang and Penefsky, 1973) and the ATPase inhibitor (Van de Stadt et al., 1973). According to Rossi et al. (1974) and Alexandre et al. (1975), ATP and ADP binding by submitochondrial vesicles as well as the ATPase activity of the latter depend on the functional state of the respiratory chain and can be controlled by respiratory inhibitors - rotenone, antimycin A, and cyanide. There is also a kind of feedback: the properties of respiratory-chain components depend on the functional state of the ATPase complex. In particular, Berden and Slater (1972) found that the presence of ATP affects the kinetics of antimycin A binding by complex III. Harmon and Sharrock (1978) observed differences in the ability of cytochrome a_3 to interact with carbon monoxide in the energized and de-energized mitochondrial membranes.

Taking into account the data of Slater's laboratory (Bertina et al., 1973) that rat-liver and beef-heart mitochondria contain equimolecular amounts of F_1 -ATPase, OSCP, and complex III (this was deduced from the results obtained by titrating mitochondria with aurovertin, oligomycin, and antimycin A), and the data on the cytochrome contents in mitochondria (Klingenberg, 1968), Ernster (1975a,b) drew the conclusion that the ATPase complex, complex III, cytochrome *c*, and complex IV are present in mitochondria in a molecular ratio of 1:1:2:2. On this basis a hypothetical scheme of the inner mitochondrial membrane was proposed in which cytochromes and ATPase formed unit assemblies.

Interaction of the ATPase complex with pyridine nucleotide transhydrogenase was studied in detail in Ernster's laboratory (Ernster, 1975a,b; Ernster et al., 1977; Nordenbrand et al., 1977). Specifically, the ATPase and transhydrogenase activities of submitochondrial vesicles were shown to be inhibited by the ATPase natural inhibitor to an equal extent though the latter activity is not rate-limiting in the ATP-dependent transhydrogenase reaction. In other experiments use was made of submitochondrial vesicles depleted of F_1 -ATPase and its inhibitor. Addition of purified factor F_1 to such vesicles brought about a parallel increase in the oligomycin-sensitive ATPase and transhydrogenase activities. The saturating F_1 concentration proved to be the same for both activities despite that the rate of ATP hydrolysis was 25 times that of transhydrogenation. These and other data allowed a suggestion that the ATPase complex and the pyridine nucleotide transhydrogenase in the mitochondrial membrane interact directly at the molecular level (i.e. within a single molecular unit) rather than indirectly, say through the electrochemical H^+-potential spread over the membrqne.

Interaction of F_1 -ATPase with the adenine nucleotide carrier has been widely discussed. It was initially supposed tnat there is no direct link between the two systems and ATP found outside mitochondria is formed from ADP and phosphate of the matrix and then transferred across the membrane in exchange for ADP from the surrounding medium. However, numerous data obtained in recent years do not comply with this point of view. Among other things, it was shown that upon addition of [^{14}C]ADP to respiring mitochondria [^{14}C]ATP appears in the medium more quickly than it could be expected assuming free mixing of [^{14}C]ADP and [^{14}C]ATP with the matrix nucleotide pool. In respiring mitochondria [^{14}C]ADP was found to be phosphorylated and converted to [^{14}C]ATP more quickly than to accumulate in the matrix. The kinetics of [^{14}C]ADP in the appearance outside mitochondria after addition of [^{14}C]ATP in the presence of 2,4-dinitrophenol indicate that the added nucleotide undergoing hydrolysis does not enter the matrix pool. These and other data, summed up in a review by Vignais (1976), prompt one to think that the ATPase complex and the adenine nucleotide carrier are in immediate proximity to each other in the inner mitochondrial membrane. It is not clear yet whether the two systems interact through a common subunit or both have direct access to the same aqueous microspace that is not in equilibrium with the matrix (Figure 12).

Concluding the section devoted to the interaction between the oligoenzyme complexes in the inner mitochondrial membrane, one may mention the idea (Bertina and Out, 1973) that electron carriers, ATPase complex, and adenine nucleotide carrier consititue an assembly (ensemble) (or a certain isolated compartment) which operates as an enzymic unit. On the other hand, the experimental data available do not rule out indirect interactions between the complexes, for instance through the gradient of H^+ ions or the transmembrane potential (Ernster, 1975a; Ernster et al., 1977).

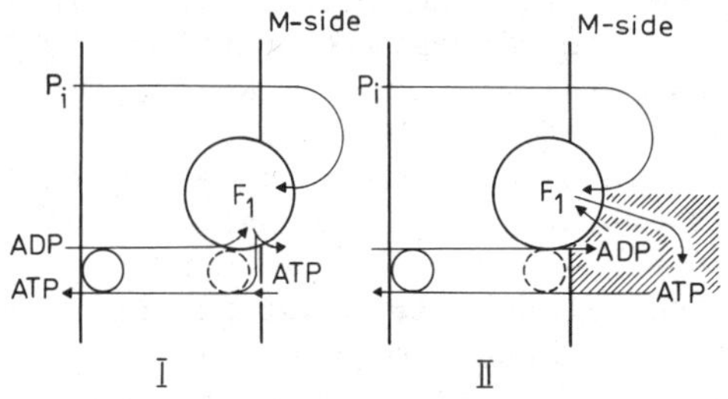

Fig. 12. Schemes of the structure-function relationship between the adenine nucleotide carrier and ATPase (F_1). Model (I) implies direct interaction between the carrier and ATPase. According to model (II), ATPase and carrier have access to the same aqueous microspace. After Vignais (1976).

Moreadith and Jacobus (1982) obtained data implying that in cardiac mitochondria the adenine nucleotide carrier and creatine kinase interact as a multistage enzymic complex. These authors showed, in particular, that ADP is produced by creatine kinase in such close proximity to the carrier as would dramatically attenuate the inhibitory effect of atractyloside on respiration. This raises the possibility of the existence of a still more complex superassembly than that mentioned above.

It must be emphasized that arguments in favor of superassemblies of oligoenzyme complexes are usually of indirect nature.

An interesting experimental approach to elucidating the possibility of formation of superensembles including several respiratory-chain complexes was devised in Hackenbrock's laboratory (Schneider et al., 1980a,b). The gist of this approach is to test how the integral functions of the respiratory chain are affected by "dilution" of the mitochondrial inner-membrane protein contents with exogenous phospholipids. It turned out that the membrane lipid bilayer could be enriched by 30-700% {sic!} in the course of controlled fusion of mitoplasts with phospholiposomes. Such fusion increases the membrane surface and lowers the density of distribution of the integral protein particles in the membrane plane, the particles being randomly scattered in the membrane and not forming any detectable packs or clusters. There is a concomitant decrease in NADH oxidase, succinate oxidase, and NADH(succinate)-cytochrome *c* reductase specific activities which is strictly proportional to the increase in the relative phospholipid content in the membrane, whereas activities of some individual enzymes (e.g., NADH-ferricyanide reductase, cytochrome *c* oxidase) are enhanced. On the basis of these data it was suggested that complexes I(II) and III do not form a structural-functional unit and that electron transfer between them takes place owing to the diffusion of coenzyme Q, either free or bound to a protein carrier (Figure 13). Unlike the integral activities indicated, duroquinol oxidase activity of mitoplasts does not fall upon phospholipid enrichment. This means either that complexes III and IV are bound into an assembly which is not split by exogenous phospholipids introduced into the membrane, or that electron transfer between the complexes performed through cytochrome *c* diffusion in the aqueous layer on the membrane surface does not limit the overall reaction rate (Figure 14).

Outwardly simple, this approach has a number of shortcomings (see Schneider et al., 1980b) which make it impossible to draw unambiguous conclusions about the actual mutual arrangement of the respiratory-chain oligoenzyme complexes in the inner mitochondrial membrane.

On the other hand, direct evidence is available that individual complexes may be bound to other mitochondrial inner-membrane proteins. This follows, in particular, from the data of Hackenbrock's group (reviewed by Hackenbrock, 1976) on the movement of protein particles in the inner mitochondrial membrane during reversible thermotropic phase transitions in the lipid bilayer. In these experiments the inner membrane-matrix fraction was first cooled to subzero temperature, which made the membrane separate into smooth lipid and particle-dense regions upon the liquid-crystalline-to-gel-state transition of the lipid bilayer. This phase separation was monitored electron-microscopically by the freeze-fracture technique. Subsequent heating of the preparation induced the reverse phase transition in the lipid bilayer with concomitant randomization of the protein particles in the plane of the membrane. When the mitochondrial inner membrane preparations were treated at -10°C (i.e. below the phase-transition temperature) with rabbit antiserum against cytochrome *c* oxidase and then with antirabbit goat IgG, the large intramembrane cytochrome *c* oxidase particles were immobilized in the immunoglobulin lattice and did not disaggregate to randomize through free translational diffusion upon subsequent heating to 25°C. At the same time, some membrane proteins migrated as usual (this pertains to the proteins on the M-side of the membrane) but some stayed with the cytochrome *c* oxidase particles; the result was that the overall number of particles inhibited in lateral diffusion was greater than expected from the calculated content of cytochrome *c* oxidase particles in the membrane. Unfortunately, the proteins displaying such affinity for cytochrome *c* oxidase have not been identified.

1.3.2.3. *Phospholipids of the Inner Mitochondrial Membrane*

As indicated in the beginning of this chapter, the inner mitochondrial membrane contains substantial amounts of phospholipid (about 30% by weight). The major components of the phospholipid fraction of liver mitochondria are phosphatidylcholine (39-44% of the total phospholipid content), phosphatidylethanolamine (25-37%), and cardiolipin (14-21%) (Parsons et al., 1967; Stoffel and Schiefer,

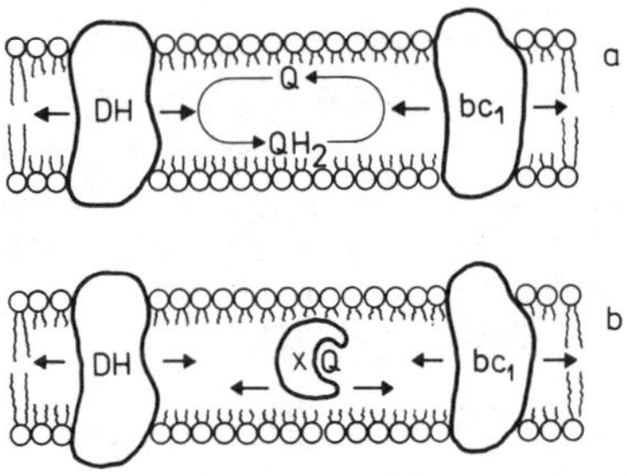

Fig. 13. Mutual disposition and interaction between dehydrogenases and complex III in the inner mitochondrial membrane. (DH) The dehydrogenase complex; (*bc*$_1$) complex III; (Q) coenzyme Q; (X) coenzyme Q-binding protein. Redrawn from Schneider et al. (1980b).

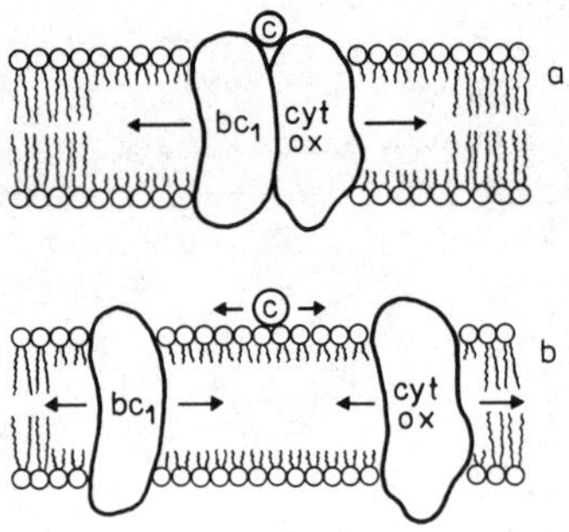

Fig. 14. Mutual disposition and interaction between complex III and cytochrome *c* oxidase in the inner mitochondrial membrane. (*bc*$_1$) Complex III; (cyt ox) cytochrome *c* oxidase; (c) cytochrome *c*. Redrawn from Schneider et al. (1980b).

1968; Lévy et al., 1969; Nachbaur et al., 1969). Phospholipids are indispensable for virtually all membrane functions: respiration (Brierly et al., 1962; Fleischer et al., 1962), oxidative phosphorylation (Petrushka et al., 1959; Burstein et al., 1971a,b), ATP hydrolysis (Kagawa and Racker, 1966a; Kopaczyk et al., 1968b; Berezney et al., 1970a; Swanljung et al., 1973), transhydrogenase reaction (Luzikov et al., 1973a; Rydström, 1975), adenine nucleotide transport (see a review by Vignais, 1976) and so on.

Unfortunately, the information on the arrangement of phospholipids in the membrane is extremely meager. On the basis of electron-microscopic data, Sjöstrand (1963) put forward a suggestion that phospholipids in the membrane form regularly distributed globular micelles about 15 Å in diameter spaced by protein molecules. Later Vanderkooi and Green (1970) proposed a similar model of the membrane (Figure 15); according to this model, the membrane framework is composed of protein globules, cavities between which are filled with phospholipids. Molecules of the latter are oriented so that their hydrophilic "heads" face the surrounding medium while their hydrophobic "tails" are buried in the interior membrane regions. This model is supported first and foremost by the fact that extraction of up to 95% of phospholipids with aqueous acetone causes no decrease in the membrane thickness or impairment of its macrostructure (Fleischer et al., 1967). In other words, the model of Vanderkooi and Green (1970) implies that the membrane structure is based on protein-protein interactions.

More recent studies have shown that this model is not true. Specifically, according to the data already mentioned (Hackenbrock, 1976), reversible thermotropic lipid phase transitions lead to the separation of smooth lipid regions and particle-dense regions in the membrane. These data imply that the inner mitochondrial membrane has a continuous lipid bilayer with protein complexes floating in it (see, e.g., Figures 13, 14).

The inner mitochondrial membrane appears to be one of the most fluid membranes of eukaryotic cells. Feinstein et al. (1975) estimated its microviscosity at 90 cP by fluorescence polarization with perylene as a probe. For comparison it may be said that microviscosities of human erythrocyte and polymorphonuclear leukocyte membranes are 180 cP and 335 cP respectively. Studying lipid phase transitions in the inner mitochondrial membrane by differential scanning microcalorimetry, Hackenbrock et al. (1976) determined that the onset temperature of the transition exotherm lies at -4°C and that of the endotherm at - 15°C, i.e. the liquid-crystalline-to-gel-state transition occurs at fairly low temperatures.

The high fluidity of the inner mitochondrial membrane is probably due to its exceptionally high relative content of unsaturated fatty acids. Specifically, Nachbaur et al. (1972) report a ratio of unsaturated to saturated fatty acids equal to 3.1 for rat-liver mitochondrial inner membrane; this ratio is 2.0 for the lysosomal membrane, 0.4 for microsomes, and only 0.2 for the plasma membrane. It must be noted that mitochondrial inner membrane fluidity varies with physiological conditions (e.g., with conditions of yeast growth), and these variations affect various processes taking place in the membrane (see Chapter Eight).

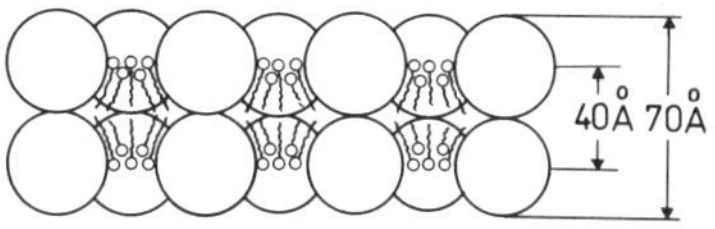

Fig. 15. Arrangement of protein particles and phospholipids in the inner mitochondrial membrane. Charged "heads" of phospholipid molecules face the surrounding medium, and their hydrophobic "tails" are submerged in the interior regions of the membrane. After Vanderkooi and Green (1970).

Though phospholipid topology in the inner mitochondrial membrane is far from being investigated in detail, still some data on that score are available. Particularly, Crane and Marinetti (1979) studied the kinetics of trinitrobenzene sulfonate interaction with phosphatidylethanolamine of mitoplasts which constitutes 35-40% of the total membrane phospholipid, and drew the conclusion that 30-40% of phosphatidylethanolamine is localized on the exterior membrane surface, 40-60% on the interior, and 10-15% is buried within the membrane. One of the reasons of the weak interaction of phosphatidylethanolamine with various reagents in mitoplasts was thought to be its screening owing to tight binding with proteins.

A certain portion of phospholipids is especially strongly bound to membrane proteins. According to Awasthi et al. (1970a), a small amount of cardiolipin (approximately one molecule per heme residue) cannot be extracted from cytochrome *c* oxidase even by drastic treatment with organic solvents (diethyl ether, hexane, isooctane, or acetone). Such enzyme preparations display low catalytic activity but are easily reactivated by adding purified phospholipids or their natural mixtures. Removal of the tightly bound cardiolipin, which can be achieved in special conditions, is accompanied by transition of cytochrome *c* oxidase to a form no longer reactivatable by added phospholipids.

Cardiolipin plays a special role in the maintenance of structural integrity not only of isolated cytochrome *c* oxidase but of the inner mitochondrial membrane as a whole. In particular, cleavage by *Naja naja* venom phospholipase A of about 70% of phosphatidylcholine and of practically all phosphatidylethanolamine in submitochondrial vesicles does not lead to any noticeable changes in the macrostructure of their membranes (Berezney et al., 1970b). But if the phospholipase treatment causes cleavage of 25-90% of cardiolipin, such changes take place: no "mushroom" formations can be seen in negatively contrasted vesicular preparations, the vesicles themselves are ruptured, the apparent thickness of their membranes increases, etc.

Burstein et al. (1971a) used *Clostridium welchii* phospholipase C to treat beef heart mitochondria and inside-out submitochondrial vesicles prepared from pretreated mitochondria, and found that a portion of mitochondrial inner-membrane phospholipids (about 30% by the phosphorus content) is relatively resistant (or inaccessible) to phospholipase C. However, these phospholipids were not analyzed. It must be emphasized that the membrane structure is in great measure determined by the fatty acid residues in phospholipid molecules. Indeed, the phosphorylating activity of submitochondrial vesicles in which up to 70% of phospholipids have been hydrolyzed by phospholipase C is almost completely restored on addition of coupling factors F_1 and F_4 (Burstein et al., 1971a). Such a restoration does not take place if vesicles treated with phospholipase C are then exposed to lipase that breaks down diglycerides remaining in the membrane after the action of phospholipase C.

SUMMARY

This chapter skims through what is known about the principles of mitochondrial structural organization. The experimental data available pertain mainly to the molecular structure of the inner mitochondrial membrane. Analysis of these data reveals two basic principles underlying the arrangement in the membrane of the protein components of the respiratory chain, energy-transducing machinery, and some transport systems. First, some of these components are localized on the M-side of the membrane; they face the matrix and are inaccessible for nonpenetrating reagents from the intermembrane space and the surrounding medium. Other components reside on the membrane C-side and have no direct contact with the matrix but are available for all reagents in the intermembrane space. Second, some polypeptides (carrying or not carrying prosthetic groups) unite into complexes capable of performing partial reactions of the respiratory chain and the ATP-synthesizing system. From this it follows that in studying the molecular aspects of mitochondrial inner membrane formation it is first of all necessary to elucidate how polypeptides made on mitochondrial or cytoplasmic ribosomes (see Chapter Five) find their place of destination in the membrane; why some of them occupy the C-side and others the M-side of the membrane while still others are buried in its interior regions; how individual polypeptides find their partners in the membrane and form catalytically active complexes, and so forth.

The peculiarities of mitochondrial inner membrane structural organization considered above are in all probability not accidental. They are necessary for the membrane to perform its catalytic functions. In particular, the asymmetric arrangement of electron carriers and components of the ATPase complex is essential for the directed transport of H^+ ions across the membrane and generation of the electrochemical potential of these ions on the membrane; formation of respiratory complexes favors the most efficient interaction between their constituents in electron transfer; integration of the ATPase complex with the adenine nucleotide carrier facilitates endogenous ADP phosphorylation.

The information available on the mitochondrial inner membrane structure is far from being complete. No details are known about the arrangement in the membrane of many components of the respiratory chain and energy-transducing machinery. We are still in the dark as to how rigorously the basic principles of the membrane design are followed for its individual elements and how frequent are the deviations from the general rules. Finally, the data on the position in the membrane of quite a few of its components (especially various translocases and biosynthetic systems) are altogether lacking.

2
General Manifestations and Molecular Mechanisms of Mitochondrial Degradation

From the preceding chapter we could see that mitochondria are organelles of elaborate structural design. They are fairly labile and quickly degrade having been isolated from cells. The "aging" of mitochondria is a process which may go one way or another depending on the source of preparation, incubation conditions, etc. Nevertheless, some degenerative alterations in mitochondria have a universal character; it is these alterations that will be dealt with in the following section.

2.1. GENERAL MANIFESTATIONS OF MITOCHONDRIAL DEGRADATION

2.1.1. Impairment of Oxidative Phosphorylation and Other Functions

Incubation of mitochondria in isotonic sucrose solutions as well as in media containing phosphate, arsenate, arsenite, pyrophosphate, Ca^{2+} ions, thyroxine, and other reagents entails a rapid loss of their ability to perform oxidative phosphorylation coupled with NAD-dependent substrate oxidation (Kielley, 1952; Slater and Cleland, 1952, 1953; Hunter and Ford, 1955; Hunter et al., 1956; Lester and Hatefi, 1958; Lehninger et al., 1959; Weinbach, 1959; Helinski and Cooper, 1960). For instance, according to Hunter and Ford (1955), in the presence of 20 mM phosphate already in 10 min at 30°C rat liver mitochondria become unable to catalyze respiration involving electron transfer via NAD or NADP. Under similar conditions succinate oxidation may be even enhanced (Cross et al., 1949; Kielley, 1952; Slater and Cleland, 1952, 1953; Raaflaub, 1953; Hunter and Ford, 1955). Introduction of excess NAD or NADP into the incubation medium completely restores respiration with such substrates as β-hydroxybutyrate, α-ketoglutarate, glutamate, pyruvate plus malate, etc.

(Kielley, 1952; Hunter and Ford, 1955; Hunter et al., 1956; Lester and Hatefi, 1958; Lehninger et al., 1959; Helinski and Cooper, 1960). After such treatment, however, the oxidative phosphorylation efficiency does not reach its initial level (Hunter and Ford, 1955; Hunter et al., 1956; Lester and Hatefi, 1958; Lehninger et al., 1959; Helinski and Cooper, 1960). The phosphorylating capacity of mitochondria subjected to prolonged incubation noticeably increases only on addition of serum albumin (Lester and Hatefi, 1958; Helinski and Cooper, 1960; see also Section 2.2.1). Thus, in most cases known, mitochondrial incubation in vitro leads in the first place to the loss of pyridine nucleotides (see below) and to degenerative changes in the energy-transducing system. It should be noted that under certain conditions the NAD-dependent oxidation of a number of substrates and the respiratory control are restored upon addition of ATP (plus Mg^{2+}), coenzyme A, and cocarboxylase to mitochondria substantially (by some 90%) depleted of NAD. Under such conditions ATP was supposed to induce NAD transfer from inactive regions to catalytic sites, possibly owing to conformational changes in the membrane (MacLennan and Tzagoloff, 1966).

Mitochondrial functions readily lost during aging are: stimulation of respiration by ADP, 2,4-dinitrophenol, and Ca^{2+} or K^+ ions in the presence of valinomycin; ATP-dependent Ca^{2+} transport and reversed electron transfer (Carafoli and Gazzotti, 1970); and partial exchange reactions ATP-$^{32}PO_4^{3-}$ and $P^{18}O_4^{3-}$-H_2O (Boyer et al., 1956; Cooper and Lehninger, 1957; Weinbach, 1959; Lehninger, 1962). At the same time, the ability to undergo ATP-induced contraction is retained by mitochondria for prolonged periods (Lehninger, 1959a,b; Carafoli and Gazzotti, 1970).

In parallel with the impairment of mitochondrial oxidative phosphorylation, there is an increase in ATPase activity (Mg^{2+}-ATPase) which causes a rapid exhaustion of the endogenous ATP pool (Chappell and Perry, 1954; Boyer et al., 1956; Weinbach, 1959). This increase can be considered another characteristic feature of organelle aging. Since there are indications that ATP is needed for the maintenance of mitochondrial high phosphorylating activity (Ernster and Löw, 1955; Hunter and Ford, 1955; Boyer et al., 1956), there are certain grounds for regarding the degradation of the mitochondrial oxidative phosphorylation system as an autocatalytic process.

2.1.2. Loss of Low-Molecular-Weight Mitochondrial Components

Siekevitz and Potter (1955) demonstrated that liver mitochondria incubated in an isotonic medium rapidly lose acid-soluble nucleotides. These included NAD, NADP, AMP, ADP, ATP, ITP, and some unidentified compounds. The process was stimulated by orthophosphate and retarded in conditions when oxidative phosphorylation took place (the incubation medium contained an oxidizable substrate, phosphate, AMP, and Mg^{2+} ions).

Especially quickly falling during mitochondrial aging is the content of pyridine nucleotides (Kielley, 1952; Hunter and Ford, 1955; Hunter et al., 1956; Lester and Hatefi, 1958; Weinbach, 1959; Hunter et al., 1959a; Kaufman and Kaplan, 1960; Brierley et al., 1964). Among the factors favoring the maintenance of a high level of these nucleotides in the organelles one may mention anaerobiosis (Hunter et al., 1956) and the presence in the incubation medium of oxidizable substrates or respiratory inhibitors (Kaufman and Kaplan, 1960). In this connection it was suggested (Kaufman and Kaplan, 1960) that nucleotide oxidation precedes their removal from mitochondria. There appears to be a direct relationship between organelle swelling and loss of pyridine nucleotides (Kaufman and Kaplan, 1960). The latter process is prevented, besides the above-mentioned agents, by EDTA, Mg^{2+} and Mn^{2+} ions, ATP, ADP, and AMP (Hunter and Ford, 1955). Exogenous NAD was shown to be incorporated into mitochondria preswelled in the presence of phosphate. Such incorporation is stimulated by ATP and ADP as well as inorganic triphosphate, UTP, and ITP (Hunter et al., 1959b). It must be emphasized that the incorporated NAD is active in oxidative phosphorylation. As suggested by Hunter et al. (1959b), the mechanism of the ATP effect is that it induces certain changes in the membrane that render it impermeable to NAD. Indeed, treatment of swelled mitochondria with ATP before the introduction of NAD into the incubation medium prevents NAD uptake by the organelles. The mechanisms of the stimulatory action of other reagents are less clear. The effect of EDTA is supposed to be associated with its ability to hinder mitochondrial swelling (Hunter et al., 1959b).

As to adenine nucleotides, a substantial portion of that remains in mitochondria after the pyridine nucleotides are already almost completely lost (Brierley et al., 1964). According to Brenner-Holzach and Raaflaub (1954), mitochondrial swelling begins when the endogenous ATP level falls to a value constituting approximately 20% of the initial. As swelling proceeds, ADP and AMP are also lost and then is coenzyme A, which comes out into the surrounding medium after dephosphorylation.

Naturally, the cofactors mentioned are not the only low-molecular-weight components that mitochondria lose during incubation. It is known, for instance, that a short-term incubation of liver mitochondria at 23-37°C causes a virtually complete loss of citrate (Schneider et al., 1956) and Mg^{2+} ions (Siekevitz and Potter, 1955; Chefurka, 1966). Mitochondrial aging is accompanied by activation of a number of latent enzymes: oxidoreductases (cytochrome *c* oxidase, glutamate dehydrogenase, malate dehydrogenase), hydrolases (acetyl-coenzyme A hydrolase, ATPase), hydrolyases (aconitate hydratase, fumarate hydratase), transferases (aspartate aminotransferase). Since in most cases known the latency is due to the inability of substrates to penetrate the membrane, such activation is to be ascribed to enhanced membrane permeability (Chappell and Greville,

1963). In this case it can be thought that mitochondrial aging is attended by a gradual loss of the substrates of all the enzymes listed.

2.1.3. Release of Soluble Proteins from Mitochondria

Besides low-molecular-weight components, mitochondria incubated in vitro lose some proteins. As shown by Lehninger and Schneider (1959), the release of proteins from mitochondria is especially marked in the presence of reduced glutathione. In this case mitochondria lose, among others, some proteins necessary for their contraction, including glutathione peroxidase and catalase (Lehninger and Gotterer, 1960; Neubert et al., 1962).

Estrada-O (1964) studied the "leakage" of several matrix enzymes from mitochondria. Incubation of liver mitochondria at 37°C was shown to lead to the appearance in solution of glutamate dehydrogenase and asparate aminotransferase activities, this process being stimulated by agents impairing oxidative phosphorylation (e.g., oligomycin or 2,4-dinitrophenol). According to later data (Estrada-O et al., 1966), leakage of soluble enzymes is associated with alterations in the mitochondrial membrane phospholipids. Indeed, free fatty acids, lysolecithin, and lecithinase intensify this process. At neutral pH the release of soluble enzymes (glutamate-oxaloacetate aminotransferase, glutamate dehydrogenase, malate dehydrogenase, and β-hydroxybutyrate dehydrogenase) is impeded by ATP and succinate. In the absence of phosphate at pH 6.0 oxidizable substrates (glutamate, succinate, β-hydroxybutyric acid) and ATP, on the contrary, are stimulatory. Electron-transfer inhibitors (antimycin A or malonate) as well as rutamycin and 2,4-dinitrophenol abolish the action of substrates and ATP. The ability of mitochondria to retain the enzymes mentioned was suggested to depend on the inner membrane state controlled by the oxidative phosphorylation mechanism (Estrada-O et al., 1970).

2.1.4 Mitochondrial Swelling

The irreversible high-amplitude swelling of mitochondria is one of the manifestations of their degradation. Indeed, there is a direct connection between mitochondrial swelling and loss of cofactors (Brenner-Holzach and Raaflaub, 1954; Hunter et al., 1956, 1959b; Kaufman and Kaplan, 1960) and some proteins (Lehninger and Gotterer, 1960; Estrada-O et al., 1970). Under certain conditions swelling is accompanied by formation of the "U-factor" (a fatty-acid mixture, see further) having a pronounced uncoupling effect on mitochondrial membranes (Lehninger and Remmert, 1959; Wojtczak and Lehninger, 1961). Various aspects of mitochondrial swelling have been treated in detail by a number of authors (see, e.g., reviews by

Lehninger, 1962; Chappell and Greville, 1963). The basic issues stemming from the numerous works devoted to this problem are briefly summarized below.

Mitochondrial swelling may be brought about by different causes depending on the nature of agents inducers this process. Wojtczak and Lehninger (1961) classified all swelling inductors into two groups: one comprises the agents which act through formation of the U-factor and are neutralized by small additions of serum albumin (Ca^{2+} ions, thyroxine, phlorizin, and the like), and the other group includes those whose action is insensitive to serum albumin (e.g., PO_4^{3-} ions, reduced glutathione).

Swelling induced by phosphate or taking place in sucrose solution is most probably associated with electron transfer in the absence of ADP. This suggestion is supported by the following experimental evidence. First, swelling is enhanced in the presence of oxidizable substrates (β-hydroxybutyrate, glutamate, α-ketoglutarate, succinate). Second, swelling is suppressed by respiratory inhibitors - amytal, malonate, antimycin A, SN 5949, azide, or cyanide - and by removing oxygen from the incubation medium (Hunter et al., 1956, 1959a; Chappell and Greville, 1958, 1960; Corwin and Lipsett, 1959; Lehninger, 1959a). Interestingly, the effect of succinate is observed in the presence of amytal (Chappell and Greville, 1958, 1960; Hunter et al., 1959a), which does not block its oxidation, but not in the presence of malonate, an inhibitor of succinate dehydrogenase. On the other hand, swelling induced by β-hydroxybutyric acid and the like substrates is impeded by amytal but not by malonate (Chappell and Greville, 1958; Hunter et al., 1959a). Third, oligomycin inhibiting β-hydroxybutyrate oxidation suppresses swelling, whereas 2,4-dinitrophenol which stimulates respiration in the presence of oligomycin also abolishes the repression of swelling caused by the latter (Chappell and Greville, 1960). As shown by Chappell and Greville (1960), in all the cases mentioned the electron acceptor is not necessarily oxygen; the function can as well be carried out by ferricyanide. In other words, swelling is evoked by electron transfer even through a limited part of the respiratory chain.

Analogous results were obtained for the agents capable of inducing the appearance of the U-factor during mitochondrial aging <u>in vitro</u>. According to Lehninger et al. (1959a), the thyroxine-provoked swelling is suppressed by antimycin A and amytal. Anaerobiosis and cyanide were shown to prevent swelling in the presence of thyroxine, Ca^{2+} ions, and other agents (Lehninger and Ray, 1957). With all this, it was pointed out that the U-factor formation is insensitive to respiratory inhibitors cyanide and azide (Lehninger and Remmert, 1959).

A special position is occupied by swelling brought about by

reduced glutathione. The action of the latter is unaffected by amytal or antimycin A but is blocked by cyanide (Lehninger and Schneider, 1959). It seems that in this case account must be taken of the peculiarities of the interaction of the swelling-inducing agent with mitochondria. For instance, reduced glutathione is known to promote, concurrently with the organelle swelling, a release therefrom of substantial amounts of protein and nucleotides.

There are experimental data in conflict with the concept implying the stimulatory role of electron transfer in mitochondrial swelling. In particular, Corwin and Lipsett (1959) showed that swelling in 0.1 M sucrose is suppressed by dithionite. Therefore the effects of anaerobiosis and respiratory inhibitors may be due not to the interruption of electron transfer but simply to reduction of some mitochondrial components (see also Lehninger and Ray, 1957; Lehninger et al., 1959). Chappell and Greville (1963) have found that the effect of oxidizable substrates on mitochondrial swelling depends on the extent of organelle preservation. Thus, the substrate (β-hydroxybutyric acid) causes only insignificant swelling of freshly prepared mitochondria, retards swelling if added after 6 h of incubation at 0°C, and produces a distinct stimulatory effect with mitochondria subjected to a 24-h preincubation.

Finally, one should linger on some general aspects of mitochondrial swelling, and in their light give a brief comment on the facts presented above. According to now accepted concepts (see a review by Brierley, 1976), mitochondrial swelling is due to the transport of ions (e.g., monovalent cations and accompanying anions) from the incubation medium into the mitochondrial metrix. Such transport may be either passive or energy-dependent. In the latter case swelling is an electrophoretic response to the negative potential building up inside mitochondria as a result of electron and proton transport across coupled membranes. Mitochondrial contraction, according to Brierley and co-workers (Brierley and Jurkowitz, 1976; Brierley et al., 1977a,b; Chávez et al., 1977), requires generation of a transmembrane gradient of H^+ concentration. A special carrier in the mitochondrial membrane performs an exchange of H^+ ions from the exterior for cations (e.g., K^+ ions) of the mitochondrial matrix, this being accompanied by neutralization of the negative charges of OH^- ions in the matrix, release into the medium of anions following the cations, and the resulting osmotic contraction of the matrix. This mechanism is operative only on condition that the membrane has low permeability for protons and metal ions. Obviously, permeability may increase owing to many events taking place in aging mitochondria, especially after addition of some inductors of aging. A weighty part in this is played by the cleavage of mitochondrial phospholipids to lysophosphatides and free fatty acids, and by lipid oxidation (see further). On the whole, mitochondria undergoing aging would lose their ability to contract as they become unable to generate the transmembrane H^+ - concentration potential.

2.1.5. Morphological Alterations in Mitochondria

Mitochondrial aging *in vitro* is attended by pronounced morphological alterations. Early studies in this field were undertaken by Harman and Feigelson (1952a,b) who ascertained the relationship between the organelle morphology and functional state. A detailed description of the types of degraded mitochondrial structures was given by Hackenbrock (1966). According to his data, after a 15-min incubation at 30°C the suspension contains (1) condensed mitochondria; (2) swollen mitochondria; (3) condensed matrix with no limiting (outer) membrane; (4) swollen matrix with no limiting membrane; and (5) mitochondria with dilated intracristal spaces appearing as matrical vesiculations. More dramatic manifestations of such alterations can be observed upon incubation of mitochondria with Ca^{2+} ions, oleate, Triton X-100, deoxycholate, or other reagents (Watson and Siekevitz, 1956; Ziegler et al., 1958; Wlodawer et al., 1966; Packer and Utsumi, 1969; Waite et al., 1969b; Butler and Judah, 1970).

The initial stage of degradation usually consists in the dilution of the matrix, increase in its volume and the subsequent straightening of the folds formed by the inner membrane. The outer membrane at this stage is frequently ruptured (Wlodawer et al., 1966). Then one may observe "vacuolarization" of the matrix (Watson and Siekevitz, 1956; Packer and Utsumi, 1969), loss of the matrix contents (Watson and Siekevitz, 1956; Ziegler et al., 1958; Wlodawer et al., 1966; Butler and Judah, 1970), total disappearance of cristae (Watson and Siekevitz, 1956; Waite et al., 1969b), and appearance in the incubation medium of small vesicles and other membranous structures (Watson and Siekevitz, 1956; Ziegler et al., 1958).

It must be emphasized that even such mitochondrial swelling as by light-scattering data is completely reversed on addition of ATP and Mg^{2+} ions actually entails irreversible morphological changes in the organelles (Wlodawer et al., 1966; Butler and Judah, 1970).

2.1.6. Sequence of Events During Aging of Isolated Mitochondria

Systematic investigations of the aging of rat liver mitochondria *in vitro* were undertaken by Parce et al. (1978, 1980). The results of these studies, summed up in Table 5, show that the degradation can be arbitrarily divided into two macroscopic stages. The first corresponds to a time interval of approximately 3 h (at 18°C) when mitochondria lose respiratory control. The latter declines far more quickly than the mitochondrial phosphorylating ability (the P/O ratio), membrane permeability for protons, or proton-translocating ability (H^+/O ratio). It is noteworthy that the attenuation of the respiratory control is due to decreasing State 3 respiration while State 4 respiration but slightly rises. This results from the leakage of adenine nucleotides from mitochondria and the impairment of

Table 5. Events Occurring During Aging of Rat Liver Mitochondria *In Vitro* (after Parce et al., 1980).
The number of arrows denotes the relative change in the property as compared to others. (↑) Increase; (↓) decrease.

Property	Change
a) Time period when respiratory control is lost:	
Endogenous adenine nucleotide content	↓↓↓
Adenine nucleotide transport	↓↓↓
State 3 respiration	↓↓↓
State 4 respiration	↑
ATPase activity	↑
Reacylation of monoacylphospholipids	↓↓↓
β-oxidative activity	↓↓↓
Proton permeability	↓↓
b) Time period when ATPase activity increases:	
Hydrolysis of membrane phospholipids	↑↑
ATPase activity	↑↑↑↑
Proton permeability	↑↑
Energy-linked fluorescent responses of 8-anilinonaphthalene-1-sulfonate	↓↓
H^+/ O ratio	↓↓
P/ O ratio	↓↓

ADP transport into the organelles. Other labile functions are β-oxidation of fatty acids and their incorporation into mitochondrial phospholipids (Parce et al., 1978). The rates of these processes fall in parallel with the respiratory control and the adenine nucleotide content. At this stage of mitochondrial degradation there is only a negligible increase in their ATPase activity.

After the loss of respiratory control the second macroscopic stage of degradation sets in, being marked by activation of mitochondrial phospholipid hydrolysis. Now the mitochondrial ATPase activity drastically increases; membrane proton permeability is enhanced; the energy-dependent responses of 8-anilinonaphthalene-1-sulfonate fluorescence, as well as the P/O and H^+/O ratios, decrease.

Thus, there are grounds for believing that the first deteriorative events taking place in rat liver mitochondria *in vitro* are in one way or another associated with the leakage of adenine nucleotides. The causes of this phenomenon are still obscure (Parce et al., 1980). Similar changes in the respiratory control were observed by Mittnacht et al. (1979) during ischemic disfunction of liver mitochondria in rats. These authors put forth a suggestion that the reason for the preferential attenuation of State 3 respiration may be either accumulation of long-chain acyl-coenzymes A, which are known as inhibitors of adenine nucleotide transport across the inner mitochondrial membrane (see a review by Vignais, 1976), or membrane de-energization. The molecular mechanisms of mitochondrial degradation in vitro and in vivo will be treated at more length below.

2.2. MOLECULAR MECHANISMS OF MITOCHONDRIAL DEGRADATION

2.2.1. Role of Endogenous Phospholipase A_2 in Mitochondrial Degradation

As mentioned above, incubation of intact liver mitochondria leads to accumulation of a certain isooctane-extractable, thermostable factor (termed U-factor by Lehninger and Remmert) displaying a pronounced uncoupling action on the oxidative phosphorylation system (Lehninger and Remmert, 1959). The factor stimulated respiration in the absence of a phosphate acceptor, inhibited the ATP-$^{32}PO_4^{3-}$ exchange, and induced an increase in ATPase activity, all these effects being abolished by serum albumin. It was suggested that the U-factor is a mixture of free fatty acids liberated from mitochondrial phospholipids under the action of a specific endogenous phospholipase.

The first works specially devoted to mitochondrial phospholipase A_2 appeared in the early sixties. Rossi et al. (1964) discovered a correlation between the uncoupling of oxidative phosphorylation and the decrease in the content of mitochondrial phospholipids (phosphatidylcholine, phosphatidylethanolamine, and phosphatidylserine) in liver mitochondria in vitro. A corollary to this was the appearance of the corresponding lysoderivatives, as well as of fatty acids either in free form or bound with amino acids. Scherphof and Van Deenen (1965) demonstrated the ability of mitochondria to catalyze the hydrolysis of exogenous phospholipids with concomitant liberation of free fatty acids. They failed to find any appreciable cleavage of endogenous phospholipids and argued that Rossi et al., had apparently observed nonenzymatic hydrolysis that had been a result of the decrease in the pH of the medium from 7.0 to 4.7 during prolonged (24 h) organelle incubation. Further works of Rossi et al. (1965), Scherphof et al. (1966), Waite and Van Deenen (1967), Nachbaur and Vignais (1968), Waite (1969), Waite et al. (1969a,b), Waite and Sisson (1971), Nachbaur et al. (1972) confirmed the presence of a specific phospholipase A in mitochondria and made it possible to assess its role in the organelle degradation.

2.2.1.1. *Specificity of Mitochondrial Phospholipase A_2*

Elucidating the question of the existence of a mitochondrial phospholipase A, it was first of all necessary to make sure that the phospholipid hydrolysis is not due to the presence of considerable amounts of other organelles in the mitochondrial preparation. To this end, Scherphof et al. (1966) and Waite and Van Deenen (1967) studied the hydrolysis of exogenous 1-[9,10-^{3}H]palmityl-2-[1-^{14}C] linoleylglycero-3-phosphoethanolamine by various rat-liver subcellular fractions. By their data, mitochondria selectively catalyze the hydrolysis of the ester bond at the C_2 atom (A_2 specificity) as the result of which the reaction medium accumulates 1-[9,10-^{3}H] palmitylglycero-3-phosphoethanolamine and 1-[^{14}C]linoleic acid. Another reaction is typical of microsomes: splitting off of [9,10-^{3}H]palmitic acid at the C_1 atom (A_1 specificity) and formation of 2-[1-^{14}C]linoleylglycero-3-phosphoethanolamine. In the presence of a mixed preparation containing mitochondria, lysosomes, and microsomes both reactions take place though the hydrolysis of the ester bond at C_1 prevails. All subcellular fractions including mitochondria display a more or less pronounced lysophospholipase activity.

According to Nachbaur et al. (1972), phospholipases of microsomes and plasma membrane have A_1 specificity whereas the mitochondrial enzyme has A_2 specificity. As to lysosomes, their phospholipase activity is determined mainly by a soluble (not membrane-bound) phospholipase active at pH 4.5; this enzyme shows mixed (A_1-A_2) specificity.

Thus, the ability to catalyze the hydrolysis of the ester bond exclusively at the C_2 atom is a characteristic feature of mitochondria. Nevertheless, the existence of the A_2-specific phospholipase activity in lysosomes may prompt one to think that phospholipid hydrolysis in mitochondria is due to lysosomes contaminating the mitochondrial preparations. Such an assertion at first seems quite plausible since mitochondria, like lysosomes, display activity in both alkaline (pH 8.0-9.0) and acidic (pH about 4.5) media (Waite et al., 1969a; Nachbaur et al., 1972). However, simple calculations based on the data of Nachbaur et al. (1972) about the phospholipase activities of mitochondria and lysosomes in acidic and mildly alkaline media show that at pH 9.0 the lysosomal phospholipase activity can account for not more than one tenth of the mitochondrial overall phospholipase activity.

In support of this one can refer to the experiments of Waite et al. (1969a) in which the mitochondrial and the lysosomal fractions were incubated separately or together, and the cleavage of endogenous phosphatidylethanolamine labeled with [^{14}C]ethanolamine was followed. The sum of monoacylglycero-3-phospho[^{14}C]ethanolamine produced during separate incubations (pH 8.8) proved to be close to that produced during joint incubation.

Finally, Nachbaur et al. (1972) believed that the most weighty argument in favor of the existence of an endogenous mitochondrial phospholipase A_2 was the detection of its activity in a purified fraction of inner mitochondrial membranes for which contamination by extraneous cytoplasmic elements was hardly probable.

2.2.1.2. *Activators and Inhibitors of Mitochondrial Phospholipase* A_2

One of the characteristic features of the mitochondrial phospholipase is its absolute dependence for its activity on Ca^{2+} ions (Nachbaur et al., 1972). Detailed information about the influence of these ions on the hydrolysis rate with endogenous and exogenous phospholipids is presented, e.g., by Waite et al. (1969a). The mitochondrial phospholipase A_2 appears to have an exceptionally high affinity for Ca^{2+}. Indeed, prolonged dialysis of organelles subjected to freezing-thawing entails only a minor decrease in the hydrolysis rate with exogenous phosphatidylethanolamine (Waite et al., 1969a). Unlike the mitochondrial one, the major lysosomal phospholipase activity with a pH optimum in the acidic region (pH about 4.0) is attenuated in the presence of Ca^{2+} ions (Waite et al., 1969a; Nachbaur et al., 1972), while the alkaline lysosomal phospholipase A is but slightly activated by Ca^{2+} ions (Nachbaur et al., 1972).

Efficient stimulators of mitochondrial phospholipase activity are free fatty acids which substantially enhance the breakdown of both exogenous and endogenous phospholipids (Figure 16). The most pronounced effect is observed with lauric (12:0), myristic (14:0), oleic (18:1), and linoleic (18:2) acids (Waite et al., 1969a). These data mean that the activating potency of a fatty acid depends rather on its molecular weight than on the degree of unsaturation. It is worthwhile to mention the data on the influence of lysophosphatides on the mitochondrial phospholipase activity. According to Waite et al. (1969a), monoacylglycero-3-phosphoethanolamine potentiates the breakdown of exogenous phosphatidylethanolamine but at the same time suppresses that of the endogenous substrate.

The phospholipase activity is also stimulated by some physical treatments that impair the organelle structural integrity. Of these one can name freezing-thawing (Waite et al., 1969a) and ultrasonic exposure (Waite and Van Deenen, 1967).

In conformity with the above, mitochondrial phospholipase activity is inhibited by EDTA binding Ca^{2+} ions. As reported by Waite et al. (1969a), ATP and ADP but not AMP significantly decrease the hydrolysis rate of exogenous phosphatidylethanolamine by competing with phospholipase A_2 for Ca^{2+} ions. In this the efficiency of the nucleotides (especially ATP) approximates that of EDTA. The effect of Mg^{2+} ions is treated analogously. It is supposed that

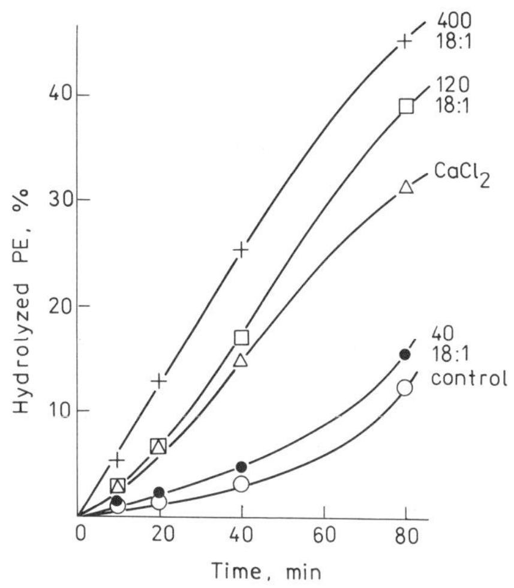

Fig. 16. Effects of $CaCl_2$ and oleic acid (18:1) on the hydrolysis of endogenous phosphatidylethanolamine (PE) in isolated rat liver mitochondria. (40, 120, 400) Nanomoles oleic acid per 10 mg mitochondrial protein. $CaCl_2$, 1 mM. Adapted from Waite et al. (1969a).

Mg^{2+} ions displace Ca^{2+} ions from the corresponding sites on the membrane and thereby attenuate the mitochondrial phospholipase activity.

Rossi et al. (1964) showed that mitochondrial aging is accompanied by a decrease in the content of mitochondrial phospholipids, and that phosphoric esters of choline, ethanolamine, and serine impede these processes, acting as competitive inhibitors towards the endogenous phospholipase A_2. Indirect support to this can be found in a later work by Estrada-O et al. (1966).

In recent years extensive use has been made of local anesthetics as inhibitors of mitochondrial phospholipase A, the most wide application being enjoyed by nupercaine (α-butoxycinchoninic acid diethylaminoethylamide chlorohydrate). In particular, inhibition of mitochondrial phospholipid hydrolysis by nupercaine was demonstrated by Scarpa and Lindsay (1972) and by Scherphof et al. (1972) (Table 6). The probable mechanism of nupercaine inhibition of mitochondrial phospholipase A_2 is that it, like other local anesthetics, competes with Ca^{2+} ions for the binding sites on the mitochondrial membranes (Scarpa and Azzi, 1968; Johnson and Schwartz, 1969; Scherphof et al., 1972; Scarpa and Lindsay, 1972; Waite and Sisson, 1972). In accordance with this, nupercaine does not suppress but rather activates the microsomal phospholipase A_1 which, unlike the mitochondrial phospholipase A_2, is inhibited by Ca^{2+} ions (Scherphof et al., 1972).

Table 6. Effect of Nupercaine on the Decrease in Respiratory Control and on Phospholipid Hydrolysis in Isolated Rat Liver Mitochondria (after Scarpa and Lindsay, 1972).

Hours of incubation	Without nupercaine			With 500 μM nupercaine		
	Respiratory control	Hydrolysis, %		Respiratory control	Hydrolysis, %	
		Phosphatidyl-choline	Phosphatidyl-ethanolamine		Phosphatidyl-choline	Phosphatidyl-ethanolamine
1	4.5	1.1	1.5	4.3	1.3	1.7
12	3.9	1.5	1.9	4.1	1.1	1.9
24	2.6	3.1	9.3	3.9	2.1	3.1
88	1.0	5.5	13.7	3.6	2.9	5.8

2.2.1.3. *Localization of Phospholipase A_2 in Mitochondria*

Nachbaur and Vignais (1968) were the first to attempt to elucidate the localization of phospholipase A_2 in mitochondria. By their data, the specific activity of this enzyme in the outer membrane fraction was almost four times that in the fraction of inner membranes with matrix. This gave grounds for suggesting that the bulk of phospholipase A_2 in liver mitochondria is associated with the outer membrane. However, the total phospholipase activity in the inner membrane fraction was too high to be accounted for by the residual outer membranes in the preparation. Therefore the authors thought it possible that phospholipase A_2 has a dual localization in mitochondria.

A similar conclusion was later arrived at by Waite (1969) who showed that the specific activities of monoamine oxidase and rotenone-insensitive NADH-cytochrome *c* oxidoreductase in the outer membrane fraction exceed those in the inner membrane fraction by more than an order of magnitude, whereas its phospholipase activity is only threefold higher. This means that the inner-membrane phospholipase activity cannot be wholly ascribed to an admixture of outer membranes. These results were undoubted proof of the dual localization of phospholipase A_2 in mitochondria. Still, they did not allow one to assess its portions associated with the outer and inner membranes. Indeed, the inner membrane preparations contained substantial amounts of matrix, as a result of which their specific phospholipase activity was underestimated.

A comparison of the specific phospholipase activities in the outer and inner mitochondrial membranes was later made by Nachbaur et al. (1972). By subjecting the inner membrane-matrix fraction to swelling in 20 mM phosphate buffer, they succeeded in removing most of the matrix proteins; thereby the specific activity of purified inner membranes reached more than 60% of the corresponding value found for the outer membrane preparation.

Additional evidence concerning the location of the enzyme was obtained in recent years. Severina and Evtodienko (1981) showed that the stimulatory action of Ca^{2+} ions on the endogenous phospholipase A_2 in rat liver mitochondria is suppressed by ruthenium red which is known (see Chapter One) to inhibit calcium transport across the inner mitochondrial membrane. From this it followed that phospholipase A_2 is activated from the matrix and is localized on the M-side of the membrane.

At the same time, according to Zurini et al. (1981), the hydrolysis of endogenous phosphatidylethanolamine in mitoplasts takes place on both M- and C-sides of the membrane. This was inferred from the results of experiments with fluorescamine which by interacting with the amino groups of phosphatidylethanolamine on the mitoplast surface (membrane C-side) partly inhibited its cleavage by the endogenous phospholipase A_2. Assuming that fluorescamine taken in low concentrations does not penetrate the membrane and that its complex with phosphatidylethanolamine is not cleaved by phospholipase A_2, the results of the experiment mentioned imply that normally phosphatidylethanolamine is broken down on both sides of the membrane while in the presence of fluorescamine the hydrolysis proceeds only on the M-side.

2.2.1.4. *Isolation of Mitochondrial Phospholipase A_2*

A final step in studying the mitochondrial phospholipase A_2 would naturally be to isolate the enzyme as an individual entity and to make a detailed description of its properties. Rossi et al. (1965) reported a purification procedure yielding a 10-fold increase in the enzyme specific activity. Waite and Sisson (1971) achieved a 169-fold increase of phospholipase activity but did not provide any data testifying that they had isolated an individual enzyme. This partially purified phospholipase A_2 preparation (constituting about 0.5% of the total mitochondrial protein) catalyzed phospholipid hydrolysis with the liberation of equimolar amounts of free fatty acids and 1-acylglycero-3-phosphoethanolamine, which indicated its specificity for the ester bond at the C_2 atom and the absence of the corresponding lysophospholipase activity. The isolated phospholipase A_2 was 50-75% inhibited by monoacylglycero-3-phosphoethanolamine (see above).

Recently De Winter et al. (1982) purified phospholipase A_2 from rat liver mitochondria to near homogeneity by combining gel filtration and column chromatography on hydroxylapatite and Matrex gel Blue. The enzyme was purified 3357-fold. Molecular-weight determinations gave values of 15,000 in sodium dodecyl sulfate-polyacrylamide gel electrophoresis and 9700 in gel filtration. The enzyme was insensitive to diisopropyl fluorophosphate and sulfhydryl reagents, but was suppressed by EDTA and an alkylating reagent p-bromophenacyl bromide. Though a single peak of phospholipase activity was observed through different stages of purification, it is still premature to say that phospholipase A_2 is common to the outer and inner mitochondrial membranes.

2.2.1.5. *Mitochondrial Alterations Associated with the Action of Endogenous Phospholipase A_2*

It is held that a certain alteration in mitochondria may be due to the endogenous phospholipase A_2 action if it evolves in parallel with phospholipid hydrolysis and is sensitive to the above-listed inhibitors and stimulators of the enzyme. A connection between phospholipid hydrolysis and the deterioration of the mitochondrial phosphorylating ability was established by Rossi et al. (1964) who also demonstrated that both phenomena are suppressed by phosphoric esters of choline, ethanolamine, and serine which are competitive inhibitors of phospholipase A_2. Also associated with the endogenous phospholipase A_2 action are the decrease of respiratory control in liver mitochondria (Chefurka, 1966; Scarpa and Lindsay, 1972), the attenuation of the 2,4-dinitrophenol-stimulated ATPase activity (Chefurka, 1966), and impairment of some other functions (Honjo et al., 1968; Scarpa and Lindsay, 1972). In the cases mentioned a protective effect was exhibited by EDTA, serum albumin, and nupercaine. Figure 17 shows how pronounced this protection can be. Obviously the most marked protective effect is exhibited by nupercaine which, simultaneously with maintaining the mitochondrial high respiratory control, notably lowers the rate of mitochondrial phospholipid hydrolysis (see Table 6). A fact worth special attention is that only 5.5% of the total mitochondrial phosphatidylcholine and 13.7% of phosphatidylethanolamine need be converted to corresponding lyso derivatives for the respiratory control to fall fourfold (Table 6).

The high-amplitude mitochondrial swelling may be a consequence of the endogenous phospholipase A_2 action. According to Waite et al. (1969b), swelling and phospholipase activity are stimulated by the same concentrations of $CaCl_2$ and oleic acid. Compared with the latter, monoacylphosphoglycerides (e.g., 1-oleylglycero-3-phosphocholine) are less effective as inductors of both swelling and phospholipase activity. A potent inhibitor of swelling, either spon-

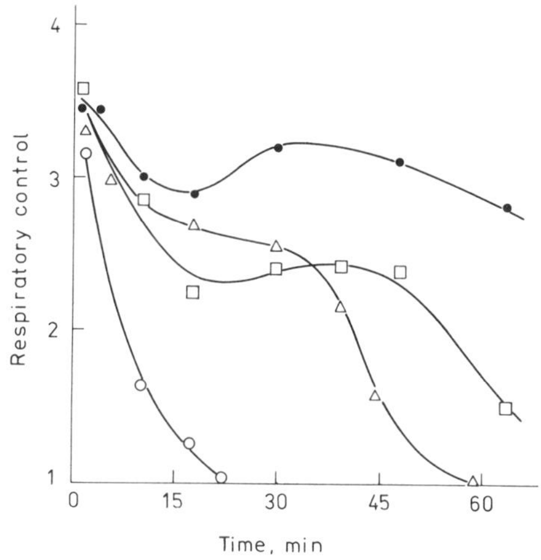

Fig. 17. Comparative efficiency of the protective action of serum albumin, EDTA, and nupercaine on rat liver mitochondria. (○) Control; (△) 3 mg/ml bovine serum albumin; (□) 0.5 mM EDTA; (●) 0.5 mM nupercaine. After Scarpa and Lindsay (1972).

taneous or induced by oleic acid and $CaCl_2$, is EDTA (Chefurka, 1966; Waite et al., 1969b). Interestingly, serum albumin abolishes the effect of oleic acid but aggravates that of Ca^{2+} ions. The latter seems to be due to the removal from the medium of the mitochondrial phospholipid hydrolysis products which can inhibit phospholipase A_2. Indeed, Chefurka (1966) showed the phospholipid breakdown rate to increase when mitochondria were incubated in the presence of serum albumin. A similar influence of serum albumin on lecithin hydrolysis by isolated phospholipase A was demonstrated by other authors (Earnshaw and Truelove, 1970). However, Honjo et al. (1968) claim that serum albumin hinders phospholipid hydrolysis in pancreatic mitochondria.

In principle, a dual character of serum albumin action is admissible: on the one hand, it may prevent the activation of phospholipase A_2 by fatty acids when their quantity is relatively small; on the other hand, binding of fatty acids may enhance phospholipid hydrolysis if the breakdown products are inhibitory towards phospholipase A_2.

Studying the thyroxine-induced mitochondrial swelling, Marzoev and Vladimirov (1977a,b) came to the conclusion that it was associated with mitochondrial phospholipid hydrolysis by the endogenous phospholipase A_2. Such swelling was not suppressed by antioxidants (α-tocopherol or β-ionol) and therefore was not related to lipid peroxidation (see further). At the same time, it did not occur in the presence of EGTA, a complexone for Ca^{2+}.

It must be noted that, according to Waite et al. (1969b), maximal mitochondrial swelling is observed when only 4% of the total mitochondrial phosphatidylcholine and phosphatidylethanolamine are hydrolyzed. This corresponds quite nicely with the above mentioned data of Scarpa and Lindsay (1972) on the relationship between phospholipid cleavage and impairment of respiratory control in mitochondria.

The probable role of the endogenous phospholipase A_2 in mitochondrial swelling and loss of catalytic functions is confirmed by the electron-microscopic observations of Aleksandrowicz et al. (1973). According to these authors, mitochondria undergo substantial alterations during a 60-min incubation at 30°C. The organelles are enlarged, their matrix is clarified, cristae are reduced and finally almost disappear. Nupercaine introduced into the incubation medium modifies somewhat the structure of freshly prepared mitochondria but prevents all the above changes attesting organelle degradation. Specifically, in the presence of the inhibitor mitochondria maintain narrow cristae and granular matrix of moderate density. The stabilizing action of nupercaine on mitochondrial structure was also described by Johnson et al. (1973).

There appears to be a direct connection between the results presented and the data of Estrada-O et al. (1966) on the possible role of the endogenous phospholipase in the increase of mitochondrial membrane permeability for "soluble" matrix enzymes. These authors found that incubation of liver mitochondria is accompanied by a release of glutamate dehydrogenase and glutamate-oxaloacetate aminotransferase from the organelles into the medium. This process was impeded by some phospholipids (phosphatidylethanolamine, phosphatidylcholine, phosphatidylserine, and sphingomyelin) as well as by phosphoric esters of choline or ethanolamine which, as already mentioned above, can inhibit phospholipase A_2. On the other hand, the loss of soluble enzymes from mitochondria was promoted by unsaturated fatty acids (oleic and linoleic) and lysolecithin. It is of interest that the effect of the latter was lessened by additions of phosphatidylcholine, phosphatidylserine, phosphatidylethanolamine, or phosphocholine. Other phospholipids (cardiolipin and phosphatidylinositol) did not affect either the spontaneous or the lysolecithin-induced leakage of enzymes from mitochondria. Thus, the mitochondrial membrane permeability for matrix enzymes is directly dependent on phospholipid integrity: a decrease in the phospholipid content and accumulation of lyso derivatives render the membrane incapable of retaining these enzymes.

2.2.1.6. *Possible Mechanisms of Mitochondrial Degradation Induced by Endogenous Phospholipase A_2*

Destructive alterations in mitochondria associated with the endogenous phospholipase A_2 may be caused by (1) the impairment of the membrane fine structure owing to phospholipid cleavage and (2) the action on the membrane of the cleavage products (free fatty acids and lysophosphatides). It was suggested that to a large extent the degradation is due particularly to free fatty acids, since their removal from the medium with serum albumin serves to preserve or even restore the mitochondrial phosphorylating ability (Weinbach, 1959; Helinski and Cooper, 1960; Hüllsmann et al., 1960; Lewis and Fowler, 1960; Wojtczak and Wojtczak, 1960; Priegnitz and Wojtczak, 1961; Honjo et al., 1968), prevents swelling (Lehninger and Remmert, 1959; Wojtczak and Lehninger, 1961; Chefurka, 1966; Waite et al., 1969b) or favors contraction of swelled organelles (Lehninger and Remmert, 1959).

However, data are available that cast doubt on this point of view. According to Waite et al. (1969b), the amount of fatty acids produced upon maximal swelling of liver mitochondria is 20-30 times less than the amount of exogenous fatty acids required to attain the same effect. From this it follows that the role of mitochondrial phospholipase A_2 consists in the breakdown of phospholipids in certain "key" sites of the membrane having paramount importance for the maintenance of its structure and functions. In all probability, at the first stages of mitochondrial degradation free fatty acids act as inductors of phospholipase activity. Indeed, EDTA inhibits phospholipase A_2 and blocks the swelling induced by small quantities of exogenously added oleic acid. When the amount of free fatty acids is large enough, they begin to exhibit a direct lytic action on mitochondrial membranes; this effect is no longer sensitive to EDTA and can only be abolished with serum albumin (Waite et al., 1969b). In any case it is obvious that mitochondrial degradation cannot be ascribed solely to the action of free fatty acids. Organelle swelling is demonstrable, for instance, even in the presence of excess serum albumin if the medium contains Ca^{2+} ions to induce phospholipase activity (Waite et al., 1969b).

Not quite clear is the role of lysophosphatides in mitochondrial degradation. On the one hand, taken in sufficient amounts they are known to produce a specific lytic effect on mitochondrial membranes (Estrada-O et al., 1966; Honjo et al., 1968; Capaldi et al., 1973; Komai et al., 1973; Rydström, 1976). On the other hand, addition of monoacylglycero-3-phosphocholine affects mitochondrial swelling to a lesser degree than that of equimolar quantities of oleic acid (Waite et al., 1969b). Besides, it was already mentioned above that lysophosphatides may be inhibitory towards the mitochondrial phospholipase A_2 (Waite and Sisson, 1971).

2.2.1.7. *Reutilization of Mitochondrial Phospholipid Hydrolysis Products*

As follows from the preceding section, there are means to activate the mitochondrial phospholipase A_2 (Ca^{2+} ions, fatty acids, disruption of mitochondrial membranes) and to inhibit it (ATP, ADP; EDTA, Mg^{2+} ions; phosphoric esters of choline, ethanolamine, and serine; nupercaine and other local anesthetics). In other words, the agents listed can in some way regulate the mitochondrial lipid composition. Another means of regulation is the reacylation of the phospholipid cleavage products.

Lehninger and colleagues (Lehninger, 1959b; Lehninger and Remmert, 1959; Wojtczak and Lehninger, 1961) showed that mitochondrial swelling induced by Ca^{2+} ions or thyroxine is accompanied by accumulation of the U-factor exhibiting a pronounced uncoupling effect on mitochondrial membranes. When mitochondria contract in the presence of ATP, the U-factor disappears at a rate close to that of the contraction (Wojtczak and Lehninger, 1961). Labeled oleate added to the incubation medium is in these conditions taken up together with other fatty acids constituting the U-factor (Figure 18), and is recovered mostly in cardiolipin and partly in the cephalin fraction. Oleate incorporation is specifically stimulated by glycero-3-phosphate, whereas phosphoric esters of choline, ethanolamine, or serine have no such ability (Wojtczak et al., 1963).

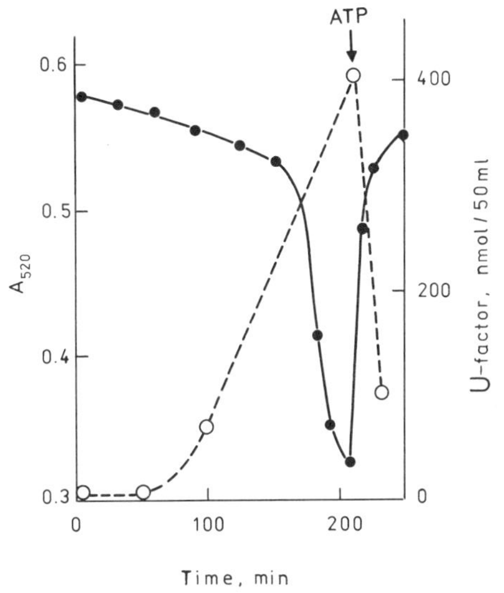

Fig. 18. Accumulation of the U-factor during spontaneous mitochondrial swelling, and its disappearance upon contraction induced by ATP and Mg^{2+} ions. (●) Absorbance at 520 nm; (○) U-factor. After Wojtczak and Lehninger (1961).

Similar investigations were carried out for the swelling induced by fatty acids (Wojtczak et al., 1963). Glycero-3-[^{32}P]phosphate proved to be incorporated into mitochondrial phospholipids during mitochondrial contraction elicited by ATP, the rates of both processes being close (Figure 19). The question arises whether the incorporation and contraction are interrelated or whether they take place concurrently but independently from each other. The first suggestion appears to hold true. Indeed, addition of glycero-3-[^{32}P]phosphate some time after introducing ATP, i.e. after the completion of mitochondrial contraction, does not result in its incorporation into phospholipids. Besides, compounds suppressing contraction (glucose, sucrose, glutathione) inhibit glycero-3-[^{32}P]phosphate incorporation as well. The incorporation level is also significantly decreased upon mitochondrial disruption by repeated freezing-thawing. Glycerophosphate incorporation depends on the presence in the incubation medium (besides ATP) of coenzyme A and fatty acids. It is negligible when mitochondrial preswelling is induced with phosphate, which does not entail a release of any appreciable amounts of fatty acids from mitochondria.

Later Marzoev and Vladimirov (1977b) observed the liberation of free fatty acids during mitochondrial swelling provoked by thyroxine, and a decrease in the free fatty-acid level upon subsequent organelle contraction in the presence of ATP.

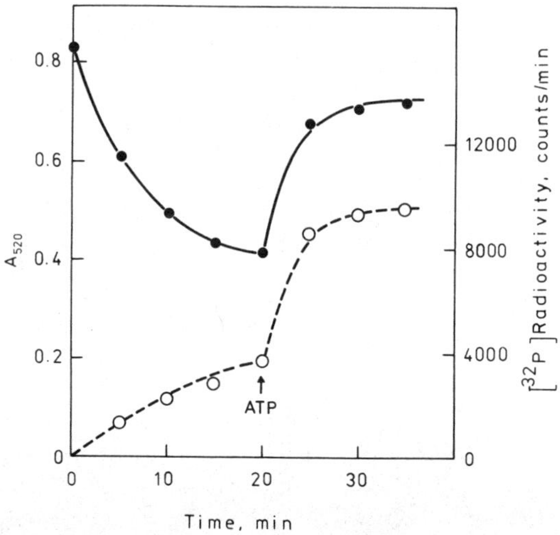

Fig. 19. Incorporation of α-glycero-[^{32}P]phosphate into mitochondrial phospholipids during oleate-induced swelling and ATP-dependent contraction. (●) Absorbance at 520 nm; (○) labeling of mitochondrial phospholipids. After Wojtczak et al. (1963).

The results presented give grounds for suggesting that the biosynthesis of some mitochondrial phospholipids is an indispensable (if not the sole) condition for mitochondrial contraction. A question arises about the identity of these phospholipids. As shown by Wojtczak and Lehninger (1961) and by Wojtczak et al. (1963), [^{14}C]oleate and glycero-3-[^{32}P]phosphate are incorporated mostly into phosphatidic acid and/or cardiolipin. The label is not found therewith in the major constituents of the mitochondtial phospholipid fraction, phosphatidylcholine or phosphatidylethanolamine. Two explanations have been advanced for the facts observed. First, mitochondrial swelling was supposed to be associated with some structural alterations in the membranes. Consequently, phospholipid resynthesis could be necessary for the reparation of the damaged membrane regions. Second, the phosphatidic acid formed during mitochondrial contraction was assumed to play an active part in the removal of surplus water from the organelles.

It must be pointed out in connection with the above that the work of Stoffel and Schiefer (1968), Nachbaur et al. (1969), Sarzala et al. (1970), Parce et al. (1978) demonstrated the ability of mitochondria to catalyze acylation of 1-acyl- and 2-acylglycero-3-phosphocholine, 1-acylglycero-3-phosphate, and 1,2-diacylglycerol. It appears from their data that the acylation reactions require ATP and take place mostly in the inner mitochondrial membrane.

2.2.2. Mitochondrial Degradation Owing to Lipid Oxidation

The mechanism of lipid peroxidation in biological membranes and its role in various pathological and normal physiological processes are treated in detail in the books by Vladimirov and Archakov (1972) and by Kozlov et al. (1972). According to these authors, the central part in lipid peroxidation is played by Fe^{2+} ions. Small amounts of Fe^{2+} (in the order of 10^{-6} M) always present in tissues suffice for lipid peroxidation to be substantially enhanced. Reactions with the participation of bivalent iron proceed at especially high rates in the presence of ascorbic acid, cysteine, or glutathione, which reduce Fe^{3+} to Fe^{2+}. Lipid oxidation in various membranes (including mitochondrial) induced by exogenous Fe^{2+} ions or reductants is inhibited by EDTA, o-phenanthroline, and other complexing compounds. Also inhibitory for this reaction are antioxidants (butyl hydroxytoluene, propyl gallate, α-tocopherol, etc.). It is of interest that excess Fe^{2+} may suppress oxidation by breaking the chain reactions. Not delving into the fine mechanism of lipid peroxidation in biological membranes, let us consider the data revealing the relationship between this process and mitochondrial degradation.

2.2.2.1. *Mitochondrial Swelling and Lysis Induced by Ascorbate, Fe^{2+} Ions, or Glutathione*

A detailed study of this type of mitochondrial degradation was made by Hunter et al. (1964a,b) who established that incubation with low concentrations of ascorbate (1-30 μm) leads, on the one hand, to accumulation of lipid peroxides and, on the other hand, to mitochondrial swelling and subsequent lysis (Figure 20). Destructive alterations were revealed electron-microscopically and by ultracentrifugation of the organelles after incubation with ascorbate. According to the results obtained, mitochondria lose about 75% of their protein content.

Lysis and lipid peroxide accumulation are impeded by natural and synthetic antioxidants (α-tocopherol, vitamins A and K_1, butyl hydroxytoluene, and propyl gallate), which indicates that the two processes are intimately connected. On the basis of these data it could be supposed that accumulation of lipid peroxides is the cause of increasing permeability and disintegration of mitochondrial membranes.

Respiratory inhibitors (cyanide, antimycin A, 2-hydroxy-3-(2-methyloctyl)-1,4-naphthoquinone, and others) prevent the ascorbate-induced mitochondrial lysis. However, they are only effective in rather large amounts exceeding by far those needed to block electron transfer through the respiratory chain. The protective effect of the compounds listed is due rather to their acting as complexones binding heavy metal ions. Lysis is also suppressed by ATP and Mg^{2+} ions,

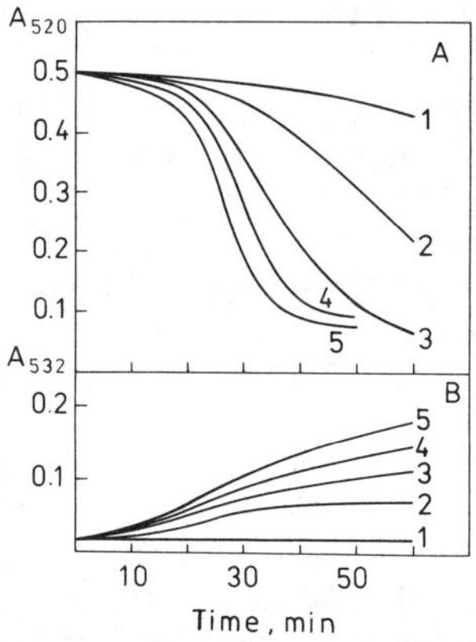

Fig. 20. Swelling-lysis (A) and lipid peroxidation (B) induced by ascorbate in liver mitochondria. (1) Control; (2, 3, 4, and 5) incubation with 1, 3, 30, and 300 μM ascorbate, respectively. From Hunter et al. (1964b).

the effect being more pronounced with both reagents present in the medium. The maximal stabilizing action is exhibited by an ATP-Mg^{2+}-serum albumin combination which is conventionally used to avert or reverse mitochondrial swelling. In other words, mitochondria in a compact form are less prone to be lyzed because of the oxidative destruction of membrane lipids. In support of this it can be added that preincubation of mitochondria at 25°C impairs their fine structure and accelerates the ascorbate-induced swelling.

Mitochondrial swelling and lysis can be induced by low concentrations (1-10 μM) of Fe^{2+} ions. In this case the destructive alterations in mitochondria also correlate with lipid peroxide formation (Hunter et al., 1963). In particular, addition of oxidized glutathione together with Fe^{2+} ions into the incubation medium was found to accelerate swelling and to simultaneously increase the rate of lipid peroxide formation. Vice versa, inhibition of swelling by antimycin is accompanied by a decrease in the peroxide level in mitochondria.

For the measure of mitochondrial injury inflicted by lipid peroxidation to be appreciated, suffice it to say that upon incubation with Fe^{2+} ions mitochondria lose 65% of total protein and 39% of lipid (McKnight and Hunter, 1966); only 33% of flavins and 55% of ubiquinone are retained, and pyridine nucleotides are completely lacking; there is a substantial loss of cytochromes aa_3, b, and especially $c+c_1$. Ferrous ions produce a significant decrease in the activities of β-hydroxybutyrate, malate, and succinate dehydrogenases and total disappearance of respiratory control and phosphorylation.

Swelling of isolated mitochondria caused by a mixture of reduced and oxidized glutathione (GSH+GSSG) is also closely associated with lipid oxidation. As in the former cases, swelling is prevented by EDTA, antioxidants (butyl hydroxytoluene, propyl gallate et seq.), and electron-transfer inhibitors (cyanide, antimycin A, 2-nonyl-4-hydroxyquinoline-N-oxide), their protective effect not being due to the impairment of the respiratory chain functions. It is noteworthy that other respiratory inhibitors such as amytal, rotenone, or malonate do not interfere with the action of the GSH-GSSG mixture on mitochondria. Swelling is completely abolished in the presence of ATP, Mg^{2+}, and serum albumin.

2.2.2.2. *Mitochondrial Degradation Linked with NADPH Oxidation*

Pfeifer and McCay (1972) demonstrated that NADPH oxidation is accompanied by destructive alterations in mitochondria which are rather similar to those described in the preceding subsection. According to these authors, NADPH oxidation entails cleavage of polyene fatty acids entering into the composition of mitochondrial membrane phospholipids. In fact, in 40 min of organelle incubation at room temperature in a medium containing NADPH and physiological amounts of Fe^{3+} the content of unsaturated fatty acids falls by half (Table 7); concomitantly an equimolar amount of malondialdehyde appears.

Deterioration of mitochondrial membranes caused by lipid oxidation manifests itself, among other things, as a gradual decrease in the rates of succinate, glutamate, and α-ketoglutarate oxidation; the mitochondrial phosphorylating ability is lost even more quickly.

Moderate heating of mitochondria or addition of p-chloromercuribenzoate inhibits NADPH oxidation and malondialdehyde production. This may mean that the lipid breakdown is due to some enzymic process. Pfeifer and McCay (1972) suppose that all membrane alterations described are associated with the functioning of mitochondrial transhydrogenase. The mechanism of the process on the whole is obscure. It must be emphasized that Fe^{3+} (or Fe^{2+}) ions are involved; this may be the reason of the inhibitory action of EDTA and excess Mn^{2+} ions which compete with Fe^{3+} for the membrane binding sites.

2.2.2.3. *Mitochondrial Degradation in the Presence of Glutathione, and the Protective Function of Glutathione Peroxidase*

As indicated above, swelling of liver mitochondria induced by glutathione (either its reduced form or a mixture of reduced and oxidized forms) is due to lipid peroxide formation. Neubert et al. (1962) have shown that in this case prevention or reversal of swell-

Table 7. Peroxidation of Mitochondrial Lipids Associated with NADPH Oxidation (adapted from Pfeifer and McCay, 1972).

Additions to the incubation medium	NADPH oxidation	Malondialdehyde production	Oxygen consumption
		nmoles/min	
None	-	2.17	0
Succinate (20 mM)	-	1.94	4341
NADPH	131.9	40.72	875
NADPH, mitochondria preheated at 65°C	23.3	3.67	0
NADPH + p-chloromercuribenzoate	5.0	6.30	84
NADPH + Mn^{2+} (1 mM)	131.9	2.17	252
NADPH + EDTA (1 mM)	117.5	3.04	236
NADPH + methylaniline (1 mM)	131.9	1.98	135

ing requires a specific factor (C-factor) comprising three components, one of which has catalase activity, the second is glutathione peroxidase, and the third has not yet been identified. It is of interest that highly purified glutathione peroxidase from erythrocytes together with crystalline catalase display all the properties of the C-factor: they reverse in the presence of ATP the glutathione-induced mitochondrial swelling, increase the efficiency of mitochondrial oxidative phosphorylation, etc. The C-factor most probably acts by removing lipid peroxides from the mitochondrial membranes owing to the glutathione peroxidase reaction (Flohé and Zimmerman, 1970). Indeed, data are available on the reduction of unsaturated-fatty-acid peroxides by glutathione with the participation of glutathione peroxidase (Christofersen, 1968; Little and O'Brien, 1968). If this is so, then mitochondria need the C-factor to put up a defense against the degradation caused by lipid oxidation.

McCay et al. (1976) supposed, however, that the protective action of glutathione peroxidase consists first and foremost in averting the free-radical attack on the membrane unsaturated lipids. By their data, during mitochondrial incubation with glutathione peroxidase the hydroxy derivatives of polyunsaturated fatty acids are not accumulated even under conditions favoring generation of substantial amounts of lipid peroxides in the membranes (e.g., in the presence of ascorbate and glutathione). Yet, it is the hydroxy derivatives that would have been the stoichiometric reaction product in the enzymic reduction of peroxide groups in polyunsaturated fatty acids by glutathione. McCay et al. (1976) believed that lipid oxidation in biological membranes proceeds at the expense of free radicals (possibly $OH^{\cdot}$ radicals) formed upon interaction of hydrogen peroxide or the superoxide anion $O_2^{\dot{-}}$ with the traces of Fe^{2+}. Glutathione peroxidase, having an affinity for H_2O_2 several orders of magnitude higher than catalase, is capable of maintaining the concentration of H_2O_2 (and consequently of $OH^{\cdot}$ radicals) at such a low level as to render the oxidation of polyunsaturated fatty acids hardly probable.

From the early data of Neubert et al. (1962) it followed that rat liver mitochondria easily lose the C-factor upon treatment with reduced glutathione or sonication. Later Green and O'Brien (1970) confirmed this. They showed that ascorbate, reduced glutathione, and oleic acid induce mitochondrial swelling and leakage of glutathione peroxidase into the medium. On the other hand, organelle incubation with phosphate or Ca^{2+} ions is not accompanied by the leakage of glutathione peroxidase. This prompts one to think that mitochondria lose the enzyme as a result of a specific membrane labilization because of lipid oxidation. Indeed, phosphate added to the incubation medium together with glutathione suppresses lipid peroxide formation and prevents the loss of glutathione peroxidase (Green and O'Brien, 1970).

Mitochondrial glutathione peroxidase has now been obtained in a purified state (see, e.g., Zakowski and Tappel, 1978). It was shown that in rat liver mitochondria it is contained in the matrix (Flohé and Schlegel, 1971) and is released into solution in an active form upon exposure of mitochondria to ultrasound (Green and O'Brien, 1970) or Triton X-100 (Sies and Moss, 1978; Zakowski and Tappel, 1978; Lötscher et al., 1979). The enzyme shows high affinity for selenium and is inhibited by iodoacetamide (Zakowski and Tappel, 1978). In Se-deficient rats glutathione peroxidase activity is within 6% of the normal one (Lötscher et al., 1979). At the same time, data are available that liver and heart mitochondria also contain a glutathione peroxidase that is inactivated by sonication or Triton; however, its activity is retained when rats are for a long time kept on a Se-free diet (Katki and Myers, 1980). This enzyme is likely to be localized in the intermembrane space, as judged by the results of mitochondrial fragmentation with digitonin.

It should be noted that rat liver mitochondria harbor about one third of the overall cell activity of glutathione peroxidase and glutathione reductase (Flohé and Schlegel, 1971). The presence of these two enzymes in such substantial amounts allows in principle a negligible level of H_2O_2 and unsaturated-fatty-acid peroxiderivatives to be maintained in mitochondria as long as the pair of enzymes is operative. It was pointed out that these enzymes are coupled with other mitochondrial metabolic systems, in particular through the pyridine nucleotide pool (Lötscher et al., 1979) according to the scheme:

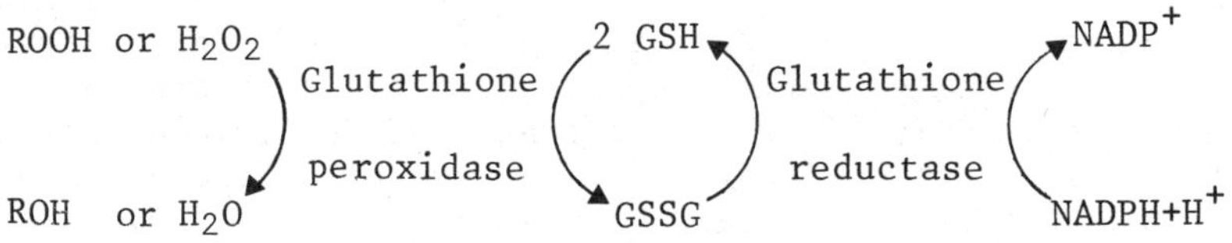

2.2.2.4. *Lipid Oxidation in Mitochondrial Membranes Induced by Chaotropic Agents*

According to Hanstein and Hatefi (1970) and Hatefi and Hanstein (1970), mitochondrial lipid oxidation by molecular oxygen can be induced with chaotropic agents (guanidine chloride, urea; SCN^-, I^-, Br^- ions, etc.) which destabilize the membrane structure. Effectiveness of these agents correlates with their ability to disturb the structure of water and increase its lipophilicity, leading to the weakening of hydrophobic interactions in the membrane and separation of its lipid and protein constituents. In some mitochondrial membrane preparations (especially in those containing large amounts of cytochromes) lipid oxidation proceeds spontaneously, without any addition of chaotropic agents. However, usually this process is slow.

Thus here we again encounter a case when the lipid oxidation rate is directly dependent on the state of mitochondrial membranes: being low in intact membranes with compact structure, it is high in those that are somehow impaired. Of particular importance in this connection are the data establishing the relationship between the membrane functional state and the susceptibility of its lipids to oxidation. Hatefi and Hanstein (1970) found the accumulation of malondialdehyde to be retarded when the submitochondrial particles performed succinate or β-hydroxybutyrate oxidation. It was supposed that the stabilizing substrate action was due to the transition of some membrane components to the reduced state. Indeed, the reduced forms of complex I (Davis and Hatefi, 1969), complex II (Baginsky and Hatefi, 1969), and complex III (Rieske et al., 1967) are more immune to the dissociating action of chaotropic agents.

We see that lipid peroxides may form in mitochondrial membranes even without any additions (such as ascorbate, Fe^{2+} ions, etc.) capable of initiating free-radical oxidative processes. Then what is the cause of lipid oxidation? According to Hanstein and Hatefi (1970), in the presence of chaotropic agents the rapid initial phase of oxidation is catalyzed by iron-sulfur proteins or by intermediate products of their breakdown. Firstly, there is a correlation between the lipid oxidation rate and the content of labile sulfides in various membrane preparations. For instance, complex I is enriched in labile sulfides as compared to complex III or submitochondrial vesicles. As follows from Table 8, it displays an accordingly higher lipid oxidation rate. Membranes of _Bacillus subtilis_ are devoid of iron-sulfur proteins, and their lipids are practically not oxidized by oxygen. Secondly, as malondialdehyde accumulates, the mitochondrial membranes lose labile sulfides (Figure 21), both processes depending on the concentration of the chaotropic agent.

However, iron-sulfur proteins are in all probability not the only initiators of lipid oxidation in mitochondrial membranes. It

Table 8. Correlation Between the Capacity of Autooxidation and the Content of Labile Sulfides in Various Membranes and Their Fragments (after Hanstein and Hatefi, 1970).

Preparation	Initial rate of oxygen uptake, nmoles/min·mg	Content of labile sulfides, nmoles/mg
Complex I	17.2	23 - 26
Submitochondrial vesicles	4.8	7.1 - 7.3
Complex III	3.5	6 - 8
B. subtilis membranes	0.0	0.5
Erythrocyte membranes	0.0	0

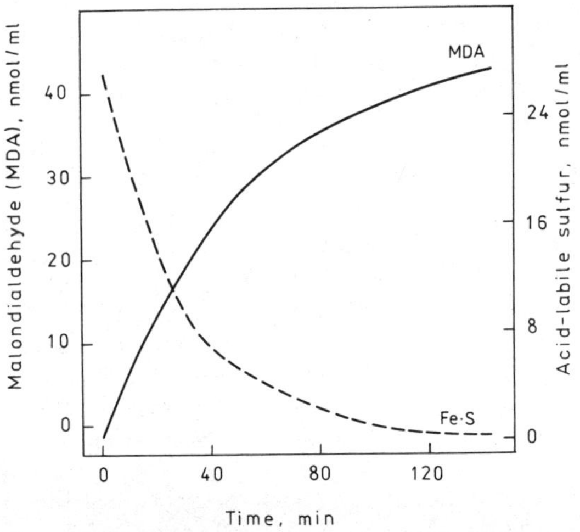

Fig. 21. Formation of malondialdehyde and decrease in the content of labile sulfides in submitochondrial vesicles exposed to perchlorate (adapted from Hanstein and Hatefi, 1970).

was shown that in preparations of complex IV, in the presence or even in the absence of chaotropic agents, lipid oxidation proceeds at a particularly high rate (Hatefi and Hanstein, 1970) though this preparation contains only trace amounts of nonheme iron and labile sulfides. Consequently, lipid oxidation in mitochondria can be triggered by cytochrome aa_3 or cuproproteins. In other words, the proneness of lipids to oxidation is determined by the very composition of mitochondrial membranes.

2.2.2.5. *Role of Superoxide Radicals in Mitochondrial Lipid Oxidation*

Weisiger and Fridovich (1973a) were the first to show that mitochondria isolated from various sources display a superoxide dismutase activity distinct from that of the cytosol. The organelles possess two types of superoxide dismutase: one contains manganese and is located in the matrix, and the other is copper- and zincum-dependent and is found in the intermembrane space (Weisiger and Fridovich, 1973b). Similar results were obtained later by Panchenko et al. (1975) and by Peeters-Joris et al. (1975). Superoxide dismutase was supposed to be needed to scavenge the excess of superoxide radicals ($O_2^{\dot{-}}$) which are highly reactive and can, among other things, induce mitochondrial lipid oxidation. As shown by Goda et al. (1973), the presence of NADPH and cytochrome *c* markedly accelerates the production of singlet oxygen by the adrenodoxin-adrenodoxin reductase system and simultaneously the accumulation of the lipid peroxidation products in adrenocortical mitochondria. Both processes are suppressed by exogenous superoxide dismutase and an antioxidant 1,4-diazobicyclo-[2,2,2]-octane.

Tyler (1975) found that incubation of beef heart submitochondrial vesicles with NADH in aerobic conditions entails a decrease in their NADH oxidase (but not succinate oxidase) activity. Since this was prevented by superoxide dismutase and a number of other agents that can annul the effect of Fe^{2+} ions, it was suggested that the inactivation results from the action of the superoxide O ion on the NADH dehydrogenase segment of the respiratory chain.

However, the results obtained later (Wang and Kimura, 1976) pointed to that the Fe^{2+}-induced lipid peroxidation in adrenocortical mitochondria is not prevented by the addition of superoxide dismutase or 1,4-diazobicyclo-[2,2,2]-octane. This allowed one to think that the superoxide ion-radicals do not appreciably contribute to mitochondrial lipid oxidation in these conditions.

A work worth mentioning in this connection is that of Kaschnitz and Hatefi (1975) who compared a number of electron-transfer proteins and superoxide radical-generating systems by their ability to initiate the oxidation of a natural phospholipid mixture (azolectin). The greatest effectiveness in this model system was found for heme-carrying compounds (hematin, peroxidase, methemoglobin, and cytochrome *c*. Iron-sulfur proteins also proved to be good oxidation initiators in the presence of chaotropic agents. At the same time, the systems generating superoxide radicals (e.g., xanthine-xanthine oxidase) turned out to be poor initiators. It is important to note that lipid oxidation elicited by ferredoxin or heme-containing compounds could not be inhibited by superoxide dismutase (erythrocuprein). It is thus quite possible that in the oxidation of natural phospholipids forming artificial liposomes, the superoxide radical is neither an initiator nor an obligatory intermediate of the reaction. It is also not excluded that the oxidation of mitochondrial membrane lipids is only in some special cases mediated by the superoxide ion-radical.

2.2.3. Mitochondrial Proteinases and Their Role in Organelle Degradation

2.2.3.1. *Amino Acid Production by Isolated Mitochondria*

The idea of the existence of specific neutral proteinases in mitochondria was initially based on the experimental fact that the organelles incubated in media with neutral pH release amino acids. This phenomenon, discovered more than two decades ago (Korner and Tarver, 1957; Penn, 1961), was then extensively studied by Bartley's group (Alberti and Bartley, 1963, 1965, 1969; Bartley and Birt, 1970; Ferdinand et al., 1973). These researchers showed that at 40°C in phosphate buffer pH 7.2 containing 0.25 M sucrose the rate of the amino acid release by rat liver mitochondria amounts to 0.144 μmoles per mg dry weight per hour. Albeit the rate falls by approximately

Table 9. Changes in the Contents of Amino Acids, Water, and Protein during Aging of Rat Liver Mitochondria (after Alberti and Bartley, 1969).

Temperature, °C	Incubation time	Decrease in protein content, % of total	Ratio of water content to to mitochondrial dry wt.	Total amino acid content, μmoles per mg dry wt.	Amino acid concentration, μM		M/S ratio	Concentration gradient, μmoles/ml
					Mitochondria (M)	Supernatant (S)		
0	0	0	2.62	0.053	12.79	0.70	18.3	12.72
	1 h	0.5	2.65	0.059	16.08	0.58	27.7	15.5
	2 h	1.6	2.66	-	-	-	-	-
	4 h	2.4	2.73	0.060	11.80	0.96	12.3	10.84
	8 h	12.0	2.96	0.082	9.78	1.67	5.9	8.11
40	0	2.5	2.47	0.037	8.4	1.2	7.0	7.2
	15 min	34.4	5.67	0.108	25.5	1.7	15.0	23.8
	30 min	42.2	6.43	0.236	76.3	3.3	23.1	73.0
	60 min	34.0	8.30	0.526	81.5	12.3	6.6	69.2

an order of magnitude when the temperature is lowered to 0°C, it is still too high for the process in total to be neglected. As follows from Table 9, the amino acids are first accumulated within the mitochondria (this is especially conspicuous at 40°C) and then liberated into the medium, which manifests itself as an increase and subsequent decrease of the M/S ratio. This can be considered the first indication that the amino acid production can hardly be attributed to the action of any exogenous proteinases.

A complete set of amino acids is found in mitochondria and in the incubation medium, the relation between them being determined by a number of factors (Ferdinand et al., 1973). The composition of the free amino acid pool in freshly isolated mitochondria does not correspond to the amino acid composition of the mitochondrial membrane and total organelle protein, which may signify the lack of appreciable proteolysis in functioning mitochondria. On the other hand, the relations between amino acids liberated by mitochondria incubated in anaerobic conditions basically match the amino acid composition of mitochondrial proteins. These data are compatible with the suggestion that incubation of isolated mitochondria is attended by the breakdown of their proteins (possibly the membrane ones). In the presence of oxygen the proteolysis is probably accompanied by metabolic conversions of some amino acids.

A fact deserving particular attention is that freshly isolated mitochondria, unlike those incubated in these or that conditions, harbor substantial amounts of sulfur-containing amino acids. Indeed, according to the amino acid analysis data the portion of cysteic acid, taurine, and glutathione reaches 50 mole %. Ferdinand et al. (1973) are of the opinion that a high glutathione level is needed to protect mitochondria from degradation owing to lipid peroxide formation (see Section 2.2.2).

The rate of amino acid liberation by isolated liver mitochondria depends on the composition of the incubation medium (Alberti and Bartley, 1963, 1965). The ability of mitochondria to release amino acids was supposed to be enhanced by their swelling. In particular, this was the explanation put forward for the inhibition of this process by EDTA and stimulation by glutathione and cysteine (Alberti and Bartley, 1963, 1965). However, later investigations showed that calcium ions and thyroxine do not produce such changes in the amino acid liberation rate as would be strictly correlated with swelling (Alberti and Bartley, 1969).

Nevertheless, the liberation rate obviously depends on the extent of mitochondrial integrity. Indeed, disruption of mitochondria causes a substantial acceleration (by 50 to 100%) of the mitochondrial protein breakdown, probably owing to increased accessibility of the substrate for endogenous proteinases.

According to Alberti and Bartley (1969), mitochondrial swelling may be directly related to amino acid production. In fact, it can be seen in Table 9 that the amino acid concentration within mitochondria is always higher than in the surrounding medium, the gradient growing more steep with increasing temperature (i.e. with increasing amino acid liberation rate). Calculations made for mitochondria incubated for 30 min at 40°C show that the contribution of the osmotic effect of intramitochondrial amino acids to the organelle swelling amounts to 60%. Thus, the proteolysis may induce spontaneous swelling in an isotonic medium accompanied by alterations in the membrane structure, breach of permeability barriers, etc.

A characteristic feature of amino acid production by isolated mitochondria is the dependence of its rate on the oxygen level in the incubation medium: in aerobic conditions the process is slower than in anaerobic (Alberti and Bartley, 1963; Ferdinand et al., 1973). It could be thought that by oxidizing endogenous substrates the mitochondria acquire some resistance to neutral proteinases. It is of special interest in this connection that isocitrate, α-ketoglutarate, succinate, malate, etc. taken in 10 mM concentrations decrease the rate of amino acid production by 20-50% (Alberti and Bartley, 1965). Unfortunately, the effect of substrates was not studied in parallel under aerobic and anaerobic conditions, which makes these data difficult to interpret. Some indirect evidence suggests that the substrate effect is not necessarily associated with electron transfer (Alberti and Bartley, 1965).

Alberty and Bartley (1969) provided experimental evidence that the amino acid production in mitochondrial suspensions cannot be ascribed to the proteinases of lysosomes which might contaminate the preparations. Firstly, in mitochondria treated with detergents (e.g., Triton X-100) or subjected to freezing-thawing the rate of amino acid liberation is only twice as high whereas the lysosomal acid phosphatase activity increases ninefold. Secondly, α-glycerophosphate, an agent protecting lysosomal membranes from rupture, does not attenuate but enhances amino acid production by mitochondrial suspensions. Thirdly, the proteolytic activity of the "light" mitochondrial fraction exceeds that of "heavy" mitochondria by 30% while their acid phosphatase activity is higher by 300%. However, there is still no unanimous opinion on this issue (see, e.g., Rubio and Grizolia, 1977).

2.2.3.2. *Some Data on Mitochondrial Proteinases*

Thus, there are grounds to think that rat liver mitochondria contain specific endogenous proteinases responsible for amino acid production. There were preliminary data on their localization in the organelles (Alberti and Bartley, 1969). Since after mitochondrial disruption and centrifugation more than 80% of the proteolytic activity (at pH 7.4) was recovered in the supernatant, the enzymes could be supposed to be not membrane-bound but localized in the matrix.

Accumulation of the trichloroacetic acid-soluble proteolysis products during mitochondrial incubation in neutral media is inhibited by 70-80% by sulfhydryl reagents iodobenzoate and p-chloromercuribenzene sulfonate. The soybean trypsin inhibitor and diisopropyl fluorophosphate are 30-40% inhibitory. These results seem to signify that the proteinases are of serine and thiol types.

There is probably a direct connection between these data of Alberti and Bartley and the recently published results of Desautels and Goldberg (1982). These authors demonstrated that the matrix of rat liver mitochondria contains a high-molecular-weight endopeptidase (mol.wt 550,000) active at alkaline pH. The proteinase is inhibited by phenylmethylsulfonyl fluoride, diisopropyl fluorophosphate, some sulfhydryl-blocking reagents, and Triton X-100. It is activated 10-40-fold by ATP at physiological concentrations, this event requiring Mg^{2+}. It is important that the polypeptide ubiquitin, which is necessary for the ATP-dependent hydrolysis of proteins in reticulocytes (Hershko and Ciechanover, 1982), is without effect on the proteolytic activity of the ATP-dependent mitochondrial endopeptidase. This points to the difference between the mitochondrial and the cytoplasmic protein-degrading systems. Desautels and Goldberg supposed that the enzyme they discovered catalyzes the initial stages of protein breakdown in rat liver mitochondria.

In addition to what was already said, mention can also be made of the report by Subramanian et al. (1975) that upon sonication of rat liver mitochondria a proteinase (or proteinases) is solubilized which is particularly active in cleaving the "structural" proteins of the mitochondrial membrane. Volodina et al. (1977) described a soluble proteinase from rat liver mitochondria which digests at a relatively high rate those proteins that are "washed out" from mitochondria upon treatment with EDTA and 2-mercaptoethanol.

Contrary to this, Rapoport et al. (1982) suggested that liberation of lysine from isolated rat liver mitochondria is associated with the action of an ATP-dependent proteolytic system localized on the outer side of the inner mitochondrial membrane. This suggestion was made on the basis of the inhibitory effect of antimycin A and carboxyatractyloside on the liberation of lysine. Both compounds had practically no influence on proteolysis when the inclubation medium was supplemented with exogenous ATP. Obviously these data cannot be considered a direct indication of the existence of a membrane-bound proteinase.

In a series of publications by Heinrich et al. (1976), Jusić et al. (1976), Haas et al. (1977), Haas and Heinrich (1979) it was found that the inner membrane fraction of rat liver mitochondria contains special proteinase and carboxypeptidase. The proteinase, identified as a histone-cleaving enzyme, was isolated and characterized by several enzymic and physicochemical properties (Heinrich et al., 1976). Later this enzyme turned out (Banno et al., 1978; Haas et al., 1978) to be apparently identical with the proteinase obtained from rat liver mitochondria by another group of researchers (Katunuma et al., 1975) and performing, as they believed, the limited proteolysis of the apoproteins of pyridoxal-dependent enzymes.

However, further studies (Haas et al., 1979) have led to the conclusion that both the proteinase and the carboxypeptidase found in the mitochondrial fraction actually originate in specific granules of mast cells, which are present in tissues of many organs and especially in liver where they may amount to 0.1% of total organ mass. From this it follows that liver and probably other mammalian organs are not suitable for studies of membrane-bound mitochondrial proteinases. It is not yet clear whether the same applies to the above mentioned "soluble" liver mitochondrial proteinases.

A recent study by Duque-Magalhaes and Regnier (1982) has shown that after treatment with the compound 48/80 which, by degranulating mast cells, presumably prevents contamination of the mitochondrial fraction by mast cell proteinases, the preparation of mitochondrial membranes retains a caseinolytic activity about 80% of the control. This activity was found predominantly in the matrix and the inner membrane fraction although the other two mitochondrial compartments also showed some ability to cleave casein. Apart from questioning the reliability of this experimental approach, one should bear in mind that as long as the proteolytic activities are assayed with such extraneous substrates as casein and the like, their physiological significance would be entirely obscure. Much more appealing in this respect are the data concerning the intramitochondrial breakdown of polypeptides made on mitochondrial ribosomes.

2.2.3.3. *Breakdown of Mitochondrial Translation Products*

Wheeldon et al. (1974) found that [^{14}C] leucine incorporated into the proteins of isolated rat liver mitochondria is gradually lost from the mitochondria in the form of products not precipitated with 10% trichloroacetic acid. In 20-30 min the acid-insoluble fraction contains only 20-40% of the label. This process is partially suppressed by N-α-tosylamidophenylethyl chloromethylketone, a specific inhibitor of serine proteinases.

The cleavage of polypeptides in mitochondria is preceded by their detachment from mitochondrial ribosomes. According to the data presented in Figure 22a, upon centrifugation of a Triton mitochondrial lysate in a sucrose density gradient a significant portion of the label incorporated into acid-insoluble products was found in the ribosomal fraction. If the lysate before centrifugation was briefly incubated with puromycin, the label disappeared from the ribosomal zone but its content in the acid-insoluble products in the upper gradient layers did not increase. In similar experiments with isolated mitochondrial ribosomes, the puromycin treatment prior to centrifugation entailed a gradual decrease of the label content in the ribosomal fraction and a corresponding increase in the free protein fraction (Figure 22b).

While puromycin, by making the nascent polypeptides detach from ribosomes, enhanced their proteolysis, chloramphenicol which hinders the transpeptidation step retarded the accumulation of trichloroacetic acid-soluble products. Cleavage of the intramitochondrially synthesized polypeptides was stimulated by ATP, which in the opinion of Wheeldon et al. (1974) was due to the acceleration of translation on mitochondrial ribosomes. The authors regarded this breakdown as the result of the action of a proteinase attacking polypeptides released from mitochondrial ribosomes and performing regulatory functions in the mitochondrial membrane formation. It was suggested that the high susceptibility to proteolysis resulted either from errors in translation or from lack of factors required for modification of nascent polypeptides or their proper insertion into the membrane.

Proteolysis of mitochondrial translation products will be discussed in more detail in Chapter Eight where we shall deal with yeast

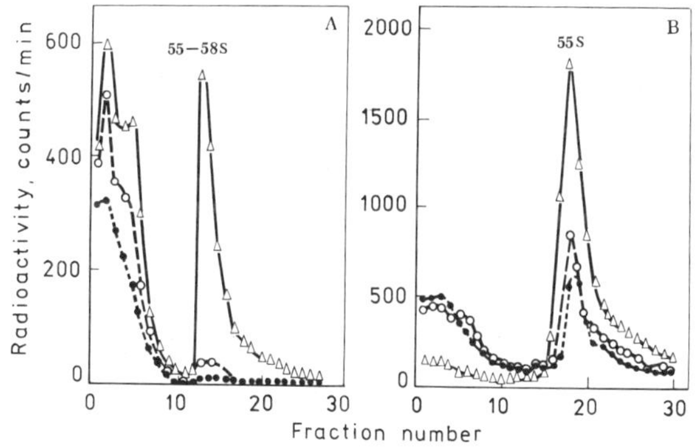

Fig. 22. Distribution of <u>in vitro</u>-labeled acid-insoluble mitochondrial translation products in sucrose density gradients upon centrifugation of (A) complete mitochondrial lysates and (B) isolated mitochondrial ribosomes. (A) (△) Pulse-labeled with [^{14}C]leucine; (○) after 10 min of chase with [^{12}C]leucine and puromycin; (●) chased with [^{12}C]leucine. (B) (△) Pulse-labeled with [^{14}C]leucine; (○) after incubation with puromycin; (●) after incubation with GTP and puromycin. After Wheeldon et al. (1974).

mitochondria. This source has an obvious advantage over liver since it permits one to work with a population of one type of cells.

SUMMARY

We see that mitochondria are extremely labile organelles. Their instability manifesting itself immediately after isolation from cells has its source in the very composition of mitochondria. It would suffice to mention the presence of phospholipase A_2 in the mitochondrial membranes, and the remarkably high content of unsaturated fatty-acid chains in mitochondrial phospholipids (see Chapter One). In immediate contact with this "combustible material" in the membrane there is a "fuze" or "primer" whose part may be played by iron-sulfur centers or heme residues. To this it must be added the fact that one of the products of mitochondrial metabolism is hydrogen peroxide (and in some cases the superoxide anion as well) which in the presence of Fe^{2+} ions yields a highly reactive $OH^{\cdot}$ radical triggering phospholipid oxidation. Finally, the organelle quite probably contains proteinases of various types and localization. Considering all this one cannot help asking, how then can mitochondria survive in the cell? Chapters Three and Four are devoted to the search of an answer to this question.

As follows from the present chapter, there are at least three more or less well-studied mechanisms of mitochondrial degradation, each of which may under certain conditions be considered prevalent. The question which of these mechanisms is primary is extremely complicated. There are grounds for thinking that the key part is played by the cleavage of mitochondrial phospholipids under the action of the endogenous phospholipase A_2. Indeed, mitochondrial swelling, loss of respiratory control, and decrease of the oxidative phosphorylation efficiency can be reversed, at least at the first stages of mitochondrial swelling, by serum albumin which is hardly capable of suppressing, let alone reversing, the hydrolysis of mitochondrial proteins or lipid oxidation.

However, it must be borne in mind that phospholipase A_2 has to be somehow activated for its effect to be noticeable. Such activation can be brought about, among other things, by alterations in the mitochondrial membrane structure because of de-energization. This issue will be discussed in a special section of the book. One cannot also exclude the involvement of endogenous proteinases in the primary steps of mitochondrial degradation. The proteinases may, for instance, de-energize the organelles by destroying the oxidative phosphorylation machinery which is particularly sensitive to proteolysis (see Chapter Three).

Speaking of the especially important role of the endogenous phospholipase A_2 in mitochondrial degradation, it should not be forgotten that this applies to mitochondria from liver (Rossi et al., 1964, 1965; Scherphof and Van Deenen, 1965; Scherhof et al., 1966; Waite and Van Deenen, 1967; Honjo et al., 1968; Nachbaur and Vignais, 1968; Waite, 1969; 1971; Waite et al., 1969a; Nachbaur et al., 1972; Scarpa and Lindsay, 1972), pancreas (Honjo et al., 1968; Ozawa et al., 1969), and probably heart (Palmer et al., 1981). As to other tissues, there the mitochondrial degradation mechanisms may be entirely different. In particular, Ozawa et al. (1969) found that brain mitochondria _in vitro_ can accumulate substantial amounts of free fatty acids, which is not accompanied, however, by any appreciable changes in the phospholipid composition. In other words, this effect is not associated with phospholipase action.

In conclusion it must be pointed out that mitochondrial degradation is a complex process. Specifically, the destructive alterations caused by phospholipase A_2 create conditions favorable for phospholipid oxidation and mitochondrial proteolysis. As shown by Green and O'Brien (1970), oleate makes mitochondria lose glutathione peroxidase necessary to prevent fatty-acid peroxide formation. On the other hand, data are available (Pappu et al., 1978) that in mitochondria isolated from livers of vitamin E-deficient rats the phospholipase A_2 activity is elevated almost twofold. This means that mitochondrial aging is an autocatalytic process in the course of which the initial degradation by one of the mechanisms triggers a whole cascade of degenerative events of various nature.

3
Mechanisms of Mitochondrial Stabilization

Cases have already been considered in the preceding chapter when destructive alterations in isolated mitochondria were retarded or even reversed. Yet, most of the stabilization means mentioned are nonphysiological by their nature; these include the protective effects of local anesthetics, heavy-metal chelators or synthetic antioxidants.

On the other hand, there is evidence dating as far back as the early fifties that mitochondrial stability depends on the extent of their energization, which may have good physiological reason. Among other things, there were indications that mitochondrial swelling is hindered if the organelles perform electron transfer in the presence of ADP or AMP (Harman and Feigelson, 1952a,b; MacFarlane and Spencer, 1953; Price et al., 1956; Di Sabato and Fonnesu, 1959; Lipsett and Corwin, 1959; Emmelot, 1960). Since this effect was abolished by 2,4-dinitrophenol if its concentration was not too high (Harman and Feigelson, 1952b; Price et al., 1956; Chappell and Greville, 1959; Di Sabato and Fonnesu, 1959; Emmelot, 1960), it could be thought to be somehow due to ADP phosphorylation coupled with oxidation of various substrates (glutamate, succinate, α-ketoglutarate). However, the action of 2,4-dinitrophenol on mitochondria is ambiguous and is not necessarily associated with the uncoupling of oxidative phosphorylation. It should be borne in mind that 2,4-dinitrophenol can stimulate the adenylate kinase reaction and thereby increase the ATP level in the medium if the latter contains ADP as the phosphate acceptor (Lipsett and Corwin, 1959). Furthermore, 2,4-dinitrophenol can itself suppress swelling in concentrations of 10^{-4}–10^{-3} M (Tapley, 1956), probably by interacting with mitochondrial membrane proteins and somehow altering the membrane structure. According to present-day concepts (see, e.g., Brierley, 1976), uncouplers impede mitochondrial contraction by destroying the gradient of H^+-ion concentration across the inner mitochondrial membrane. It is interesting to note in this context that the protective effect of an oxidizable substrate and ADP during the phosphate-induced mitochondrial swelling is not abolished by an H^+-ATPase inhibitor, oligomycin (Chappell and Greville, 1961; Connely and Hallstrom, 1967). At the same time, contraction of swollen mitochondria upon addition of ATP is suppressed by oligomycin (Lehninger, 1962).

There are also other lines of evidence for the relationship between mitochondrial stability and energy transformation. According to Siekevitz and Potter (1955), the loss of acid-soluble nucleotides by mitochondria is retarded in the presence of an oxidizable substrate, phosphate, AMP, and Mg^{2+} ions. Estrada-O et al. (1970) showed that succinate oxidation or ATP hydrolysis by coupled mitochondria prevent the loss of some soluble enzymes. Slater and Cleland (1952, 1953) demonstrated the stabilization of the mitochondrial α-ketoglutarate oxidation system in the presence of the reaction mixture used to measure oxidative phosphorylation.

In most of the works mentioned mitochondrial aging was judged by their swelling, which certainly is an oversimplification. Suffice it to say that during prolonged incubation of mitochondria with thyroxine (Lehninger, 1959a) they retain the ability to contract in the presence of ATP, whereas oxygen consumption, to say nothing of oxidative phosphorylation, is irreversibly lost. More precise data were obtained in the seventies on the connection between the stability of isolated mitochondria and their energetic state.

3.1. ROLE OF OXIDATIVE PHOSPHORYLATION IN STABILIZATION OF ISOLATED MITOCHONDRIA

Romani et al. (1970, 1973, 1974) and Goldblatt and Romani (1977) showed that animal and plant mitochondria can maintain their respiratory control for 3-4 days at 25°C if the incubation medium contains sucrose, phosphate, α-ketoglutarate, NAD, thiamine pyrophosphate, Coenzyme A, ADP, and serum albumin; incubation conditions must necessarily be aerobic (Figure 23). The need for catalytic amounts of ADP for maximal mitochondrial preservation was considered proof that organelle survival requires "metabolically supplied energy." It is noteworthy that, according to Romani and colleagues, ATP by itself does not mimic the action of the substrate-ADP mixture, suggesting that the entire oxidative phosphorylation system should function to stabilize mitochondria. This is supported by the results of electron-microscopic observations. It must be pointed out that the maintenance of mitochondrial integrity *in vitro* does not seem to require mitochondrial protein synthesis. Indeed, all the above results were obtained in the presence of chloramphenicol, which blocks mitochondrial translation.

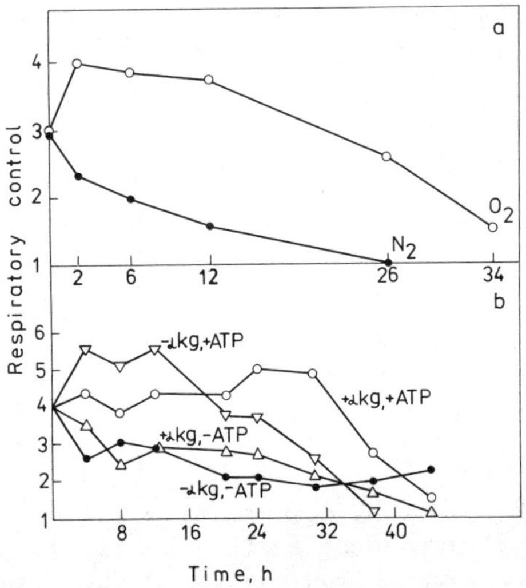

Fig. 23. Dependence of the stability of isolated avocado fruit mitochondria on their functional state. (a) Effect of anaerobiosis on mitochondrial survival in vitro: (o) aerobic incubation in a medium containing sucrose, phosphate, α-ketoglutarate, NAD, thiamine pyrophosphate, ADP, $MgCl_2$, coenzyme A, bovine serum albumin, and chloramphenicol; (•) the same in anaerobic conditions (under nitrogen). (b) Effect of different combinations of ATP and α-ketoglutarate (αkg) on mitochondrial survival: (+) compound present, (-) compound absent. Adapted from Romani and Ozelkök (1973).

According to Parce et al. (1978), a crucial point in the maintenance of the integrity of isolated mitochondria is the intramitochondrial ATP level. As it decreases during organelle aging, they lose their ability to reacylate lysophosphatides and to perform fatty-acid β-oxidation. When the respiratory-control ratio falls to unity and the ATP content to 1 nmole per mg protein, there is a drastic increase in the amounts of monoacylglycerophosphoethanolamine (lysophosphatidylethanolamine) and free fatty acids (Figure 24), which marks the onset of the irreversible phase in mitochondrial degradation. On the other hand, ATP added into the incubation medium increases or even restores to the initial values (if the aging has not been too long) the rates of reacylation and fatty-acid β-oxidation.

Figure 24 demonstrates that anaerobiosis rapidly decreases the respiratory control and the intramitochondrial ATP content, the lag in lysophosphatidylethanolamine accumulation being therewith dramatically shortened. The same effect is brought about by an uncoupler

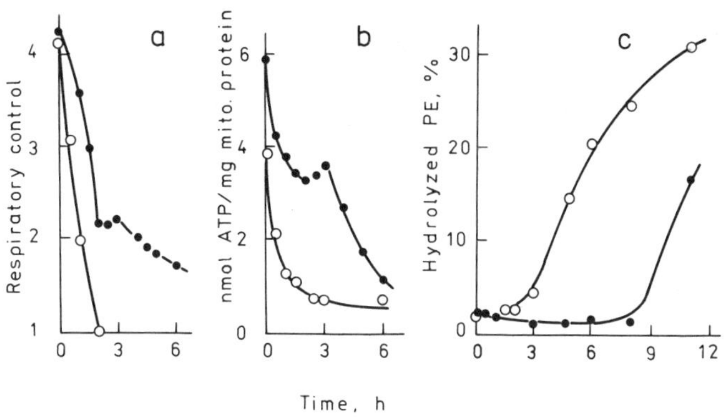

Fig. 24. Effect of anaerobiosis on some characteristics of rat liver mitochondria. Freshly isolated mitochondria were incubated at 18°C in 0.25 M sucrose under (●) aerobic or (o) anaerobic conditions; (a) respiratory control; (b) ATP content; (c) hydrolysis of phosphatidylethanolamine (PE). After Parce et al. (1978).

(carbonylcyanide p-trifluoromethoxyphenylhydrazone) introduced into the incubation medium. Anaerobiosis also attenuates the mitochondrial capacity of reacylation and especially of fatty-acid oxidation.

These data undoubtedly mean that oxidative phosphorylation in mitochondria (at least in rat-liver) is an indispensable prerequisite to preventing their deterioration, since it is the only means of maintaining a high ATP level in the organelles. The question is merely the mechanism of the ATP stabilizing action. One of the functions of ATP is to decrease the mitochondrial free-fatty-acid level at the expense of the ATP-dependent lysophosphatide reacylation and fatty-acid β-oxidation. The explanation, however, is not only in the necessity of ATP for these processes to counteract those leading to accumulation of fatty acids which produce uncoupling and lytic effects on mitochondrial membranes. In fact, the mitochondrial reacylating activity disappears long before any appreciable hydrolysis of phosphatidylethanolamine is observed. On the other hand, Parce et al. (1978) reported that as long as mitochondria contain ATP, their phospholipase activity is not elevated even in the presence of excess Ca^{2+}, so that the membrane phospholipid components are preserved. This may be explained by the existence of a special ATP-dependent structural state of the inner mitochondrial membrane in which either the exogenous calcium is not available to the endogenous phospholipase A_2 or the membrane phospholipids are hardly accessible to the phospholipase. The latter possibility will be dealt with in a special section of this chapter.

Similar ideas were developed by Bragin et al. (1975, 1977a,b) who studied the degradation of liver mitochondria under anoxia. According to their scheme, the events during anoxia proceed in the following order: mitochondrial de-energization → release of Ca^{2+} ions from the organelles → phospholipase A_2 activation → hydrolysis of membrane phospholipids → impairment of the energy-dependent functions of oxidative phosphorylation.

It is easy to see that in all the works mentioned the key moment in the degradation of whole mitochondria *in vitro* is considered to be the disturbance to the energy-transducing system.

3.2. STUDIES ON THE MOLECULAR MECHANISMS OF MITOCHONDRIAL STABILIZATION IN MODEL SYSTEMS

Important information about the mechanisms of the oxidative-phosphorylation-linked mitochondrial stabilization was obtained in model experiments with exogenous phospholipase A_2 and neutral proteinases. These experiments were run to study the action of lytic enzymes on the oligoenzyme complexes of the respiratory chain, on the reconstituted respiratory chain, and on coupled and uncoupled submitochondrial vesicles under conditions when these objects did or did not perform their functions in electron transfer and energy transformation.

3.2.1. Action of Exogenous Phospholipase A_2 on the Mitochondrial Oxidative Phosphorylation System

Enzymes most resembling the mitochondrial phospholipase A_2 are those found in the venoms of some snakes (*Naja naja*, *Crotalus adamanteus*, *Crotalus terrificus*, and others). These phospholipases catalyze the hydrolysis of the ester bond at the second carbon atom (Van Deenen and De Haas, 1964; Wu and Tinker, 1969), have an activity optimum at pH 7.0-8.0, and are activated by calcium ions (Saito and Hanahn, 1962; Wu and Tinker, 1969; Salach et al., 1971). Furthermore, phospholipases A_2 from *Naja naja* and *Crotalus adamanteus* cleave in the mitochondrial membranes mostly phosphatidylethanolamine and phosphatidylcholine, the first being broken down at a higher rate (Awasthi et al., 1970b). It must be recalled that a similar pattern was observed in the changes in the phospholipid composition of aging mitochondria: the most labile component was also phosphatidylethanolamine (Waite et al., 1969b; Scarpa and Lindsay, 1972). Thus snake venom phospholipase A_2 appears to attack in the first place the same mitochondrial membrane phospholipids that the endogenous phospholipase A_2 does.

In the inner mitochondrial membrane exposed to exogenous phospholipase A_2 (*Naja naja* venom), substantial functional disorders are elicited long before there are any detectable changes in its composition and structure (Nygaard and Sumner, 1953; Awasthi et al., 1970b). Especially sensitive to phospholipases A_2 from various sources is the energy-transducing machinery. In fact, the phosphorylating ability falls substantially more rapidly than the respiratory activity when mitochondria (Petrushka et al., 1959; Burstein et al., 1971b) or ultrasonic submitochondrial vesicles (Burstein et al., 1971a) are treated with cobra venom phospholipase A, so that the P/O ratio decreases very quickly.

3.2.1.1. *Energy-Transducing Machinery*

The most detailed study was made of the degradation of the energy-transducing machinery under the action of phospholipase A from Crotalus terrificus venom (Luzikov et al., 1973a). A characteristic feature of this process is the substantial variance in the rates with which different energy-dependent functions are impaired in the ultrasonic submitochondrial vesicles. The "vulnerable" functions include the ATP-dependent NAD reduction by succinate (reversed electron transfer) and the ATP-dependent and succinate-dependent transhydrogenase reaction (NADH + NADP → NAD + NADPH). The phosphorylating ability of the vesicles (phosphorylation coupled with succinate oxidation) and the ATP-$^{32}P_i$ exchange (Figure 25) are affected at much lower rates. This means that the ATP-synthesizing system proper is relatively resistant to Crotalus terrificus venom phospholipase A_2.

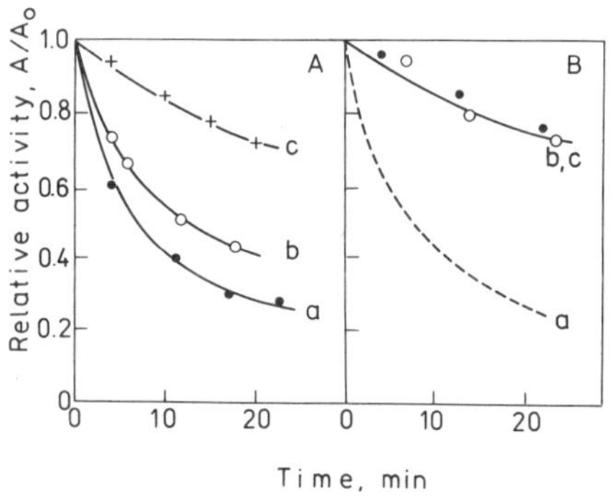

Fig. 25. Suppression of some functions of submitochondrial vesicles by phospholipase A from Crotalus terrificus venom. (A) Energy-dependent functions: (a) ATP-dependent NAD reduction by succinate (reversed electron transfer); (b) ATP-dependent transhydrogenation; (c) ATP-$^{32}P_i$ exchange. (B) Energy-independent functions: (b) NADH-fumarate reductase reaction and (c) energy-independent transhydrogenation; (a) is the same as in (A) for comparison. From Luzikov et al. (1973a).

However, data are available that treatment of mitochondria (Burstein et al., 1971b) and submitochondrial vesicles (Burstein et al., 1971a) with phospholipase A_2 from Naja naja venom quickly decreases the rate of ATP-$^{32}P_i$ exchange. This discrepancy may be due to some difference in the substrate specificity of phospholipases A_2 from the two sources.

It must be pointed out that all the above-described effects of phospholipase A_2 on the energy-transducing machinery cannot be ascribed to the uncoupling or lytic action of the hydrolysis products of the mitochondrial membrane phospholipids, because the energy-dependent functions were always assayed in the presence of excess serum albumin. Therefore, the impairment of the functions concerned is caused by the changes in the lipid composition and/or the accompanying changes in the fine structural organization of certain membrane regions.

3.2.1.2. *Respiratory Chain*

The activities first hampered by phospholipase A_2 in the respiratory chain of mitochondria or submitochondrial vesicles are NADH oxidase and succinate oxidase. In fact, succinate oxidase activity of rat liver mitochondria treated with phospholipase A_2 from Crotalus terrificus venom is altogether lost when cytochrome *c* oxidase and succinate dehydrogenase activities are not yet appreciably affected (Nygaard and Sumner, 1953). Naja naja venom phospholipase A_2 completely inactivates mitochondrial succinate oxidase and succinate-cytochrome *c* reductase, producing no appreciable effect on succinate-ubiquinone reductase (Cerletti et al., 1966). Similarly, reconstituted NADH oxidase is more sensitive to cobra venom phospholipase A_2 than its constituent complexes I+III (NADH-cytochrome *c* reductase) and complex IV (cytochrome *c* oxidase) (Luzikov et al., 1967).

This prompts one to think that there are two sites in the respiratory chain most susceptible to phospholipase A_2: one is between the cytochrome *c*-reducing and the cytochrome *c*-oxidizing parts of the respiratory chain (i.e. at the level of cytochrome *c*), and the other is shared by NADH-cytochrome *c* reductase and succinate-cytochrome *c* reductase segments (i.e. lies within complex III). Another thing worth mentioning is that in submitochondrial vesicles treated with Naja naja venom phospholipase A_2 there is a relatively rapid decrease in NADH-ubiquinone reductase activity while NADH-ferricyanide reductase activity is not changed (Awasthi et al., 1970b). This suggests a labile link preceding coenzyme Q (ubiquinone) in the respiratory chain.

The general pattern of injury to the respiratory chain can be confirmed and supplemented by the results of spectrophotometric measurements. Nygaard (1953) reported that phospholipase A_2 prevents reoxidation of reduced cytochrome *b*, having no effect on its reduction by NADH or succinate in cardiac-muscle and liver preparations; a conclusion was made that the phospholipase-sensitive site is between cytochromes *b* and *c*. According to the later data of Minakami et al. (1964), Naja naja venom phospholipase A_2 interrupts electron transfer in submitochondrial vesicles between the NADH dehydrogenase flavoprotein and coenzyme Q and also between cytochromes *b* and c_1.

Additional information on the phospholipase A_2-sensitive respiratory-chain sites was furnished by steady-state kinetic analysis of electron transfer (Luzikov et al., 1971a). Phospholipase A_2 proved to elicit pronounced alterations in the cytochrome reduction levels in the steady state of succinate aerobic oxidation. The pattern of these alterations indicated the existence of two sites in the succinate oxidase chain showing particularly high susceptibility to phospholipase A_2: at the level of cytochrome *c* and between cytochromes *b* and c_1. All that has been said about the sites of phospholipase A_2 action on the respiratory chain in summarized in the scheme: (ph.A:phospholipase A_2)

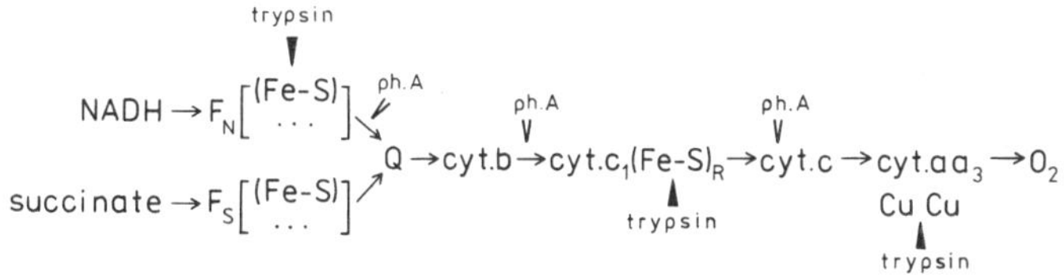

The respiratory chain destruction by phospholipase A_2 in mitochondria or submitochondrial vesicles appears to be due first of all to phospholipid breakdown and cannot be attributed solely to the action of the hydrolysis products. Thus Edwards and Ball (1954) showed that the minimal amount of oleate needed to completely inhibit succinate oxidase is about three times that of free fatty acids actually liberated from the membrane preparation. Still, the phospholipid hydrolysis products probably make a certain contribution to the inactivation of respiratory-chain components; this particularly applies to the NADH-ubiquinone reductase segment (Salach et al., 1967).

3.2.2. Increasing Resistance of the Oxidative Phosphorylation System to Exogenous Phospholipase A_2

Taking into account work dealing with the role of the endogenous phospholipase A_2 in mitochondrial degradation (see Chapter Two), one can easily predict that the damage done by an exogenous phospholipase A_2 to the oxidative phosphorylation system would be alleviated in the presence of serum albumin, which binds the "toxic" products of phospholipid hydrolysis (free fatty acids and lysophosphatides). It is also obvious that phospholipids added to phospholipase-treated mitochondria or submitochondrial particles must have some reactivating effect. Such results were indeed obtained for the respiratory chain and the energy-transducing machinery (see, e.g., Fleischer et al., 1964; Machinist and Singer, 1965; Berezney et al., 1970a; Burstein et al., 1971a,b). Especially interesting, however, are the data on the relationship between the functional state of these systems and their resistance to phospholipase A_2.

3.2.2.1 *Energy-Transducing Machinery*

As has been shown above, different energy-dependent functions of ultrasonic (inside-out) submitochondrial vesicles are not equally affected by Crotalus terrificus venom phospholipase A_2 (see Figure 25). The highest sensitivity is displayed by the ATP-dependent NAD reduction by succinate, and by the ATP-dependent (or succinate-dependent) transhydrogenase reaction (NADH + NADP → NAD + NADPH). Under the same conditions the rate of the ATP-$^{32}P_i$ exchange undergoes a markedly lesser decrease.

The suppression of reversed electron transfer is retarded if, together with phospholipase A_2, the incubation medium is supplemented with an NADH-generating system (e.g., NAD-ethanol-alcohol dehydrogenase-semicarbazide). As the protective effect is observable only in aerobic conditions, it can be thought to be due to electron transfer. Reduced NAD has no protective effect in the presence of carbonylcyanide m-chlorophenylhydrazone or potassium oleate (Figure 26) which, while fully effective in uncoupling, do not directly influence either the NADH oxidase activity of the vesicles or the sensitivity of the above energy-dependent functions to phospholipase A_2. Adenosine 5'-triphosphate (but not ADP) also retards the suppression of reversed electron transfer during the exposure of submitochondrial vesicles to phospholipase A_2, its effect being abolished by uncouplers (Luzikov et al., 1973a). These results give grounds for suggesting that the protective action of NADH and ATP is somehow associated with the functioning of the energy-transducing machinery.

Rather similar results were obtained on studying the degradation of the ATP-dependent transhydrogenase in submitochondrial vesicles treated with phospholipase A_2. In this case the energy-transducing machinery was also somewhat stabilized by NADH oxidation or ATP hydrolysis. At the same time the ATP-$^{32}P_i$ exchange, which was much less sensitive to phospholipase A_2, was not protected by vesicle energization.

3.2.2.2. *Respiratory Chain*

Degradation of NADH oxidase reconstituted from complexes I+III and IV in the presence of cytochrome *c* was shown (Luzikov et al.,

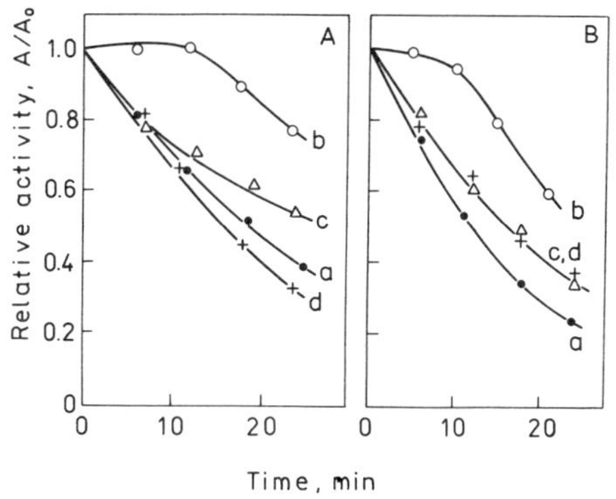

Fig. 26. Stabilization of the energy transformation system during the exposure of submitochondrial vesicles to phospholipase A from Crotalus terrificus venom. ATP-dependent NAD reduction by succinate (reversed electron transfer) was measured. (A) Protective action of NADH: (a) incubation without NADH; (b) with NADH (aerobic); (c) with NADH (anaerobic); (d) with NADH (aerobic) plus potassium oleate. (B) Protective action of ATP: (a) incubation without ATP; (b) with ATP; (c) with ATP and carbonylcyanide m-chlorophenylhydrazone; (d) with carbonylcyanide m-chlorophenylhydrazone. In all cases the incubation medium contained phospholipase A. Assays were carried out in the presence of serum albumin and EDTA to stop the phospholipase action. From Luzikov et al. (1973a).

1967) to be retarded when the incubation medium contained, besides phospholipase A_2, an oxidizable substrate (NADH). The protective effect of the latter is elicited only in aerobic conditions (Figure 27). In other words, the requirements for NADH oxidase stabilization are an electron donor and an acceptor. The same data were obtained for succinate oxidase reconstituted from complexes II, III, IV, and cytochrome *c*: it was stabilized by succinate in the presence of oxygen (Luzikov et al., 1967). On the basis of these results it was suggested that inactivation of complexes I+III+IV+cytochrome *c* and complexes II+III+IV+cytochrome *c* by phospholipase A_2 is retarded when electron transfer takes place from NADH or succinate to oxygen.

A phenomenon worth special attention is the "cross-protective" effect of substrates (Luzikov et al., 1968). This is that NADH oxidase of complexes I+II+III+IV+cytochrome *c* is less easily inactivated by cobra venom phospholipase A_2 not only when the incubation medium contains NADH, but also when the sole oxidizable substrate is succinate, i.e. when electrons are transferred via complexes II+III+IV. Analogously, succinate oxidase activity of the whole respiratory chain is preserved in the presence of NADH, i.e. when complexes I+III+IV are operative. It should be noted that the protective

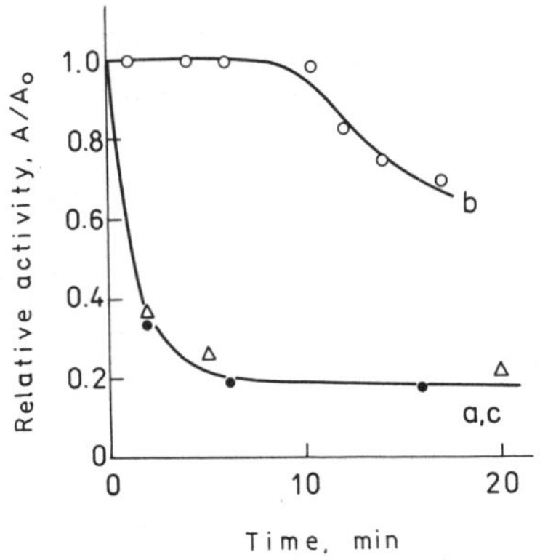

Fig. 27. Protective effect of the oxidizable substrate (NADH) on the inactivation of reconstituted NADH oxidase (complexes I+III+IV + cytochrome *c*) by cobra venom phospholipase A. Incubation (a) without NADH; (b) with NADH (aerobic); (c) with NADH (anaerobic). From Luzikov et al. (1967).

effect of succinate on the NADH oxidation system is suppressed by tenoyl trifluoroacetone which inhibits complex II, and the effect of NADH on succinate oxidase inactivation is not observed in the presence of amytal, a specific inhibitor of complex I.

The facts presented allow the following conclusions to be made. Firstly, NADH and succinate can exhibit protective effects while not interacting directly with the oxidase system under consideration. Secondly, their protective action takes place only in aerobic conditions and is prevented by compounds known as specific respiratory-chain inhibitors. From this it follows that the effects of both substrates can hardly be ascribed to the formation of stable enzyme-substrate complexes or to nonspecific interactions of NADH and succinate with the oxidase systems, and that a plausible cause of the stabilization of the respiratory chain is electron transfer through it.

Similar results were obtained for the succinate oxidase system of submitochondrial vesicles (alkaline or ultrasonic). This system was protected by succinate from cobra venom phospholipase A_2, the substrate being effective only in aerobic conditions (Luzikov et al., 1969). It has earlier been mentioned that phospholipase A_2 induces substantial changes in the extents of cytochrome steady-state reduction indicative of the existence of "weak" sites at the level of cytochrome *c* and between cytochromes *b* and c_1. If, besides phospholipase A_2, the incubation medium contains succinate and oxygen, such changes do not occur, as shown in Table 10 (Luzikov et al., 1971a).

Table 10. Effect of Succinate (in Aerobic Conditions) on the Inactivation of the Succinate Oxidase System of Ultrasonic Submitchondrial Vesicles by Cobra Venom Phospholipase A (from Luzikov et al., 1971a).

Incubation conditions	Succinate oxidase, μg-atoms oxygen per min per mg protein	Cytochrome steady-state reduction levels, % of total substrate-reducible		
		cyt. *b*	cyt. $c_1 + c$	cyt. aa_3
Control	1.4	29±1	27±2	15±1
With cobra venom	0.9	49±3	34±3	7±1
With cobra venom and succinate	1.4	30±1	27±2	15±2

3.2.3. Proteinase Action on the Mitochondrial Oxidative Phosphorylation System

3.2.3.1. *Energy-Transducing Machinery*

The mitochondrial energy-transducing machinery shows an exceptionally high sensitivity to exogenous neutral proteinases. According to the data of Racker (1962), submitochondrial vesicles treated with trypsin rapidly lose their phosphorylating ability even when the amounts of the proteinase and vesicles relate as 1:300 on a weight basis. Since subsequent addition of coupling factor F_3 stimulates the ATP-$^{32}P_i$ exchange, this factor can be thought to be the "weak" site of the energy-transducing unit. Other weak sites are the ATPase protein inhibitor (Horstman and Racker, 1970), factor F_C (Bulos and Racker, 1968), and factor F_6 (Fessenden-Raden, 1972; Kanner et al., 1976). Data are available favoring the idea that the susceptibility of a particular energy-transducing component to proteinases depends on the way it is arranged in the membrane. To give a case in point, isolated factor F_C is readily destroyed by trypsin, whereas in submitochondrial vesicles it is inaccessible to the proteinase (Bulos and Racker, 1968). Likewise, removal of factor F_1 from the vesicles markedly enhances the tryptic cleavage of factor F_0, and binding of F_1 to the membrane suppresses this process (Racker, 1965).

A peculiarity of trypsin action on the energy-transducing machinery is that various energy-dependent functions of submitochondrial vesicles are affected at the same rate (Luzikov et al., 1971b). In particular, Figure 28 depicts such data for phosphorylation coupled with NADH or succinate oxidation, ATP-$^{32}P_i$ exchange, and ATP-dependent NAD reduction by succinate (reversed electron transfer). Similar effects are produced on submitochondrial vesicles by oligomycin (Ernster et al., 1967). This similarity may mean that trypsin (like oligomycin) impairs the ATP-synthesizing system proper. It differs in this respect from Crotalus terrificus venom phospholipase A_2 which hampers the energy-dependent functions listed at different rates (see Figure 25).

3.2.3.2. *Respiratory Chain*

The respiratory chain is somewhat more resistant to trypsin and chymotrypsin than the energy-transducing machinery. Thus rat liver mitochondria treated with trypsin immediately lose respiratory control and phosphorylating ability but retain respiration with succinate and NAD-linked substrates (Weinbach and Garbus, 1968). Similarly, in beef-heart ultrasonic submitochondrial vesicles exposed to trypsin the phosphorylating ability (P/O ratio), ATP-$^{32}P_i$ exchange, and ATP-dependent reversed electron transfer are lost when neither NADH oxidase nor succinate oxidase activity undergoes any changes (Luzikov et al., 1971b).

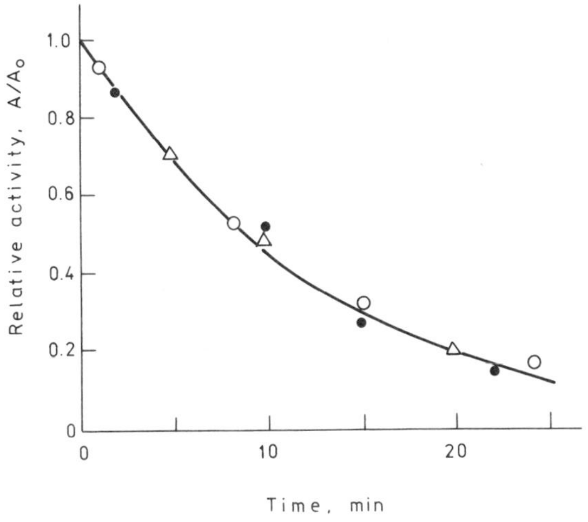

Fig. 28. Trypsin action on different energy-dependent functions of inside-out submitochondrial vesicles. (●) Oxidative phosphorylation (P/O ratio with NADH as substrate); (○) ATP-$^{32}P_i$ exchange; (△) ATP-dependent NAD reduction by succinate (reversed electron transfer). From Luzikov et al. (1971b).

For the mechanism of respiratory-chain inactivation by proteinases to be understood, information is needed about its components most susceptible to proteolysis. Trypsinization of high-molecular-weight NADH dehydrogenase was shown to be accompanied by a decrease in the iron-sulfur protein content per mole flavin (Watari et al., 1962). According to Baum et al. (1967), the most trypsin-sensitive of the components of isolated complex III is an iron-sulfur protein which is broken down more quickly than cytochromes *b* or c_1 or the core protein. Ball et al. (1977) have later shown that during treatment of complex III with trypsin the cleavage of the iron-sulfur protein and impairment of electron transfer proceed at comparable rates.

These data agree with the results obtained for alkaline submitochondrial vesicles. Indeed, when the latter are treated with chymotrypsin, the first to undergo destruction are the iron-sulfur centers giving in the reduced state an epr signal with a g-factor of 1.94 (Luzikov, 1973a).

3.2.4. Stabilization of the Oxidative Phosphorylation System Against Exogenous Proteinases

3.2.4.1. *Respiratory Chain*

Inactivation of reconstituted NADH oxidase (complexes I+III+IV+cytochrome *c*) and succinate oxidase (complexes II+III+IV+cytochrome *c*) has been shown to be retarded when the incubation medium contains NADH or succinate (Luzikov et al., 1967). Quite similarly, the NADH oxidase system of alkaline or ultrasonic submitochondrial vesicles (Luzikov et al., 1969) acquires resistance to trypsin or chymotrypsin in the presence of NADH, its protective effect being exhibited only in aerobic conditions. A phenomenon of "cross-stabilization" by substrates was discovered (Luzikov et al., 1969): the NADH oxidase system of the submitochondrial vesicles is protected from tryptic attack not only by NADH but also by succinate which does not directly interact with NADH oxidase. The effect of succinate is abolished by tenoyl trifluroacetone which inhibits the succinate-ubiquinone reductase segment of the respiratory chain.

On the other hand, data are available that NADH analogs that cannot serve as electron donors for NADH oxidase do not protect it from proteolysis even if they efficiently bind to the enzymic system (Luzikov and Romashina, 1972). These results together with the substrate cross-stabilizing effects indicate that the formation of a complex with the substrate and oxygen can hardly be the cause of the enhanced resistance of the NADH oxidase system to proteinases, and that electron transfer is imperative for its stabilization.

This suggestion is favored by the dependence of the extent of the substrate protective effect on the rate of electron transfer (Luzikov and Romashina, 1972). In these experiments, different inhibitors of NADH oxidase were used. One of them (Zn^{2+} ions) hindered electron transfer mostly between cytochromes *b* and c_1 (Skulachev et al., 1967; Kleiner and Von Jagow, 1972); other (NAD, deamino NAD, and thio-NAD) competed with the substrate for the enzyme active site.

As shown in Figure 29a, addition of Zn^{2+} ions to the incubation medium attenuates the protective action of NADH during the inactivation of NADH oxidase in submitochondrial vesicles. The effect is completely abolished when the electron-transfer rate falls by 70%. Zincum ions as such did not change in these experimental conditions the affinity of NADH oxidase for NADH and did not affect its own resistance to chymotrypsin, but produced noticeable alterations in the cytochrome steady-state reduction levels (Luzikov and Romashina, 1972).

Similar results were obtained with competitive inhibitors of NADH oxidase (Luzikov and Romashina, 1972). These compounds also

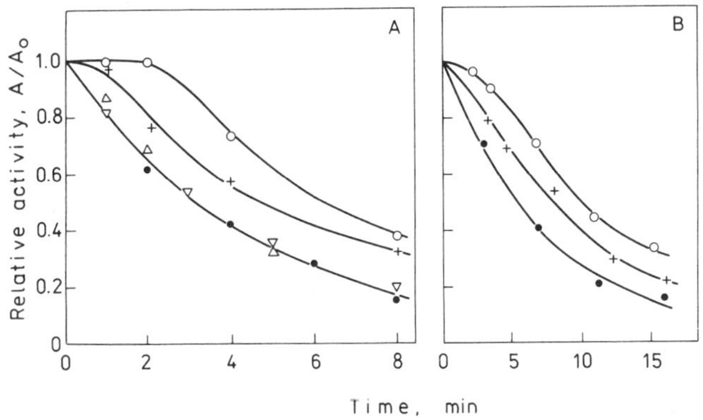

Fig. 29. Dependence of the stability of the respiratory chain on the rate of electron transfer: effect of Zn^{2+} ions and deamino-NAD on the stabilizing action of NADH during the destruction of the NADH oxidase system in submitochondrial vesicles by chymotrypsin or trypsin. (A) Effect of Zn^{2+}: (●) incubation without NADH (control); (○) with NADH; (+) with NADH and Zn^{2+} (40% inhibition of electron transfer; (△) with NADH and Zn^{2+} (70% inhibition); (▽) with Zn^{2+}, without NADH. (B) Effect of deamino-NAD: (●) without NADH (control); (○) with NADH; (+) with NADH and deamino-NAD (30% inhibition of electron transfer). In all cases the incubation medium was aerobic and contained trypsin or chymotrypsin. From Luzikov and Romashina (1972).

lower the NADH protective efficiency but do not by themselves affect the rate of the breakdown of the NADH oxidase system (Figure 29b). In contrast to Zn^{2+} ions, they do not appreciably change the cytochrome steady-state reduction levels during NADH oxidation. This gives grounds for suggesting that for the substrate protective effect to be attenuated it is enough to slow down the electron transfer without any other alterations in the NADH oxidase system.

It must be specially emphasized that deamino NAD and Zn^{2+} ions, despite the basically different inhibition mechanisms, interfere identically with the substrate protective action being taken in amounts needed to suppress electron transfer to the same extent (Luzikov and Romashina, 1972). That these two agents produce similar effects though one does while the other does not substantially change the cytochrome reduction levels, testifies to the fact that the role of electron transfer in the respiratory-chain stabilization cannot be restricted to mere maintenance of a certain relation between the reduced and the oxidized states of electron carriers. Experiments with NADH oxidase competitive inhibitors, in which the only apparent cause of attenuated protective effect of NADH was the lower rate of electron transfer, prompt one to think that for the respiratory chain to be protected from proteolysis it is important that the individual acts of electron transfer occur frequently enough to preclude the proteinase from productively binding to the "weak spots" of the system.

3.2.4.2. *Energy-Transducing Machinery*

Unlike the respiratory chain, the energy-transducing machinery is not protected from tryptic attack by respiratory substrates. Indeed, the P/O ratio decreases at the same rate whether or not an oxidizable substrate is present in the incubation medium (Figure 30). Adenosine 5'-diphosphate by itself also has no protective effect. The proteolysis of the ATP-synthesizing system noticeably slows down only when the incubation medium simultaneously contains ADP, phosphate, an oxidizable substrate (NADH or succinate), and oxygen. Adenosine 5'-triphosphate also protects the ATP-synthesizing system from proteolysis. From this it follows that the resistance of the ATP-synthesizing machinery to trypsin is enhanced in all probability owing to its functioning, i.e. ATP synthesis or hydrolysis. Similar results were obtained in assays of the P/O ratio, ATP-$^{32}P_i$ exchange, and ATP-dependent reversed electron transfer (Luzikov et al., 1971b; Luzikov, 1973a).

To illustrate the necessity of the functioning of the ATP-synthesizing system for its protection from proteinases, data can be

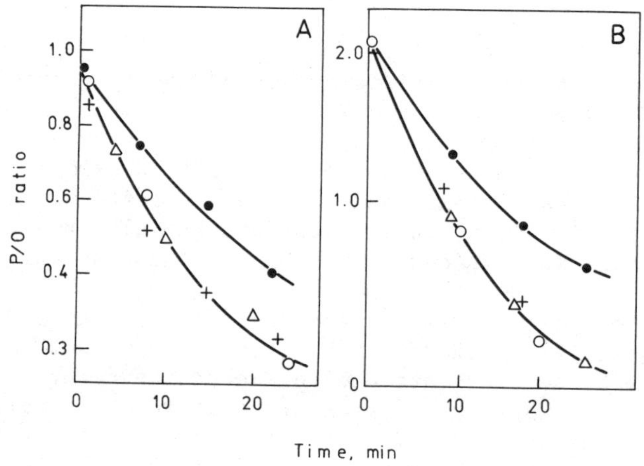

Fig. 30. Protection of the phosphorylating system of submitochondrial vesicles from trypsin. (A) Phosphorylation coupled with oxidation of NADH (supplied by an NADH-generating system). (○) Control; (△) with oxidizable substrate; (+) with ADP; (●) with both ADP and substrate. The incubation medium was always aerobic and contained phosphate and trypsin. From Luzikov et al. (1971b).

cited about the influence of uncouplers on the protective effect of ADP in the presence of phosphate and oxidizable substrate. Potassium oleate taken in small enough amounts was shown (Luzikov et al., 1971b) to suppress the protective effect, producing by itself no changes in the rate of tryptic inactivation of the ATP-synthesizing system or in NADH oxidase and succinate oxidase activities of submitochondrial vesicles. In other words, even when the incubation medium contains the complete set of components necessary for ADP phosphorylation or ATP hydrolysis, and when electron transfer or ATPase reaction take place but the system is uncoupled, the ATP synthetase acquires no resistance to trypsin.

3.2.5. Some Issues Arising from Model Experiments

It follows from the material presented in the preceding subsections that when mitochondrial inner membrane preparations (inside-out submitochondrial vesicles) perform oxidative ADP phosphorylation or coupled electron transfer they pass into a special state marked by enhanced resistance to exogenous lytic enzymes - phospholipase A_2 and neutral proteinases. Membrane stabilization can hardly be identified with its energization. On the contrary, the experimental data available testify to the specificity of the stabilizing action of various energizing agents. In particular, it can be mentioned that oxidizable substrates on the one hand, and ATP on the other, display different effects on the stability of the oxidative phosphorylation system to trypsin in coupled submitochondrial vesicles (Luzikov et al., 1971b).

It has been suggested above that endogenous ATP may stabilize aging isolated mitochondria through maintaining the inner membrane in such a state as would make the mitochondrial phospholipase A_2 incapable of cleaving membrane phospholipids at an appreciable rate. The results of the model experiments just described confirm this suggestion.

The very fact of decreasing sensitivity of various mitochondrial inner-membrane energy-dependent functions to trypsin and phospholipase A_2 implies a change in the membrane structure upon its energization. Mention should be made in this context of a recent publication by Hatefi et al. (1982) which showed that partial uncoupling of submitochondrial vesicles by adding an uncoupler or valinomycin with nigericin in the presence of K^+ ions decreases the apparent V_{max} and increases the apparent K_m for corresponding substrates in four energy-dependent processes: oxidative ADP phosphorylation, NTP-$^{32}P_i$ exchange, ATP-dependent NAD^+ reduction by succinate, and transhydrogenase reaction. These results were interpreted as indicating that membrane energization brings about a conformational change in the membrane-bound enzymes manifesting itself as enhanced enzyme-substrate affinity and reaction rate.

Unfortunately the conformational changes taking place in the inner mitochondrial membrane cannot yet be discussed in terms of intramembrane coordinates of molecules and intramolecular coordinates of atoms. It can only be said thus far that the trypsin-sensitive and the phospholipase A_2-sensitive membrane regions become inaccessible to the lytic enzymes upon energization, apparently owing to altered mutual arrangement and orientation of protein molecules in the membrane. The latter suggestion arises, among other things, from the cooperative character of the stabilization. In fact, isolated cytochrome *c* oxidase (Luzikov and Romashina, 1972) or complexes I+III (Luzikov, 1973a) were shown not to be protected from phospholipase A_2 despite their functioning. At the same time, they acquire resistance to phospholipase A_2 when they form the operative NADH oxidase (complexes I+III+IV).

Finally, though the increasing resistance to trypsin of the uncoupled respiratory chain performing electron transfer may have no bearing on the problem of mitochondrial stabilization, this phenomenon is of certain interest for enzymology. It has long been known that many enzymes upon forming a complex with their substrate or cofactor go into a state resistant to various denaturing factors and proteinases. Such data are available for glyceraldehyde-3-phosphate dehydrogenase (Racker and Krimsky, 1958; Elődy and Szabolczi, 1959; Bolotina et al., 1967), alcohol dehydrogenase (Okunuki, 1961; Di Sabato and Kaplan, 1965), aldolase (Szabolczi and Biszku, 1961), creatine phosphokinase (Nan Sen Tseng and Cunningham, 1966), tryptophan pyrrolase (Schimke et al., 1965), isoleucyl-tRNA synthetase (Baldwin and Berg, 1966), RNA polymerase (Khesin et al., 1967), and others. Stabilization of the respiratory chain carrying out electron transfer can be considered a dramatic demonstration of the existence of a special "working" state of a multienzyme assembly, quite distinct from its fully reduced or fully oxidized states which show high sensitivity to trypsin and phospholipase A_2.

3.3. PROTECTION OF MITOCHONDRIAL LIPIDS FROM PEROXIDATION

It has been already mentioned that polyunsaturated fatty-acid residues in mitochondrial membrane phospholipids can be protected from peroxidation by two matrix enzymes: glutathione peroxidase and glutathione reductase. The first one catalyzes lipid hydroperoxide reduction to alcohols (or simply eliminates traces of hydrogen peroxide), and the other reduces oxidized glutathione (see Chapter Two). This section will deal with another two mechanisms physiological by their nature. They imply a relationship between the resistance of mitochondrial lipids to oxidation and the functional state of the oxidative phosphorylation system. Before considering these mechanisms, it can be noted that mitochondrial protection by the glutathione peroxidase - glutathione reductase couple is also indirectly dependent on the organelle functional state, since swelling that

Table 11. Relation Between Malondialdehyde Production and the Level of Ubiquinone Reduction in Submitochondrial Vesicles in the Presence or in the Absence of ADP·Fe^{3+} (taken from Takayanagi et al., 1980).

Additions	Ubiquinone reduction (% of total)		Malondialdehyde formed (nmoles per 10 min per ml)	
	−ADP·Fe^{3+}	+ADP·Fe^{3+}	−ADP·Fe^{3+}	+ADP·Fe^{3+}
NADH	43±2	43±1	0.1	6.0±0.6
NADH + rotenone	1±1	1±1	0.1	38.1±0.9
NADH + antimycin A	80±1	78±2	0	1.6±0.6
NADPH	4±2	1±1	0.1	29.3±0.5
NADPH + rotenone	1±1	0	0.1	30.9±1.3
NADPH + antimycin A	59±2	0	0.1	26.3±1.4

usually accompanies mitochondrial de-energization leads to the leakage of these protective enzymes.

3.3.1. Lipid Peroxidation Induced by Fe^{2+} Ions

Exogenous Fe^{2+} or agents reducing endogenous Fe^{3+} to Fe^{2+} are known (see Chapter Two) to induce peroxidation of polyunsaturated fatty-acid chains in mitochondrial phospholipids. This process can be evoked in beef heart submitochondrial vesicles by NADH or NADPH in the presence of a chelate complex ADP-Fe^{3+} (Takayanagi et al., 1980). Using inhibitors of electron transfer through the respiratory chain - rotenone and antimycin A - these authors showed that malondialdehyde accumulation is inversely correlated with the extent of ubiquinone reduction: the higher the ubiquinol content, the less malondialdehyde is found in the submitochondrial vesicles (Table 11). A peculiarity of NADPH in this respect was that in the presence of the ADP-Fe^{3+} complex this substrate failed to maintain a noticeable ubiquinol level even if antimycin A was also added (the sites of inhibitor action on the respiratory chain are shown in the Ohnishi's scheme, Section 1.3.1). In accordance with the above, succinate decreased the malondialdehyde content, its effect being potentiated by antimycin A and cyanide but attenuated by malonate which inhibits succinate dehydrogenase and thereby impedes ubiquinone reduction (Table 12). Administration of ubiquinol-2 into the incubation medium virtually abolished lipid oxidation, whereas ubiquinone-2 was totally ineffective (Table 12). Finally, duroquinol which is known to donate electrons directly to cytochrome *b* also exhibited no protective effect.

Table 12. Effects of Electron-Transfer Inhibitors and Succinate on NADH- and NADPH-Dependent Formation of Malondialdehyde in Rotenone-Treated Submitochondrial Vesicles (taken from Takayanagi et al., 1980).

	Malondialdehyde formation (nmoles in 10 min per mg protein)	
	with NADH	with NADPH
Control	4.54	4.28
+ succinate	0.81	0.73
+ succinate + malonate	2.63	2.23
+ succinate + antimycin A	0.02	0.02
+ succinate + KCN	0.09	0.05
+ duroquinol + antimycin A	4.18	3.92
+ ubiquinol-2 + antimycin A	0.11	0.04
+ ubiquinone-2 + antimycin A	3.98	3.91

On the basis of these and some other data Takayanagi et al. (1980) proposed the following scheme:

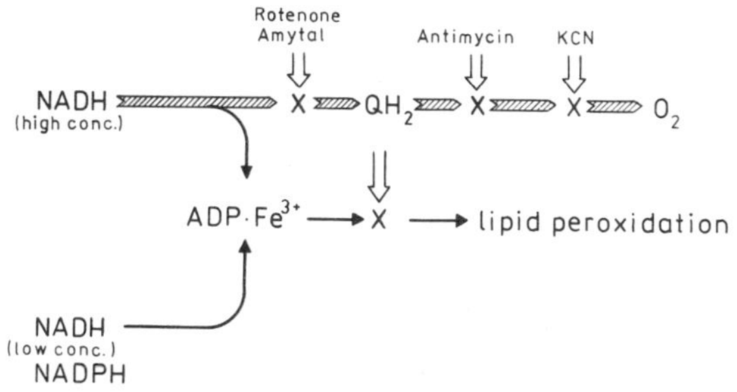

It can be seen that NADPH, or NADH taken in small amounts, at low electron-transfer rates reduces the ADP-Fe^{3+} chelate at the site positioned before the rotenone-sensitive link of NADH dehydrogenase, and the ADP-Fe^{2+} formed further acts as inducer of lipid peroxidation. At high NADH concentrations the electron-transfer rate is sufficient to produce a high steady-state level of ubiquinol which acts as an antioxidant capable of "extinguishing" lipid free radicals. In this case there is no peroxidation, and no malondialdehyde is accumulated in the submitochondrial vesicles. An inefficient electron donor such as NADPH cannot provide a high ubiquinol content but can reduce the ADP-Fe^{3+} chelate. As a result, the ubiquinone is mostly in the oxidized state, and lipid peroxidation proceeds even in the presence of antimycin A or cyanide.

The merit of the data of Takayanagi et al. (1980) is not only in that they establish the ability of ubiquinol to prevent polyunsaturated-fatty-acid peroxidation in mitochondrial phospholipids, but also in the implication that for this purpose mitochondria have to be in a metabolically active state. The rate of electron transfer through the respiratory chain has to be such as to ensure a high steady-state ubiquinol level. Obviously, this is only possible - if one is dealing with coupled mitochondria - when a high rate of oxidative ADP phosphorylation is maintained by the organelles.

3.3.2. Lipoxygenase Action on Mitochondrial Phospholipids

Maturation of red blood cells is known to be attended by inactivation and subsequent disappearance of mitochondria. As shown in Rapoport's laboratory (Schewe et al., 1977; Rapoport et al., 1979), this process, being particularly rapid during bleeding anemia, is induced by a special cytoplasmic enzyme, lipoxygenase, which is a metalloprotein having two atoms of iron per a molecule of some 78,000 daltons. The results of experiments in which rat liver mitochondria or submitochondrial vesicles were treated with lipoxygenase isolated from reticulocytes evidenced that mitochondrial membranes are selectively subject to the action of this enzyme, resulting in malondialdehyde formation and suppression of electron transfer through the respiratory chain. The latter is in all probability due to the destruction of iron-sulfur centers in the respiratory-chain complexes I and II by intermediary radicals produced in lipid oxidation (Schewe et al., 1981). Exposure of mitochondria to lipoxygenase also results in the leakage of matrix enzymes and impairement of the organelle ultrastructure (Krause and Halangk, 1977).

Such destructive alterations are averted by ATP or by ADP with succinate (Krause and Halangk, 1977). In particular, it follows from Table 13 that in both cases the release of malate dehydrogenase from mitochondria is retarded. Mitochondrial resistance to lipoxygenase is unlikely to be directly linked to their conformational state. Indeed, degradation proceeds even in a hypertonic sucrose solution (see Table 13) where mitochondria assume a condensed form like that in the presence of ATP or succinate and ADP. Though the mechanism of the protective effect is not clear yet, it is quite certain that mitochondria performing energy transformation are less susceptible to lipoxygenase than nonfunctioning ones.

Summing up the material of the last two chapters, it is easy to see that mitochondria possess everything that is necessary for both "self-destruction" and "self-defense." Notwithstanding that we do not yet know all the details of mitochondrial degradation <u>in vitro</u> or the precise mechanisms of their stabilization, we can state with certainty that mitochondrial stability is directly dependent on their metabolic activity, i.e. on how quickly and efficiently they perform oxidative phosphorylation of ADP and coupled reactions. If isolated mitochondria are provided with the required set of substrates and cofactors, a phosphate acceptor, etc., they maintain in aerobic conditions a high respiratory control and morphological integrity for several days at room temperature, whereas in the absence of a single necessary factor of the environment (e.g., oxygen) the life-span of their energy-transducing form is shortened to several hours. Thus the mitochondrion must be regarded as a dynamic rather than a static system.

Table 13. Release of Malate Dehydrogenase from Intact Rat Liver Mitochondria Upon Exposure to Lipoxygenase (adapted from Krause and Halangk, 1977).

Incubation conditions	Release of malate dehydrogenase, %	Structural alterations
In 0.3 M sucrose	25	++
with ADP and succinate	3	-
with ATP	5.5	-
with succinate and DNP*	9	+

* DNP, 2,4-dinitrophenol.

Having considered in Chapter One the principles of mitochondrial compartmentation, we came to the conclusion that they correspond to the concept of maximal functional expedience. In other words, the mitochondrial compartmentation is such as would maximally facilitate their main function which is to phosphorylate ADP by utilizing pyruvate, fatty acids, and a number of other substrates. Actually, the components of the mitochondrial matrix (including the enzymes of the tricarboxylic-acid cycle and fatty-acid β-oxidation together with the appropriate set of cofactors and substrates) feed substrates to the respiratory chain in the inner mitochondrial membrane where electron transfer takes place, the transmembrane electrochemical proton potential is generated, ADP is phosphorylated, and other energy-generating and energy-dependent processes are carried out. Were there no functioning inner membrane, the matrix components would dissipate in the surrounding medium (Brenner-Holzach and Raaflaub, 1954; Siekevitz and Potter, 1955; Estrada-O et al., 1970). The adenosine triphosphate produced is transported to the intermembrane space where with the help of a number of enzymes (creatine phosphokinase, nucleoside diphosphate phosphotransferase, adenylate kinase, etc.) it is quickly converted back to ADP which is again subject to phosphorylation. The lack of intermembrane-space enzymes would inevitably result in a deficiency of ADP, attenuating thereby the intensity of the mitochondrial energy metabolism (Jacobus and Lehninger, 1973; Liberman et al., 1980). This makes obvious that for oxidative phosphorylation to proceed the outer mitochondrial membrane is indispensable as a barrier to confine the necessary enzyme set in the immediate proximity to the inner membrane.

All this amounts to the idea that the known pattern of distribution of the mitochondrial components among the compartments is the only viable one.

4
Mitochondrial Degradation in the Cell

The molecular mechanisms of mitochondrial degradation and stabilization in vitro having been considered, it is only natural to ask: how does all this bear on the reality of the living cell? In an endeavor to answer this question, it is reasonable to deal first of all with two problems: (1) what are the actual mechanisms of mitochondrial degradation in the cell? and (2) is oxidative phosphorylation necessary for the maintenance of mitochondrial integrity in vivo, as could be expected from the results of model experiments?

4.1. MECHANISMS OF INTRACELLULAR MITOCHONDRIAL DEGRADATION

4.1.1. Mode of Mitochondrial Turnover

The early electron-microscopic observations of Ashford and Porter (1962), de Duve and Wattiaux (1962), Novikoff and Essner (1962), Swift and Hruban (1964) showed that in the cell mitochondria may be digested by lysosomes. In the years that followed these notions were confirmed and developed. In particular, Travis and Travis (1972) and Topping and Travis (1974) presented data on the participation of lysosomes in mitochondrial degradation in aging cardiac muscle.

These observations are in line with the biochemical evidence obtained by a number of authors. For instance, Hanes and Tappel (1971) discovered in purified liver lysosomal preparations unusual hemochrome pigments with main absorption bands at 428 and 562 nm. Incubation of lysosomes with mitochondria at pH 5.0 gave rise to products with spectral characteristics corresponding to those of the lysosomal hemochromes. More than a score years ago Fletcher and

Sanadi (1961) reported that the half-lives of the main rat-liver mitochondrial components (soluble and insoluble proteins, and lipids) are rather close to each other. These data led them to the suggestion that a mitochondrion in the cell turns over as a unit with a half-life of about 10 days. Rather similar half-lives were further estimated for mitochondrial proteins, phospholipids, and DNA by other authors (Table 14). However, interpretation of the experimental data available encounters substantial difficulties. Firstly, it is usually not individual proteins or phospholipids that are considered but their mixtures including numerous components which may have widely differing half-lives. Secondly, the estimated half-lives rarely reflect the actual breakdown rates since the assay techniques do not always account for the fact that labeled amino acids, fatty acids, nucleotides, and other compounds released upon breakdown of proteins, lipids, nucleic acids, etc., may be reutilized in the biosynthesis of these cell components. Indeed, noticeably shorter half-lives (4-6 days) were found for liver mitochondrial proteins in more recent work (Table 14). These results were obtained by labeling the protein with [^{14}C-guanido]arginine and the heme with δ-amino[2,3-^{3}H]levulinic acid. The label from the former compound is known to be quickly excreted from mammalian cells with urea and therefore not recycled in protein synthesis. It must be admitted that the half-lives of six rat-liver mitochondrial protein fractions measured with [^{14}C-guanido]arginine proved to be rather similar to each other and also close to those estimated for cytochromes aa_3, b, and c by labeling their hemes with δ-amino[2,3-^{3}H]levulinic acid (Druyan et al., 1969; Ashenbrenner et al., 1970; Ip et al., 1974). This fits into the hypothesis of Fletcher and Sanadi; however, the corrected half-lives of liver mitochondrial protein components - crude protein fractions (Swick et al., 1968; Ashenbrenner et al., 1970; Gear et al., 1974) as well as individual proteins (Druyan et al., 1969; Ashenbrenner et al., 1970; Ip et al., 1974) - differ almost twice from the values found for mitochondrial lipids and DNA (Gear et al., 1974).

More detailed studies of the renewal of mitochondrial protein and phospholipid components showed that the hypothesis of Fletcher and Sanadi oversimplifies the actual mechanism of mitochondrial degradation. First of all it can be mentioned that, according to Swick et al. (1968), the half-lives of two mitochondrial enzymes (leucine aminotransferase and ornithine aminotransferase) are 0.73 and 1.03 days, i.e. five times less than the corresponding value for the pooled protein fractions; the half-life of δ-aminolevulinic acid synthetase is as short as 72 min (Marver et al., 1966). The proteins of the outer and inner mitochondrial membranes were reported to be renewed at different rates (Brunner and Neupert, 1968; Beattie, 1969; De Bernard et al., 1969; Lipsky and Pedersen, 1981). Rajwade et al. (1975) found that significant differences in the half-lives of some mitochondrial protein fractions (from 8 to 12 days) may evolve because of hormonal regulatory disorders in rats. Finally, just

Table 14. Half-Lives of Some Rat Liver Mitochondrial Components.

Fraction or component	Half-life (days)	Method of determination	Reference
Soluble protein	10.8	Decrease in ^{35}S content after pulse of [^{35}S]-methionine	Fletcher and Sanadi (1961)
Insoluble protein	10.1		
Cytochrome *c*	9.7		
Lipids	10.6	Decrease in ^{14}C content after pulse of [^{14}C]-acetate	
Synthetase of δ-aminolevulinic acid	about 70 min	Decline in enzymic activity after addition of puromycin	Marver et al. (1966)
Total mitochondrial protein	8.39	Decrease in ^{14}C content after pulse of [^{14}C]leucine	Beattie et al. (1967)
Water-soluble protein	8.56		
Cytochrome *c*	8.56		
Contractile protein	8.37		
Structural protein	8.77		
Insoluble protein	9.0	Decrease in ^{35}S after pulse of [^{35}S]-methionine	Bailey et al. (1967)
Phospholipids:		Decrease in ^{32}P content after pulse of [^{32}P]phosphate	
component I	1.6		
component II	10.0		
Water-soluble protein	4.2–5.1	Decrease in ^{14}C content after pulse of [^{14}C-guanidino]-arginine	Swick et al. (1968)
Protein soluble in 0.9% KCl (cyt. *c*)	4.5–5.0		
Protein soluble in 0.6 M KCl	4.6–5.3		
Protein precipitated with 13%-saturated ammonium sulfate	4.3–5.3		
Protein precipitated with 35%-saturated ammonium sulfate	3.9–4.8		
Residue	4.7–8.8		
Alanine aminotransferase	0.73	Decline in enzymic activity after switching animals from 60% to 12% protein diet	Swick et al. (1968)
Ornithine aminotransferase	1.03		
Total outer-membrane protein	4.2	Decrease in ^{35}S content after pulse of [^{35}S]methionine	Brunner and Neupert (1968)
Total inner-membrane protein	12.6		
Soluble protein	11.6		

Table 14. (continued)

Fraction or component	Half-life (days)	Method of determination	Reference
Outer-membrane protein	7.0	Decrease in ^{14}C content after pulse of [^{14}C]leucine	Beattie (1969)
Protein of the "inner membrane-matrix" fraction	8.4		
Subfractions of the inner membrane:			
water-soluble protein	8.2		
soluble in 0.9% KCl	6.9		
insoluble in 0.9% KCl	10.2		
Soluble protein fraction	6.8		
Total mitochondrial protein	6.8	Decrease in ^{14}C after pulse of [^{14}C-guanidino]arginine	Arias et al. (1969)
Total mitochondrial protein	9.0	The same after pulse of randomly labelled [^{14}C]arginine	
Cytochrome *b*	5.5	Decrease in ^{3}H content after pulse of δ-amino-[^{3}H]-levulinic acid	Druyan et al. (1969)
Cytochrome *c*	6.1		
Cytochrome b_5	4.4		
Mitochondrial DNA	9.4	For details see reference	Gross et al. (1969)
Phospholipids:			
cardiolipin	11.5		
lecithin	10.0		
phosphatidyl-ethanolamine	9.9		
sphingomyelin	11.4		
Cytochrome *c*, Heme *a*, Total protein	5.5-6.2	Decrease in ^{3}H and ^{14}C after pulse of δ-amino[2,3-^{3}H] levulinic acid and [^{14}C-guanidino]-L-arginine	Ashenbrenner et al. (1970)
Ornithine amino-transferase	1.9	Decrease in ^{14}C content after pulse of $NaH^{14}CO_3$	Ip et al. (1974)
Cytochrome *c* oxidase	5.7		
Outer-membrane protein	2.8	Decrease in ^{14}C content after pulse of $NaH^{14}CO_3$	Lipsky and Pedersen (1981)
Inner-membrane protein	3.8		
Matrix protein	3.9		

recently Hare and Hodges (1982a) with the use of a double-labeling technique have determined the half-lives of the 33 most abundant polypeptides of the inner mitochondrial membrane resolved in a bi-directional polyacrylamide gel system. In nonproliferating cells of a hepatoma monolayer culture these half-lives were found to vary from 20 to 100 h. It is of interest that four polypeptides entering into the composition of the cytochrome bc_1 complex (e.g., apocytochrome b and core protein II) have a half-life of about 40h whereas the value for the other polypeptides of the complex (including apocytochrome c_1) is about 100 h. This means that neither the inner mitochondrial membrane nor its oligoenzyme complexes turn over as a unit. Thus, the more information becomes available about the renewal times of individual mitochondrial proteins, the more obvious it is that these values vary in a wide range, in contradiction to the hypothesis of Fletcher and Sanadi.

This is still more graphic for mitochondrial phospholipids. As shown by Bailey et al. (1967), rat-liver mitochondrial phospholipids contain two components displaying half-lives of 1.6 and 10 days. The more stable second component appears to be cardiolipin (Taylor et al., 1967). It is of special interest that, according to these authors, the acyl groups of the phospholipid molecules are renewed far more rapidly than their glycerophosphate core. Substantial differences in the half-lives of cardiolipin and other phospholipids in brain mitochondria were reported by Cuzner et al. (1966).

All this gives grounds for regarding the mitochondrion in vivo as a flexible structure whose protein, lipid, and other constituents are continuously degraded and replaced. As to the Fletcher-Sanadi hypothesis that the mitochondrion in the cell is degraded as a unit, it seems to apply to "dead" rather than to normally metabolizing organelles. From this point of view, the rather similar maximal half-lives of various mitochondrial components may reflect just the rate of elimination of such defunct organelles from the total mitochondrial pool, their trapping and digestion by lysosomes. In other words, the lysosomal mechanism of mitochondrial degradation is not likely to be the primary one (see the concluding section of Chapter Two).

According to the calculations of Grisolia et al. (1980), not less that 50% of mitochondrial protein is degraded by mechanisms other than those involving autophagic vacuoles.

To date, quite a number of cases are known when mitochondrial degradation in the cell does not involve lysosomes. One example of this is the degeneration of the tadpole's (Xenopus larvae) tail (Weber, 1964). Novikoff (1963) studied lymphocyte degradation in the thymus after cortisone treatment or X-ray irradiation, and found pronounced destructive alterations in mitochondria that were not related directly to lysosomal action. Finally, it was shown in Rapoport's laboratory (Rapoport et al., 1979; Müller et al., 1980; Dubiel et al., 1981a,b) that mitochondrial breakdown and elimination during reticulocyte maturation are at the first stages associated with the action of lipoxygenase which induces peroxidation of mitochondrial membrane phospholipids, and only then completed by the cytoplasmic proteolytic system. These events will be dealt with in more detail below in a special section of this chapter.

Even during joint incubation of isolated mitochondria and lysosomes, hydrolysis of mitochondrial proteins is preceded by functional and structural alterations in mitochondria. Mellors et al. (1967) showed that lysosomes induce high-amplitude mitochondrial swelling, uncoupling of oxidative phosphorylation, and activation of ATPase. Such swelling can be ascribed neither to lipid peroxide formation nor to the action of proteolytic enzymes. Indeed, it could not be suppressed by antioxidants, chelators, or proteinase inhibitors (diisopropyl fluorophosphate and 6-aminocaproic acid). At the same time, the lysosome effect on mitochondria was completely abolished by serum albumin. On the basis of these and some other data it was suggested that under the action of an endogenous phospholipase A the lysosomes liberate free fatty acids which trigger mitochondrial degradation. It was also thought that mitochondrial membranes could be directly destroyed by the lysosomal phospholipase.

4.1.2. Role of Mitochondrial Phospholipase A in Organelle Degradation *In Vivo*

In the light of the molecular mechanisms already considered in Chapter Two, of special interest is the question of the part played by mitochondrial phospholipases in organelle degradation *in vivo*. That phospholipase A may probably participate in mitochondrial breakdown is suggested by the data obtained in studies of various ischemic organs. Boime et al. (1970) showed that liver ischemia is marked by accumulation of free fatty acids in mitochondria accompanied by decreasing respiratory control and profound morphological alterations. The initial stages of mitochondrial degradation are reversible: the damage is repaired on addition of serum albumin to isolated mitochondria. The authors propose that the bulk of free fatty acids found in the mitochondrial fraction is of microsomal origin, and their accumulation is due to the action of either microsomal phospholipase A_1 or lysosomal phospholipase A_1+A_2. With all this, an important part, especially in the irreversible mitochondrial injury, is apparently played by the endogenous mitochondrial phospholipase A_2, since there is a gradual decrease in the cardiolipin content in mitochondria. Loss of cardiolipin, which is localized exclusively in the inner mitochondrial membrane, was also observed by other authors (Ekholm et al., 1968; Vasdev et al., 1980) in ischemic heart.

According to Ozawa et al. (1969), the interruption of oxidative phosphorylation in ischemic pancreas is the result of a decrease in the content of acyl residues in mitochondrial phospholipids and accumulation of lysolecithin, as in isolated aging mitochondria.

Thus, there are grounds for believing that a certain part in the degradation of mammalian mitochondria, at least in some pathological conditions, is played by the mitochondrial phospholipase A_2.

4.1.3. Role of Mitochondrial Proteinases in Mitochondrial Degradation *In Vivo*

Studying the effects of a hypocholesterolemic drug clofibrate (ethyl α-p-chlorophenoxyisobutyrate) in rats, Gear et al. (1974) found that it induced a considerable increase in the overall mitochondrial mass in liver with a concomitant decrease in the mitochondrial neutral proteolytic activity. These and some other data were used to suggest that the content of the mitochondrial material in liver cells may be controlled by endogenous mitochondrial proteinases. The regulatory functions of these enzymes were ascribed to their putative ability to selectively digest a group of rapidly synthesized membrane proteins constituting a minor portion of the total mitochondrial protein but playing an exceptionally important role in mitochondrial formation (see Chapter Six). This point of view is in line with the already mentioned data of Wheeldon et al. (1974) about the rapid degradation of nascent polypeptides *in vitro* which was assumed to involve mitochondrial proteinases. It must be mentioned, however, that according to a recent report of Lipsky and Pedersen (1982) the increase in the content of the mitochondrial material in the cells of clofibrate-treated animals is more probably due to enhanced synthesis of most mitochondrial proteins.

Rajwade et al. (1975) established that the half-life of water-insoluble rat liver mitochondrial proteins appreciably increases (by 27-37%) in thyroidectomized rats. Simultaneously the mitochondrial proteolytic activity fell approximately by half while liver cathepsin activity remained almost constant. It was hence suggested that endogenous proteinases may adjust the turnover rate of mitochondrial proteins. Further, thyroidectomy affected the half-lives of some mitochondrial proteins (see above) but not of others (the soluble mitochondrial protein fraction); so mitochondrial proteinases could be thought to be also responsible for the synchronization of protein turnover in the organelles, since partial suppression of these enzymes evoked a divergence in the half-lives of some proteins.

It is easy to see that all the data presented are only indirect evidence for the participation of mitochondrial proteinases in the breakdown of mitochondrial proteins *in vivo*. This question will be treated in more detail in Chapter Eight.

4.1.4. Role of Lipid Oxidation in Mitochondrial Degradation *In Vivo*

Some pathological conditions (malignant proliferation, radiation injury, E-avitaminosis, etc.,) are known to be attended by accumulation of hydroperoxides in the mitochondrial lipid fraction (Turovetskii et al., 1972) and partial impairment of oxidative phosphorylation (Tappell, 1973).

One of the most dramatic examples of mitochondrial degradation *in vivo* because of lipid oxidation is reticulocytosis upon bleeding anemia. According to Rapoport et al. (1979), in such conditions a large amount of lipoxygenase is induced in reticulocytes, which then catalyzes selective oxidation of mitochondrial phospholipids (as compared to, e.g., plasma membrane lipids), impairing the respiratory chain function and mitochondrial structural integrity (see Chapter Three). Such degraded mitochondria are the main target for the cytoplasmic proteolytic system (Dubiel et al., 1981b) which under certain conditions can digest up to 50% of the total stroma protein within only 2h. In accordance with the above, proteolysis in reticulocyte lysates is suppressed by a lipoxygenase inhibitor salicyl hydroxamate or by anaerobiosis (Table 15).

Hydrolysis of mitochondrial proteins in reticulocytes is ATP-dependent. As follows from Table 16, it is suppressed in intact cells by a respiratory inhibitor antimycin A or by an uncoupler 2,4-dinitrophenol. In reticulocyte lysates the proteolysis takes place only on addition of ATP or in the presence of an operative ATP-generating system, which may be the mitochondrial oxidative phosphorylation system. It should not be overlooked that lipoxygenase, on the one hand, triggers the proteolysis by rendering the substrate (mitochondrial proteins) accessible to intracellular proteinases but, on the other hand, hinders it by exhausting the ATP reserve in reticulocytes and impairing oxidative phosphorylation. In this respect the lipoxygenase influence on the proteolysis must resemble that of antimycin and 2,4-dinitrophenol (Table 16).

Of special interest is the question how widely the lipoxygenase-linked mitochondrial degradation is distributed through various tissues and cells. Thus far it can only be said that lipoxygenase is certainly not ubiquitous though the list of lipoxygenase-containing objects is by no means restricted to reticulocytes (Table 17).

4.2. ROLE OF OXIDATIVE PHOSPHORYLATION IN MITOCHONDRIAL SURVIVAL *IN VIVO*

4.2.1. Effects of Anaerobiosis on Yeast Mitochondria

The objects of choice for studying the role of oxidative phosphorylation in the maintenance of mitochondrial integrity *in vivo* are

Table 15. Effects of Salicylhydroxamate (SHAM) and Anaerobiosis on Protein Breakdown in Rabbit Reticulocyte Lysates in the Presence of ATP (taken from Dubiel et al., 1981a).

Reticulocytes, %	Protein breakdown, %		Reticulocytes, %	Protein breakdown, %	
	−SHAM	+SHAM		Aerobiosis	Anaerobiosis
39.5 ± 41.0	17.8 ± 2.7	5.4 ± 1.9	30.5 ± 32.0	14.0 ± 1.6	2.0 ± 1.2

Table 16. Energy Dependence of Protein Degradation in Intact Reticulocytes (taken from Müller et al., 1980).

Reticulocytes, %	Protein degradation, %		Inhibitor, concn.
	Control	With inhibitor	
48.0	49.8	2.5	AntimycinA, $5 \cdot 10^{-6}$ M
48.7	43.4	2.1	2,4-Dinitrophenol, $5 \cdot 10^{-5}$ M

Table 17. Lipoxygenase Activity in Different Rabbit Cells and Tissues (taken from Rapoport et al., 1979).

Type of cells	Lipoxygenase activity, $\Delta A_{234} \cdot min^{-1} \cdot ml\ cells^{-1}$	Immunological reactivity
Erythrocytes	0	<0.05
Reticulocytes	0.33	1
Bone marrow	0.003	<0.05
Peritoneal macrophages	0.012	<0.05
Lung	0.15	<0.05
Brown-Pearce tumor	0.30	<0.05

facultative anaerobes, which can normally grow and develop in the absence of oxygen. Such organisms include some species of yeast. As shown by Meisel' et al. (1964), switching of yeast from aerobic growth to alcohol-yielding fermentation is accompanied by reduction of mitochondrial cristae. Still, the organelles never completely disappear under anoxia. Chapman and Bartley (1968) followed the changes in enzymic activities in baker's yeast homogenates in the course of cell adaptation to anaerobiosis. In these experiments the yeast grown aerobically in a medium with a moderate glucose concentration (0.9%) were transferred into a deaerated medium containing much glucose (10%). In other words, the anaerobic adaptation was superimposed on glucose repression, which made it difficult to discern mitochondrial alterations caused by anaerobiosis as such. Under the conditions specified, an increase was observed in the specific activities of some glycolytic enzymes (e.g., pyruvate kinase and lactate dehydrogenase). At the same time the activities of some Krebs-cycle enzymes (α-ketoglutarate dehydrogenase, malate dehydrogenase, and isocitrate dehydrogenase) decreased.

The influence of anaerobiosis on yeast mitochondria was studied in detail in our laboratory (Zubatov et al., 1971; Zubatov, 1972; Luzikov et al., 1971c, 1973b). Figure 31 shows how anaerobiosis affects the respiratory system of aerobically grown Saccharomyces cerevisiae. Deaeration of the medium in the onset of the stationary phase (i.e. at the moment when culture density, cell respiration, and mitochondrial respiratory activity reach their highest values) entails a substantial fall in mitochondrial NADH oxidase and succinate oxidase activities, while their succinate-ferricyanide reductase, NADH-ferricyanide reductase, cytochrome *c* oxidase, and other partial respiratory-chain activities decrease at a lower rate. Since this decrease was observed under conditions when glucose had been largely consumed and the ethanol accumulated could not be utilized in the absence of oxygen, it might be thought that the alterations in the respiratory system are a result of cell starvation rather than the specific sequelae of anaerobiosis. However, the data presented below

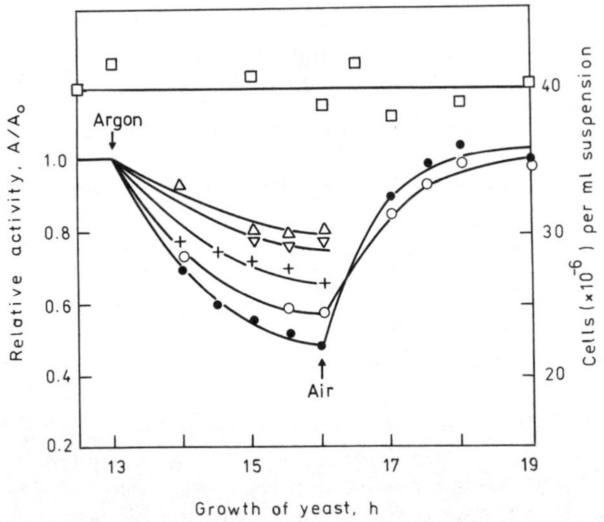

Fig. 31. Changes in the activities of some mitochondrial enzymes during deaeration and subsequent aeration of aerobically grown Saccharomyces cerevisiae. The yeast were deaerated in the end of the exponential phase of aerobic growth on glucose, and at the times indicated aliquots of suspension were taken to isolate mitochondria and assay the following specific activities: (●) succinate oxidase; (○) NADH oxidase; (+) cytochrome *c* oxidase; (▽) succinate-cytochrome *c* reductase; (△) NADH-cytochrome *c* reductase. (□) Concentration of the cell suspension during the experiment. From Luzikov et al. (1971c).

(see Figure 32) show that yeast deaeration in the early exponential phase, when there is still enough glucose, brings about an even more pronounced deterioration of the respiratory system.

The decrease in the mitochondrial respiratory activity under the conditions specified was not due to the suppression of the synthesis of mitochondrial proteins in the absence of oxygen. Such a possibility could not be ruled out a priori, since glucose had for the most part been consumed by the beginning of the stationary phase, and interruption of oxidative phosphorylation could rapidly exhaust the cell energy resources. Nevertheless, introduction of the inhibitors of mitochondrial and cytoplasmic protein synthesis (chloramphenicol and cycloheximide, respectively) in the beginning of the stationary phase of aerobic growth does not inactivate the respiratory system (Figure 33a) as deaeration does.

Interpretation of these data is somewhat complicated by the data of Michel et al. (1976) that in Neurospora crassa cycloheximide

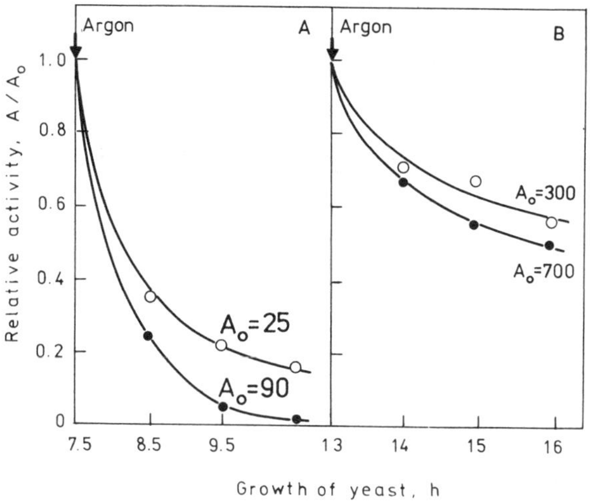

Fig. 32. Degradation of the mitochondrial respiratory system in Saccharomyces cerevisiae deaerated in the early (A) and late (B) exponential phases of aerobic growth on glucose. Cells were harvested, thickened to suppress cell division, and incubated under argon. At moments indicated mitochondrial fractions were isolated and (•) NADH oxidase and (o) succinate oxidase activities were measured. The A_0 values are the initial activities expressed as nmoles of substrate oxidized per min per mg mitochondrial protein at 30°C. From Luzikov et al. (1973b).

suppresses the synthesis of a short-lived proteinase efficiently cleaving mitochondrial proteins. Cycloheximide also hinders the degradation of the Saccharomyces cerevisiae respiratory system during glucose repression (see further), presumably by inhibiting the synthesis of phospholipase D responsible for the degradation (Grossman et al., 1973). In this connection it could be thought that mitochondria were preserved in the presence of cycloheximide and chloramphenicol (Figure 33a) just because the destructive machinery was inoperative. However, both inhibitors added at the beginning of the stationary phase did not prevent the respiratory system degradation induced by anaerobiosis (Figure 33b).

It should be noted that, in contrast to Neurospora, in the yeast Schizosaccharomyces pombe the breakdown of mitochondrial proteins is not prevented by cycloheximide added into the growth medium (Michel et al., 1976). As to the induction of phospholipase D synthesis, it could be characteristic of glucose repression but hardly probable in yeast deaerated in the early stationary phase.

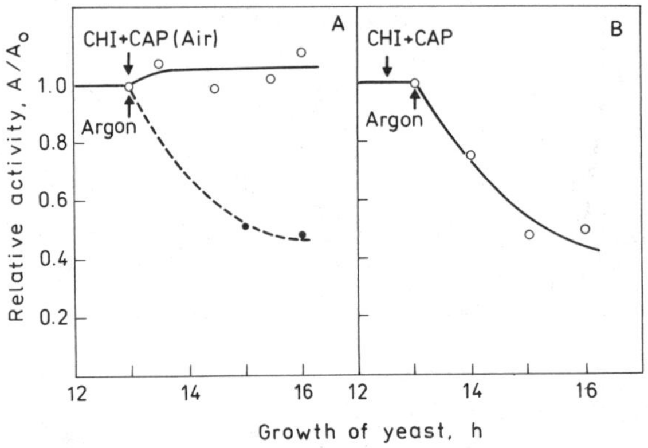

Fig. 33. Effect of protein-synthesis inhibitors on the stability of the mitochondrial NADH oxidase system in Saccharomyces cerevisiae under aerobic and anaerobic conditions. (A) (●) Effect of anaerobiosis (same as Figs. 31, 32); (○) aerobic incubation with cycloheximide (CHI) and chloramphenicol (CAP). (B) Deaeration of aerobically grown yeast in the presence of CHI + CAP.

The effect of anaerobiosis also cannot be attributed to a decrease in the content of unsaturated fatty acids and ergosterol which are necessary for the formation and maintenance of mitochondrial membranes (Jollow et al., 1968; Wallace et al., 1968; Damsky et al., 1969; Watson et al., 1970); indeed, introduction of these compounds into the growth medium does not impede the decline of the mitochondrial specific respiratory activity in deaerated yeast (Luzikov et al., 1971c).

The most plausible cause of decreasing mitochondrial respiratory activity in anaerobic conditions is the destruction of inner mitochondrial membranes proceeding via an unknown mechanism. This suggestion is favored by electron-microscopic data obtained in this laboratory (Luzikov et al., 1971c, 1973b).

The mitochondrial degradation described in this section has several peculiarities. One of them is that the functional and morphological alterations are completely reversible, and at least the initial stages of restoration, for which it is enough to aerate the yeast in the same medium (Figure 31), are not suppressed by inhibitors of mitochondrial (chloramphenicol) or cytoplasmic (cycloheximide) protein synthesis (Luzikov et al., 1971c). In this respect the restoration of the mitochondrial respiratory system after relatively short anaerobiosis differs from glucose derepression or respiratory adaptation of anaerobically grown yeast. By the data of a number of authors (Henson et al., 1968a; Linnane et al., 1968;

Rouslin and Schatz, 1969; Ball and Tustanoff, 1970; Lin et al., 1974), the last two processes are hindered by chloramphenicol and cycloheximide because they require synthesis of cytochromes and other respiratory components.

The anaerobic degradation of yeast mitochondria in the early stationary phase is also marked by the lack of changes in the cell cytochrome content (Luzikov et al., 1971c) though mitochondrial NADH oxidase and succinate oxidase activities decrease considerably and the organelles undergo pronounced degenerative structural alterations. Hence one may think that the restoration of the respiratory system would not require synthesis of cytochromes and probably other mitochondrial membrane components. This explains why the protein-synthesis inhibitors have no effect on the restoration of the respiratory machinery which must apparently be regarded as reorganization of partly disorganized membranes.

The above-described case of respiratory system degradation without loss of cell cytochromes is not unique. Turner and Lloyd (1971) found that in nonproliferating cells of a protozoan Tetrahymena pyriformis chloramphenicol causes a considerable decrease in the respiration rate with α-ketoglutarate and succinate as substrates, loss of respiratory control, and other alterations. These, however, were not accompanied by any decrease in the cytochrome content either per cell or per unit mitochondrial protein.

4.2.2. Degradation of Yeast Mitochondria Induced by Respiratory Inhibitors and Uncouplers

Bartley and Tustanoff (1966) showed that respiratory inhibitors (4-hydroxy-2-nonylquinoline-N-oxide and antimycin A) hinder the development of respiration during the respiratory adaptation of anaerobically grown Saccharomyces cerevisiae. According to Kováč et al. (1970), yeast cells grown in a glucose medium containing cyanide, antimycin A, 2,4-dinitrophenol, or oligomycin do not possess a complete set of cytochromes characteristic of the normal respiratory chain. This was explained as aggravation of glucose repression in the presence of the above-listed compounds.

However, from the later results of Kováč (1972) it follows that antimycin A and oligomycin prevent the normal respiratory chain development even when yeast are grown in a galactose-containing medium, i.e. under nonrepressive conditions. The abnormalities manifest themselves at least as increased content of cytochrome *c*, markedly decreased (3-4-fold) content of cytochrome aa_3, and lowered cell respiration rate. It is regrettable that Kováč (1972) provided no data on the functional peculiarities of the respiratory apparatus of the yeast grown in the presence of oligomycin and/or antimycin A. In particular, nothing was said about its sensitivity to cyanide or other respiratory inhibitors, its coupling to the energy-transducing machinery, etc.

Thus it appears that the normal respiratory chain is not formed in yeast in the presence of the inhibitors of respiration and oxidative phosphorylation, even when there is no catabolite repression and the cells have a necessary energy source for their development. In such a case it seems reasonable to assume that the cause of the abnormal respiratory chain formation lies in the inability of the assembled mitochondrial structures to perform the usual set of functions characteristic of the oxidative phosphorylation system, in the destabilization of such structures and their partial degradation. An alternative to this would be a suggestion that normal respiratory chain development requires an intramitochondrial energy source and that respiration is needed to regulate mitochondrial formation through a feedback mechanism (i.e. through heme synthesis, creation of a template for further mitochondrial membrane differentiation, maintenance of a certain level of NAD reduction, etc.,) (Perlman and Mahler, 1974). The only argument put forward by these authors to support their suggestion is that the respiratory inhibitor antimycin A suppresses the increase in the activities of some dehydrogenases and cytochrome *c* oxidase and in the content of cytochromes *b*, c_1, and aa_3 during yeast respiratory adaptation.

Having analyzed the data of Kováč and colleagues together with the similar results of Perlman and Mahler, it is natural to ask how the compounds capable of suppressing oxidative phosphorylation affect the completely assembled mitochondria in mature yeast cells. The answer to this question is given by the work of Luzikov et al. (1970), in which yeast was grown aerobically until the beginning of the stationary phase; then the suspension was supplemented with respiratory inhibitors (cyanide, antimycin A) or uncouplers (2,4-dinitrophenol, carbonylcyanide m-chlorophenylhydrazone), and the changes in the respiratory activity of the mitochondrial fraction or the morphological state of the organelles were studied. Cyanide and antimycin A in aerobic conditions proved to affect the mitochondria in just the same way as anaerobiosis (Figure 34). Moreover, the respiratory system is degraded during yeast incubation in the presence of 10^{-6} M carbonylcyanide m-chlorophenylhydrazone and 10^{-4} M 2,4-dinitrophenol when initially the cell respiration is somewhat stimulated. It is especially important that though respiratory inhibitors, uncouplers, and anaerobiosis impair oxidative phosphorylation in different ways, the mitochondrial respiratory chain is in all cases degraded at practically the same rate. The similarity of the effects of anaerobiosis and uncouplers was demonstrated not only functionally but morphologically as well (Luzikov et al., 1973b).

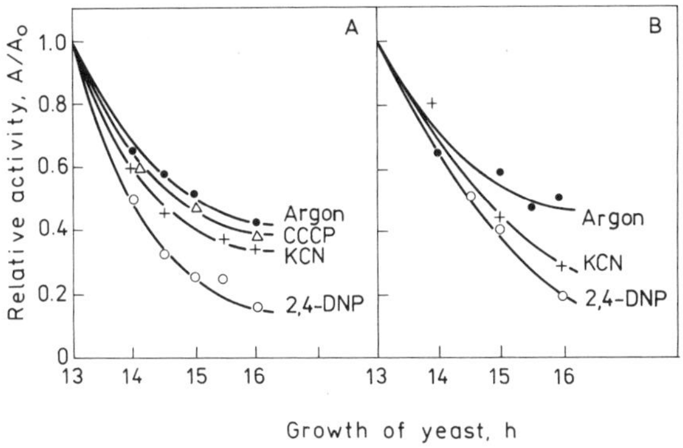

Fig. 34. Changes in NADH oxidase (A) and succinate oxidase (B) activities of mitochondria during deaeration of aerobically grown Saccharomyces cerevisiae (●) and aerobic incubation with potassium cyanide (+) or uncouplers (○,△). (2,4-DNP) 2,4-Dinitrophenol; (CCCP) carbonylcyanide m-chlorophenylhydrazone. From Luzikov et al. (1970b).

4.2.3. **Degradation of Yeast Mitochondria During Glucose Repression**

Transfer of yeast cells into a glucose-rich medium is well known to entail mitochondrial degradation. According to Jayaraman et al. (1966), the degradation manifests itself as decreasing mitochondrial content related to the cell protein, decreasing cell respiration rate, cytochrome breakdown, decreasing activities of a number of mitochondrial respiratory enzymes, impairment of respiratory control and mitochondrial structural integrity. Similar results were reported by Utter et al. (1967).

Later Dharmalingam and Jayaraman (1971, 1972) found that phospholipase D activity is enhanced in yeast cells undergoing glucose repression. During subsequent derepression this activity is lowered in parallel with increasing oxygen uptake rate. Yeast grown on maltose, which has no repressive action, are characterized by low phospholipase activity. It is interesting that blocking of mitochondrial protein synthesis with chloramphenicol is also attended by an increase in phospholipase activity. From these data it was suggested that mitochondrial membrane breakdown during glucose repression is due to the increasing content (induction of synthesis) of phospholipase D in yeast cells.

The possible role of phospholipase D in yeast mitochondrial degradation during glucose repression was studied in more detail by Grossman et al. (1973). It was shown that administration of glucose into the incubation medium (final concentration 6%) is accompanied by a decrease in the cell respiration rate, in mitochondrial succinate dehydrogenase activity, and in the content of covalently protein-bound flavin; there is a parallel increase in phospholipase D activity in mitochondria. All the changes indicated are suppressed by cycloheximide which inhibits cytoplasmic translation. This allows a suggestion that in the presence of high glucose concentrations phospholipase D synthesis is induced in the yeast cytoplasm, and that this synthesis can be considered a trigger mechanism in mitochondrial degradation (Figure 35).

Glucose repression also brings about other alterations in mitochondria which in principle could cause the impairment of oxidative phosphorylation in yeast. Among other things, they include the increase in the relative content of saturated fatty acids in mitochondrial phospholipids (see Chapter Eight). It is known (see, e.g., Marzuki et al., 1975a) that such changes lead to the loss of mitochondrial phosphorylation and in certain conditions even of respiratory ability.

4.2.4. Changes Induced in Yeast Mitochondria by Inhibitors of Mitochondrial Protein Synthesis

Growth of yeast (Clark-Walker and Linnane, 1966, 1967; Kellerman et al., 1969; Mahler and Perlman, 1971), fungi (Howell et al., 1971; Von Jagow and Klingenberg, 1972; Wakiyama and Ogura, 1972), and other cell types including protozoans, algae, and animal cells in the presence of specific inhibitors of mitochondrial protein synthesis (chloramphenicol or erythromycin), ethidium bromide, etc., results in pronounced alterations in the mitochondrial population manifesting

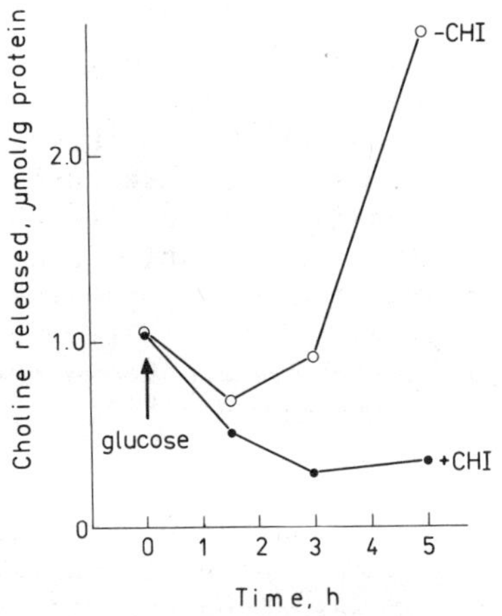

Fig. 35. Induction of phospholipase D in aerobic yeast transferred to a medium containing 6% glucose. (CHI) cycloheximide. Adapted from Grossman et al. (1973).

themselves as decreasing activities of the respiratory enzymes and impairment of the coupling mechanism (Clark-Walker and Linnane, 1967; Mahler and Perlman, 1971; Turner and Lloyd, 1971; King et al., 1972; Von Jagow and Klingenberg, 1972), decreasing content of cytochromes aa_3 and *b* (Clark-Walker and Linnane, 1966, 1967; Kellerman et al., 1969; González-Cadavid et al., 1970; Mahler and Perlman, 1971; Wakiyama and Ogura, 1972), reduction of cristae and other deviations from normal organelle structure (Clark-Walker and Linnane, 1967; Kellerman et al., 1969; King et al., 1972). The question arises as to whether these alterations are due to the upset balance between the synthesis and breakdown of mitochondrial proteins (and as a consequence to the decreasing content of certain membrane protein components) or to a 'disordered' accumulation of cytoplasmically made proteins and formation of defective organelles under conditions of blocked mitochondrial protein synthesis.

Important information which to a certain extent enables one to choose between these alternatives is furnished by the work of González-Cadavid et al. (1970) who showed that chloramphenicol induces formation of functionally deficient organelles in regenerating liver, which is marked by enchanced rates of biosynthetic processes, and has no effect on mitochondria in normal liver. Hence it might be concluded that the above-mentioned changes in mitochondria are due to formation of 'monster' organelles rather than to a noncompensated breakdown of mitochondrial translation products in the presence of chloramphenicol. Also, Stone and Wilkie (1975) presented evidence for the degradation (possibly proteolytic) of cytochrome *c* oxidase during Saccharomyces cerevisiae growth in the presence of erythromycin or chloramphenicol.

4.2.5. **Some General Aspects of Mitochondrial Degradation in the Cell**

4.2.5.1. *Trigger Mechanisms of Degradation*

As follows from the material presented in this chapter, mitochondrial degradation in the cell may be triggered by (1) induction of synthesis or activation ("delatentization") of destructive enzymes, or (2) destabilization of mitochondria owing to their deenergization.

The first mechanism can be exemplified by the synthesis of phospholipase D during yeast glucose repression (Grossman et al., 1973) or by lipoxygenase superinduction in reticulocytes in bleeding anemia (Rapoport et al., 1979). Another probably pertinent case is proteinase activation during yeast sporulation (Betz and Weiser, 1976) when the cells lose up to four fifths of the total cytochrome *b*, c_1, and aa_3 contents while the activities of aminopeptidase and proteinases A and B increase severalfold. Degradation of this type is usually suppressed by protein-synthesis inhibitors.

On the other hand, mitochondrial degradation induced in an early-stationary yeast culture by anaerobiosis (Luzikov et al., 1971c, 1973b) or respiratory inhibitors and uncouplers (Luzikov et al., 1970, 1973b) most probably results from their being destabilized because of de-energization. This is evidenced by the close, if not identical, rates of respiratory system disorganization under such influences and by the ineffectiveness of the inhibitors of cytoplasmic and mitochondrial translation in retarding this process (see above). It must be admitted that the actual molecular mechanism of degradation in this case is still obscure.

It is quite possible that mitochondrial degradation sets in with their de-energization in various ischemic organs. There are grounds for supposing that in such cases mitochondria become more susceptible to endogenous and exogenous A-type phospholipases (Ekholm et al., 1968; Ozawa et al., 1969; Boime et al., 1970; Vasdev et al., 1980). The mechanism of such destruction was discussed at length in the two preceding chapters.

4.2.5.2. *Depth of Mitochondrial Degradation*

It must be emphasized that the term "degradation" does not mean complete "dismantling" of mitochondria. As shown by Meisel' et al. (1964), Criddle and Schatz (1969), Plattner and Schatz (1969), Watson et al. (1970), and Groot et al. (1971), mitochondrial structures are found in the cells of facultative anaerobes grown in the absence of oxygen. These structures ("promitochondria") have outer and inner membranes, contain mitochondrial DNA, and can perform a number of functions characteristic of normal mitochondria (for instance, they display ATPase, NADH-ferricyanide reductase, and succinate dehydrogenase activities, catalyze the ATP-$^{32}P_i$ exchange and carry out ion transport). Mitochondrial structures are never completely lost during glucose repression (Jayaraman et al., 1966; Neal et al., 1970). Suppression of mitochondrial protein synthesis by chloramphenicol added to the yeast growth medium (Linnane et al., 1968) or as a result of cytoplasmic petite mutations (Yotsuyanagi, 1962; McClary and Bowers, 1968; Packer et al., 1973) also does not lead to the disappearance of mitochondrial structures from the cells.

With all this, the mitochondrial structures found in the cells of various organisms under the above conditions markedly differ from normal organelles. Indeed, they lack many respiratory chain components, their specific enzymic activities are low (e.g., the rate of the ATP-$^{32}P_i$ exchange in promitochondria is only about one fifth that in mature yeast mitochondria), the number of cristae and the overall inner-membrane area are reduced, etc.

Thus, mitochondrial degradation in the cell implies first and foremost their transition to a more primitive level of organization.

Such a 'demotion' may manifest itself simply as impairment of integral enzymic functions (e.g., coupling of respiration and phosphorylation, electron transfer through the respiratory chain, and the like). However, degradation may proceed further, involving breakdown of certain proteins and phospholipids and a decrease in the total mitochondrial mass. Jayaraman et al. (1966) reported that the amount of protein in the mitochondrial fraction falls more than by half during glucose repression. On the other hand, Criddle and Schatz (1969) believe that the protein content in the promitochondrial fraction depends on yeast growth conditions and may be close to that in the mitochondrial fraction.

These approaches cannot be considered quantitative for many reasons, in particular because neither the purity of mitochondrial or promitochondrial fractions nor the completeness of their isolation from the cells were controlled (no standards were used to which the experimental estimates could be compared). A better assessment seems to have been made by Betz and Weiser (1976) for the changes in the content of some mitochondrial proteins in sporulating Saccharomyces cerevisiae. According to these authors, the "unit size" of a yeast mitochondrion, i.e. the relative membrane surface area per mitochondrial genome, decreases during ascospore formation to less than one fourth of the value determined for growing derepressed yeast cells.

The depth of mitochondrial degradation is to a considerable extent determined by the 'age' of the yeast culture. According to our data (Luzikov et al., 1973b), mitochondria isolated from yeast in the lag period have a very low NADH oxidase activity (20-30 nmoles NADH oxidized per min per mg protein at 30°C); this cyanide-insensitive activity does not decrease upon subsequent incubation of yeast without oxygen. In the early exponential phase of yeast growth their mitochondria display higher respiratory activity (up to 90 nmoles NADH oxidized per min per mg protein at 30°C) which is almost completely inhibited by cyanide. Anaerobic incubation of such cells for 3 h entails complete inactivation of the respiratory system, so that mitochondrial NADH oxidase and succinate oxidase activities are lost altogether; meanwhile the partial respiratory chain activities (cytochrome *c* oxidase and succinate-ferricyanide reductase) only slightly decrease (Figure 32). As the respiratory system develops, its sensitivity to anaerobiosis in vivo is lowered. In particular, when yeast are deaerated in the early stationary phase, mitochondrial NADH oxidase and succinate oxidase activities decline markedly more slowly than in the 'young' yeast (Figure 32).

It is obvious that in all cases the rate and depth of mitochondrial degradation depend on two basic things: (1) the lability of mitochondria in given experimental conditions (i.e. their potential susceptibility to lytic enzymes, oxidative processes, etc.), and (2) the activity of the apparatus responsible for mitochondrial degradation (this, among other things, will be dealt with in Chapter Eight which considers the proteolysis of mitochondrial translation products). Therefore, it is quite conceivable that even upon impairment of oxidative phosphorylation the destabilized mitochondria may survive in vivo for a long time without any substantial alterations. Further, it follows from general considerations that degradation of mitochondria that do not perform oxidative phosphorylation can be retarded if they are energized in the cell at the expense of energy supplied externally, e.g., in the form of ATP produced in glycolysis, photophosphorylation, and other processes.

A case in point can be given with the cells of rice coleoptiles. As shown by Vartapetian et al. (1971, 1972), mitochondria undergo no morphological alterations during a 24-h deaeration of coleoptiles. Still, the mitochondria of 14-day coleoptiles grown without oxygen show signs of destructive changes (reduced number of cristae, clarification of matrix, etc.). Such mitochondria have 2-3 times less cytochromes and oxidize Krebs-cycle substrates at markedly lower rates than the mitochondria of the aerobic variant (Vartapetian et al., 1975). From electron-microscopic observations it follows that the anaerobic degradation of mitochondria in rice coleoptiles as well as in squash and rice roots can be prevented by adding glucose to the growth medium (Vartapetian et al., 1976, 1977). This was supposed to be due to an increase in the cellular ATP content owing to enhanced glycolysis.

If the sole crucial point in mitochondrial stabilization is their energization by this or that means, it can be assumed that functioning of separate components of the energy-transducing system would favor the survival of the mitochondrion as a whole, including its inoperative constituents. According to Zvyagil'skaya et al. (1977), in the obligately aerobic yeast Endomyces magnusii grown in the presence of antimycin A only the first coupling site is functional. In this case the cell content of cytochrome aa_3 does not decrease, and coupling site III remains potentially active albeit idle in vivo for the lack of an appropriate electron donor. The data of Zvyagil'skaya et al. (1977) might be regarded as evidence in support of the above suggestion. It must be commented, however, that the experimental data presented cannot yet be considered sufficient for stating that during Endomyces magnusii growth with antimycin A the cytochrome *c* oxidase segment of the respiratory chain was completely switched off.

A similar situation may occur in cells where ATP synthesis takes place coupled to NADH oxidation by fumarate under anoxia or in the presence of the inhibitors of the terminal respiratory-chain segment. For example, this process was well studied in cyanide-treated rat-heart submitochondrial vesicles (Wilson and Cascarano, 1970).

SUMMARY

As shown in Chapters Two and Three, mitochondria can be regarded as rather labile structures, the lability being inherent in the very composition of the organelles. Indeed, mitochondria harbor specific phospholipases and proteinases which under certain conditions can promptly break them down, and reactive centers initiating phospholipid oxidation.

A prominent if not decisive role in the maintenance of mitochondrial integrity *in vivo* is played by oxidative phosphorylation. There are grounds for believing that any impairment of energy transformation in mitochondria *in vivo* entails their degradation. This conclusion is in general accordance with the predictions made on the basis of the *in vitro* experiments described in the preceding chapter. The intensity and scope of destructive processes taking place in mitochondria *in vivo* depend not only on the level of metabolic activity of the oxidative phosphorylation system but also on a number of other factors: the amount of nonmitochondrially made ATP available, the state of the machinery (mitochondrial or cytoplasmic) responsible for the degradation of whole mitochondria or their individual fragments, and so forth. The information presented in this and preceding chapters prompts one to suggest a direct connection between the content of the mitochondrial material in the cell and the availability of substrates used to generate energy in mitochondria. In other words, organelle formation may be regulated at the substrate level.

Another suggestion issuing from the last two chapters concerns the possible mechanism of control over mitochondrial assembly. The gist of it is that defective, nonfunctioning structures formed during mitochondrial assembly are eliminated in the course of mitochondriogenesis. This issue will be considered in detail in Chapter Eight. The dependence of mitochondrial stability on their metabolic activity may serve as grounds for a hypothesis that *any* mutation in yeast resulting in impairment of oxidative phosphorylation would be pleiotropic because of the lability of the afflicted mitochondrial structures. In other words, a genetically determined defect in any site of the oxidative phosphorylation system would be accompanied by phenotypic degenerative alterations in its other regions. This hypothesis will be discussed in Chapter Eight.

The basic rule that nonfunctioning mitochondria do not usually survive in the cell appears also to have other consequences. The reader is left at liberty to speculate in this field.

5
Topography of the Synthesis of Mitochondrial Proteins

As follows from the first chapter, mitochondria are elaborately organized structures in which various enzymic systems are arranged in a strictly defined way. The general principles of the structural organization of the inner mitochondrial membrane have been elucidated. It is established that the protein components of the membrane are asymmetrically distributed between its inner and outer sides. There are grounds for believing that within the membrane, individual polypeptides and proteins carrying catalytically active groups are subintegrated into complexes which can be isolated upon mild fragmentation of the membrane.

The aim of the present chapter is to give a general picture of the topography of the synthesis of mitochondrial protein components. Bidimensional polyacrylamide gel electrophoresis showed that mitochondria from various sources comprise from 120 to 150 types of polypeptides (Henslee and Srere, 1979). Some of these are coded for by nuclear DNA and expressed through the cytoplasmic translation system; others are encoded in the mitochondrial genome and are products of mitochondrial translation. The peculiarities of the mitochondrial protein-synthesizing machinery and its interrelation with the nucleus have been extensively discussed in multiple reviews by quite a number of authors (see, e.g., Roodyn and Wilkie, 1968; Nass S., 1969; Borst and Kroon, 1969; Ashwell and Work, 1970; Rabinowitz and Swift, 1970; Borst and Grivell, 1971; Borst, 1972; Linnane et al., 1972; Schatz and Mason, 1974; Saccone and Quagliariello, 1975; Attardi et al., 1975; Mahler et al., 1975; Butetow and Wood, 1978; Tzagoloff et al., 1979; Tzagoloff, 1982), so there seems to be little need to go over this again. Still, for the reader to be prepared for the next chapter dealing with mitochondrial assembly _in vivo_, it is necessary to consider the material on the sites of translation of particular mitochondrial proteins. First brief information will be given about the criteria applied in ascertaining the origin of a given polypeptide (for details see Schatz and Mason, 1974).

5.1. CRITERIA OF THE ORIGIN OF MITOCHONDRIAL PROTEINS

5.1.1. Composition and Properties of Mitochondria from Cells Grown with Inhibitors of Mitochondrial Protein Synthesis

Cycloheximide is known to selectivily inhibit amino acid incorporation into the protein of eukaryotic cells (see, e.g., Siegel and Sisler, 1965) whereas chloramphenicol has no such effect (So and Davie, 1963). The latter, like other antibacterial antibiotics (erythromycin, aureomycin, teramycin), inhibits protein synthesis in prokaryotes (Gale and Folkes, 1953, Matthaei and Nierenberg, 1961) by binding to the 50 S ribosomal subunit (Vazquez, 1964) and blocking the peptidyltransferase reaction (Cundliffe and McQuillen, 1967; Denslow and O'Brien, 1974). It was also found that chloramphenicol, erythromycin, linkomycin, spiramycin, carbomycin, and oleandomycin prevent the incorporation of amino acids into protein by mitochondria (Mager, 1960; Wintersberger, 1965; Lamb et al., 1968; Kroon and De Vries, 1971; Kislev and Eisenstadt, 1972; Mockel, 1972; Maddaiah et al., 1973). Unlike these compounds, cycloheximide neither inhibits the protein-synthesizing systems of bacterial origin nor affects protein synthesis in isolated mitochondria (Ashwell and Work, 1968; Lamb et al., 1968; Kislev and Eisenstadt, 1972; Mockel, 1972; Maddaiah et al., 1973).

In accordance with this, retention of a certain protein component in mitochondria despite completely blocked mitochondrial protein synthesis in cells growing in the presence of chloramphenicol or erythromycin signifies that this protein is a product of the cytoplasmic protein-synthesizing machinery. Conversely, disappearance of a mitochondrial protein under such conditions of cell growth may be regarded as an indication of its mitochondrial origin. The latter issue is not so persuasive as the former. Indeed, as will be seen below, when mitochondrial protein synthesis is blocked the binding to the membrane of some polypeptides synthesized on cytoplasmic ribosomes is hampered for the lack of necessary "partners." In such a case the absence of a certain protein from mitochondria can be due either directly to the cessation of its synthesis on mitochondrial ribosomes or to its impaired binding to the membranes. Another difficulty in interpreting the results of such experiments crops up from the fact that the very existence of some proteins in mitochondria is frequently judged by the absence or presence of corresponding peaks in the absorption spectra or by corresponding enzymic activity. However, in the presence of protein-synthesis inhibitors incomplete forms of the proteins may appear having no such properties.

Assessing the results of long-term experiments with antibacterial antibiotics, one must not neglect their side effects such as suppression of growth and gradual loss of viability of obligate aerobes, uncoupling of oxidative phosphorylation, hindered amino acid transport into the cells, and so forth (Schatz and Mason, 1974).

5.1.2. Effects of Protein-Synthesis Inhibitors on Amino Acid Incorporation in Mitochondria *In Vivo*

As mentioned above, protein synthesis on cytoplasmic ribosomes in eukaryotes is inhibited by cycloheximide [as well as by anisomycin (Lizardi and Luck, 1972), emetine (Perlman and Penman, 1970), or pederine (Brega and Baglioni, 1971)], while the synthesis on mitochondrial ribosomes is sensitive to chloramphenicol (Mager, 1960), erythromycin, and linkomycin (Lamb et al., 1968). From this it follows that incorporation of radiolabeled amino acids into a certain mitochondrial protein upon short-term incubation of the cells with the label and chloramphenicol would testify to the synthesis of this protein on cytoplasmic ribosomes, and amino acid incorporation in the presence of cycloheximide to its synthesis on mitochondrial ribosomes. To rule out possible errors resulting from the inhibitor side effects, the sensitivity of the synthesis of the mitochondrial component under study is usually tested with both types of inhibitors. The site of translation of each protein is held to be more or less clear if amino acid incorporation is suppressed by chloramphenicol but not by cycloheximide, or vice versa.

5.1.3. Mitochondria of Yeast Cytoplasmic Petite Mutants

Cytoplasmic petite mutants of yeast are known to partly or completely lack the normal genetic information in mitochondrial DNA (Mounolou et al., 1966; Goldring et al., 1970; Nagley and Linnane, 1970) and to have no functional protein-synthesizing system (Kužela and Grečná, 1969; Schatz and :, 1969; Tzagoloff and Akai, 1972); the latter is a consequence of loss of either all or some of the cistrons for mitochondrial rRNAs (see references in Whittaker, 1979). It is therefore natural that all the proteins found in the mitochondrial structures of cytoplasmic petites are of nuclear origin. It could be thought that the absence of a certain protein from the defective mutant mitochondria is evidence for its synthesis on mitochondrial ribosomes in the corresponding wild type. However, interpretation of such experimental data is not quite unequivocal. Indeed, it is always possible that the protein is synthesized in the cytoplasm but its binding to mitochondria is impeded by the lack of another component made on mitochondrial ribosomes. Similarly, since cytoplasmic petites are deficient in mitochondrial DNA and rRNAs, one may fail to find in the defective mutant mitochondria the proteins that are usually associated with these nucleic acids (ribosomal proteins, initiation and elongation factors, rifampicin-sensitive RNA polymerase).

5.1.4. **Protein Synthesis in Isolated Mitochondria**

In isolated mitochondria radiolabeled amino acids are incorporated only into proteins synthesized on mitochondrial ribosomes. The lack of labeling of a protein in this case may indicate its cytoplasmic origin. However, such data per se are not very convincing, since a failure to detect the label in a certain protein in mitochondria does not necessarily mean that it is not made on mitochondrial ribosomes. In those cases when the polypeptides of mitochondrial origin are constituents of catalytic complexes, the label incorporated into individual polypeptides may be not recovered in isolated complexes for the simple reason that the labeled polypeptides were not integrated into these complexes. To illustrate this point, one may cite the early data of Beattie et al. (1970), according to which liver mitochondria in vitro do not incorporate amino acids into cytochrome *c* oxidase, cytochrome *b*, or oligomycin-sensitive ATPase, although to date it is known for certain that these mitochondrial inner-membrane components are partially or wholly made on mitochondrial ribosomes.

The most nearly perfect means of detecting the labeling of a certain mitochondrial protein is its precipitation with a monospecific immunoglobulin (and a secondary antiserum to the latter) from mitochondrial lysates, with a subsequent assay of radioactivity in the immunoprecipitate. Still, it should be borne in mind that solubilization of a really integral membrane protein may require such drastic conditions as would hinder the ensuing immunoprecipitation.

While negative results obtained for amino acid incorporation into some proteins in isolated mitochondria are often difficult to interpret, the very fact of labeling of a polypeptide can be considered proof it its mitochondrial origin. For the final conclusion to be made, however, it must be tested whether label incorporation is inhibited by chloramphenicol or erythromycin.

It must be specially noted that the polypeptides synthesized by mitochondria in vitro are normal translation products identical to that formed in vivo. Indeed, polyacrylamide gel electrophoresis in the presence of sodium dodecyl sulfate or other protein-disaggregating reagents yields identical patterns of radioactivity distribution for the proteins labeled in isolated mitochondria or in vivo in the presence of cytoplasmic-translation inhibitors (Sebald et al., 1969a,b; Coote and Work, 1971; Ibrahim et al., 1973a; Lederman and Attardi, 1973).

5.1.5. **Genetic Criteria**

Genetic methods can be applied to elucidating the site of synthesis of mitochondrial proteins, assuming that a protein coded for by nuclear DNA is synthesized on cytoplasmic ribosomes and a protein coded for by mitochondrial DNA is synthesized on mitochondrial ribosomes. Here direct genetic analysis is possible of crosses between lines carrying different forms of the protein under study, as well as its combination with various biochemical approaches.

Analysis of polymorphic types in initial lines, as well as in the series of generations obtained from crosses between these lines and between the first-generation hybrids and parents, was performed for a number of mitochondrial enzymes: malate dehydrogenase of *Neurospora crassa* (Munkres et al., 1965; Munkres, 1968) and maize (Longo and Scandalios, 1969), and glutamate-oxaloacetate aminotransferase (Davidson et al., 1970) and β-hydroxybutyrate dehydrogenase of *Paramecium aurelia* (Tait, 1968). For all these enzymes, the polymorphic variants were inherited according to Mendelian laws, indicating that they are coded for by nuclear DNA.

An example of other genetic approaches is the work of Knowles and Tait (1972) in which an original method for elucidating the genetic control of mitochondrial proteins was proposed and demonstrated for fumarase. Mitochondria of a *Paramecium aurelia* strain resistant to erythromycin and possessing a certain form of mitochondrial fumarase were injected into the cells of an erythromycin-sensitive strain having another form of the enzyme. Injected cells were incubated in a medium with erythromycin to select those that had acquired resistance to the antibiotic. Starch gel electrophoresis showed that such hybrids have the fumarase variant characteristic of the host cell. This means that mitochondria are not involved in the genetic control of this enzyme.

Gross et al. (1968) reported that the mitochondrial leucyl-tRNA synthetase in *Neurospora crassa* is most probably controlled by a nuclear gene, since the mitochondria of a nuclear point mutant lacked leucyl-tRNA synthetase activity.

5.2. TOPOGRAPHY OF THE SYNTHESIS OF PROTEINS OF DIFFERENT MITOCHONDRIAL COMPARTMENTS

5.2.1. Outer Mitochondrial Membrane

All the proteins of the outer mitochondrial membrane are synthesized on cytoplasmic ribosomes. Beattie et al. (1967b), Neupert et al. (1967), and Neupert and Ludwig (1971) were the first to show that upon incorporation of radiolabeled amino acids into isolated liver mitochondria only negligible radioactivity (not more than 5% of the total mitochondrial) is recovered in the outer-membrane fraction. As could be expected, this incorporation was inhibited by chloramphenicol but insensitive to cycloheximide (Beattie et al., 1967b). Later it was found for mammalian cells

(HeLa), yeast (Saccharomyces cerevisiae), and fungi (Neurospora crassa) that in vivo incorporation of labeled amino acids into mitochondrial outer-membrane proteins is almost completely (by 97-98%) suppressed by cycloheximide and pederine, inhibitors of cytoplasmic translation (Brega and Baglioni, 1971; Neupert and Ludwig, 1971; Bandlow, 1972). Under the same conditions labeling of the inner mitochondrial membranes constitutes 20-30% of control.

In the hepatocyte outer mitochondrial membrane 6-8 polypeptides are found, two of which (molecular weight 31,000, pore protein; and molecular weight 15,000, cytochrome b_5) account for 20-30% of the total membrane protein. Cycloheximide totally inhibits [^{35}S] methionine incorporation into all these polypeptides (Gellerfors and Lindén, 1981).

It can be added that electron-microscopic observations (Yotsuyanagi, 1962; McClary and Bowers, 1968; Packer et al., 1973) show that yeast cytoplasmic petites possess mitochondrial structures with outer membranes. The same applies to the yeast grown in the presence of chloramphenicol.

5.2.2. **Inner Mitochondrial Membrane**

5.2.2.1. *NADH Dehydrogenase and Succinate Dehydrogenase*

As shown by Linnane et al. (1968) and Kellerman et al. (1969), yeast growth in the presence of chloramphenicol is accompanied by a substantial decrease in NADH dehydrogenase and succinate dehydrogenase activities of the cell homogenates. Under the same conditions the cells completely lose cytochromes *a* and a_3 and display no cytochrome *c* oxidase activity. The chloramphenicol effect on the behavior of the dehydrogenase activities is conditioned by quite a number of factors: glucose concentration, lipid content in the growth medium, duration of inhibitor treatment, etc. (Vary et al., 1969, 1970; Mahler and Perlman, 1971; Mahler et al., 1971). In certain cases (e.g. during yeast respiratory adaptation in the presence of lipids) chloramphenicol does not at all affect the rise in succinate dehydrogenase activity whereas cycloheximide blocks it entirely (Vary et al., 1969, 1970). Similar results were obtained by other authors on studying the development of NADH dehydrogenase and succinate dehydrogenase activities in yeast (Mahler and Perlman, 1971; Kim and Beattie, 1973) and mammalian cells (Kleitke and Wollenberger, 1968; Fettes et al., 1972). It was shown that ethidium bromide, while inducing a decrease in the contents of cytochromes aa_3 and *b*, in cytochrome *c* oxidase, oligomycinsensitive ATPase, and succinate-cytochrome *c* reductase activities in Dictyostelium discoideum mitochondria, does not affect their NADH dehydrogenase and succinate dehydrogenase activities (Stuchell et al., 1975).

In the immunoprecipitates obtained from beaf-heart mitochondrial lysates with an antiserum against NADH dehydrogenase, Hare and Hodges (1982b) failed to find any mitochondrially made polypeptide belonging to this complex.

Thus, both dehydrogenases are in all probability built not of products of the mitochondrial protein-synthesizing system but of proteins synthesized on cytoplasmic ribosomes. With all this, evidence is available for a certain dependence of the dehydrogenase formation on biosynthetic processes taking place in mitochondria (Vary et al., 1970).

5.2.2.2. *Respiratory Chain bc_1 Segment*

Lin et al. (1974) and Rouslin (1977) showed that formation of complex III in yeast is suppressed by cycloheximide as well as by chloramphenicol. This prompts one to think that expression of ubiquinol-cytochrome *c* reductase activity requires components synthesized both on cytoplasmic and on mitochondrial ribosomes.

To date, synthesis of apocytochrome *b* is held to take place on mitochondrial ribosomes. In the early work of a number of authors who studied the influence of the inhibitors of mitochondrial protein synthesis on the development of the respiratory system in yeast, fungi, algae, and mammalian cells, cytochrome *b* synthesis was found to be partially (Linnane et al., 1968; Mahler and Perlman, 1971; Soslau and Nass, 1971; Von Jagow and Klingenberg, 1972; Sato et al., 1973) or completely suppressed (Clark-Walker and Linnane, 1967; Marchant and Smith, 1968; Davey et al., 1969; Von Jagow and Klingenberg, 1972; Wakiyama and Ogura, 1972). In these studies the inhibitors were introduced into the growth media for extended periods, and the cytochrome *b* content was estimated spectrophotometrically; as mentioned above, such an approach cannot be considered entirely reliable.

In recent years the experiments with short-term labeling in vivo in the presence of chloramphenicol or cycloheximide furnished more definite information on the site of cytochrome *b* synthesis in yeast cells. As pointed out in Chapter One, the question of how many *b*-type cytochromes there are in mitochondria and which polypeptides are their apoproteins has not been finally solved. The most plausible point of view to date is that apocytochrome *b* is a polypeptide with a molecular weight of about 30,000. According to Weiss and Ziganke (1974), chloramphenicol prevents pulse incorporation of $[^3H]$leucine into cytochrome *b* in Neurospora crassa whereas cycloheximide has no such effect. These authors isolated cytochrome *b* as a protein with a minimal molecular weight of 30,000 from cultures labeled in the presence or in the absence of protein-synthesis inhibitors and determined its radioactivity. The isolation procedure included gel filtration in the presence of bile acid salts and was rather complicated. A protein of molecular weight 11,000 was observed as an admixture to the main component.

A similar work was later carried out by Lin et al. (1978) who isolated cytochrome *b* from yeast mitochondria (labeled in vivo with [^{3}H]leucine) with the use of an antiserum to its purified preparation containing one major polypeptide (apparent molecular weight 31,000). Incorporation of [^{3}H]leucine into cytochrome *b* was inhibited by chloramphenicol but not by cycloheximide. Besides, cytochrome *b* was not found in a cytoplasmic petite unable to perform mitochondrial protein synthesis.

Finally, Katan et al. (1976) designed a technique for isolating from yeast mitochondria the cytochrome bc_1 complex comprising seven different polypeptides. By their data, chloramphenicol (or acriflavine) suppresses in vivo amino acid incorporation into a single polypeptide which by its molecular weight can be identified as apocytochrome *b*.

The data on the influence of chloramphenicol on the synthesis of cytochrome c_1 obtained by growing various organisms with this inhibitor were contradictory. It has been reported that chloramphenicol completely or partially suppresses the formation of a component having the spectral properties of cytochrome c_1 (Clark-Walker and Linnane, 1967; Firkin and Linnane, 1968; Davey et al., 1969; Kellerman et al., 1969; Sato et al., 1973), that it has no effect on this process (Mahler and Perlman, 1971; Fettes et al., 1972), and that it stimulates the increase in cytochrome c_1 contents (Soslau and Nass, 1971; Von Jagow and Klingenberg, 1972; Wakiyama and Ogura, 1972).

The site of cytochrome c_1 synthesis was finally ascertained in the experiments of Ross and Schatz (1976b) who labeled yeast with [^{3}H]leucine (1) in the absence of protein-synthesis inhibitors, (2) in the presence of acriflavine, and (3) in the presence of cycloheximide. After labeling, the mitochondrial fraction soluble in sodium cholate was treated with an antiserum to highly purified cytochrome c_1. The immunoprecipitates were subjected to sodium dodecyl sulfate-polyacrylamide gel electrophoresis to identify the cytochrome c_1 band. The label proved to be incorporated into this polypeptide (molecular weight 31,000) in the presence of acriflavine but not of cycloheximide. Furthermore, Ross and Schatz found that a cytoplasmic petite lacking mitochondrial protein synthesis still contains cytochrome c_1 tightly bound to mitochondrial membranes and present in amounts close to that characteristic of the initial wild-type yeast strain.

As mentioned above, of the seven polypeptides constituting the bc_1 complex of yeast mitochondria only one appearing to be cytochrome *b* is synthesized on mitochondrial ribosomes (Katan et al., 1976). This means that the other polypeptides corresponding to two core proteins, antimycin-binding protein, Rieske iron-sulfur protein, and cytochrome c_1 (see Chapter One) are made on cytoplasmic ribosomes.

5.2.2.3. *Cytochrome c*

Sherman et al. (1966) demonstrated that the primary structure of iso-1-cytochrome *c* of baker's yeast is coded for by the nuclear genome. As cytochrome *c* synthesis in various organisms (including yeast, fungi, protozoans, and mammals) is not suppressed by chloramphenicol or ethidium bromide (Clark-Walker and Linnane, 1967; Evans and Lloyd, 1967; Marchant and Smith, 1968; De Vries and Kroon, 1970; King et al., 1972; Von Jagow and Klingenberg, 1972; Wakiyama and Ogura, 1972), it can be held to take place on cytoplasmic ribosomes. In support of this it can be added that isolated mitochondria do not incorporate labeled amino acids into cytochrome *c* (Roodyn et al., 1962).

The same conclusion was arrived at in experiments where cytochrome *c* was pulse-labeled with [^{14}C]lysine or δ-amino[^{14}C]levulinic acid, and its radioactivity was measured in different subcellular fractions. Small amounts of cytochrome *c* (not more than 10% of its total cell content) were found in the microsomal fraction. Soon after the addition of label the specific radioactivity of cytochrome *c* in this fraction was markedly higher than in mitochondria (González-Cadavid and Campbell, 1967; Pennial and Davidian, 1968; Kadenbach, 1969). As it decreased with time in microsomes and increased in mitochondria, it was suggested that cytochrome *c* is synthesized in the endoplasmic reticulum and then transported to mitochondria.

Opinions vary on whether cytochrome *c* is completely formed on microsomes or only its apoprotein is synthesized there. According to Kadenbach (1970), cytochrome *c* formation is a multistage process starting in microsomes with the synthesis of "precytochrome *c*" which is then transported to mitochondria where it binds heme. On the other hand, Pennial and Davidian (1968) and Davidian et al. (1969) believed that all stages of cytochrome *c* 'assembly' take place in the endoplasmic reticulum. From the data of de Córdova et al. (1977) it follows that all the enzymes necessary for protoporphyrin synthesis from δ-aminolevulinic acid, incorporation of iron into the protoporphyrin to form the heme, and covalent attachment of the heme to apocytochrome *c* are contained in the cytosol of liver cells.

The most convincing information on the synthesis of apocytochrome *c* and its conversion to holocytochrome *c* is presented by Korb and Neupert (1978). With the use of antibodies specific for each of these two forms they showed that the conversion of the apo-form to cytochrome *c* occurs only in mitochondria. In particular, these authors found that formation of holocytochrome *c* can be observed upon incubation of the postribosomal supernatant containing apocytochrome *c* (synthesized in Neurospora crassa homogenate) with isolated mitochondria. In their absence no conversion took place.

5.2.2.4. *Cytochrome c Oxidase*

The results of numerous studies demonstrate that cytochrome *c* oxidase formation involves both cytoplasmic and mitochondrial protein-synthesizing systems. Thus it was shown that chloramphenicol, erythromycin, and ethidium bromide suppress the development of cytochrome *c* oxidase activity (Kroon and Jansen, 1968; De Vries and Kroon, 1970; Heilporn and Limbosch, 1971; Howell et al., 1971; Mahler and Perlman, 1971; Fettes et al., 1972; King et al., 1972) and cytochrome aa_3 formation (Clark-Walker and Linnane, 1966, 1967; Evans and Lloyd, 1967; Firkin and Linnane, 1968, 1969; Marchant and Smith, 1968; Davey et al., 1969; Kellerman et al., 1969; De Vries and Kroon, 1970; Mahler and Perlman, 1971; Soslau and Nass, 1971; Dixon et al., 1972; Von Jagow and Klingenberg, 1972; Wakiyama and Ogura, 1972) in the cells of various organisms (yeast, fungi, protozoans, mammals). Chloramphenicol also hinders cytochrome *c* oxidase synthesis during yeast respiratory adaptation (Chen and Charalampous, 1969; Vary et al., 1969). However, the same effect is produced by cycloheximide (Chen and Charalampous, 1969; Vary et al., 1969).

As indicated in Chapter One, the most generally acknowledged point of view is that cytochrome *c* oxidase from the mitochondria of Saccharomyces cerevisiae and Neurospora crassa comprises seven polypeptides. Three larger polypeptides (I, II, and III, see Chapter One) are by the data obtained in many laboratories (Mason and Schatz, 1973; Rubin and Tzagoloff, 1973b; Sebald et al., 1973; Lansman et al., 1974) synthesized on mitochondrial ribosomes, since incorporation of labeled amino acids into these subunits is suppressed by chloramphenicol or erythromycin but not by cycloheximide. The four smaller subunits (IV, V, VI, and VII) are not mitochondrial translation products. They are synthesized in the presence of erythromycin (Mason and Schatz, 1973) but not in the presence of cycloheximide (Weiss et al., 1971; Mason and Schatz, 1973; Rubin and Tzagoloff, 1973b; Sebald et al., 1973). Analogous results were later obtained by Koch (1976) for the culture of toad (Xenopus laevis) oocytes and by Hare et al. (1980) for HeLa cells. Rascati and Parsons (1979) demonstrated $[^3H]$leucine incorporation into the three larger subunits of cytochrome *c* oxidase in isolated rat liver mitochondria.

5.2.2.5. *ATPase Complex*

According to the data summarized in Chapter One, the ATPase complex includes at least 10 types of polypeptide. Of these, well identified are the five polypeptides constituting the soluble F_1-ATPase, the ATPase protein inhibitor, the oligomycin-sensitivity-conferring protein, and the dicyclohexyl carbodiimide-binding protein. The functions performed by each subunit constituting the "membrane sector" of the complex have not yet been exactly determined.

Now the F_1-ATPase is known to be synthesized on cytoplasmic ribosomes. It was found in the mitochondria of yeast cytoplasmic petites (Kováč and Weissova, 1968; Schatz, 1968; Packer et al., 1973) and yeast grown in the presence of chloramphenicol (Schatz, 1968). According to Tzagoloff (1969), chloramphenicol does not affect the increase in the total ATPase activity of yeast cells upon their transfer from high-glucose to low-glucose medium. Under the same conditions cycloheximide suppresses the increase in ATPase activity both in mitochondria and in cytosol. Special experiments (Tzagoloff, 1969; Tzagoloff et al., 1972) showed that in all cases the ATPase activity belonged to factor F_1. Finally, Tzagoloff and Meagher (1972) found that cycloheximide blocks the incorporation of labeled leucine into F_1-ATPase during yeast glucose derepression.

The protein conferring sensitivity to oligomycin upon F_1-ATPase (OSCP) is also synthesized in the cytoplasm. Chloramphenicol does not prevent the formation of this protein during yeast derepression (Tzagoloff, 1970), and it is accumulated in the soluble cell fraction (postribosomal supernatant) and is probably responsible for F_1 binding to mitochondrial membranes. No such accumulation is observed in the presence of cycloheximide.

According to Ebner and Maier (1977), yeast cytoplasmic ribosomes make a low-molecular-weight basic protein which can inhibit the mitochondrial ATPase. Some properties of this protein (in particular, its thermal stability and high sensitivity to proteinases) as well as the fact that it is not found in appreciable amounts in a nuclear petite (pet 936) devoid of F_1-ATpase and in a cytoplasmic petite that among other mitochondrial translation products lacks the "membrane sector" polypeptides, give grounds for regarding it as a bona fide component of the ATPase complex.

Of other H^+-ATPase constituents, especially interesting is the proteolipid (ninth subunit) that probably acts as the proton channel. This polypeptide having high affinity for N,N'-dicyclohexyl carbodiimide is synthesized on mitochondrial ribosomes in yeast (Tzagoloff and Meagher, 1972; Esparza et al., 1981; Somlo et al., 1982); however, it is made on cytoplasmic ribosomes in rat liver (De Jong et al., 1980; Kužela et al., 1980), fungi Neurospora crassa and

Botryodiplodia theobromae (Sebald and Jackl, 1975; Wenzler and Brambl, 1981), and Xenopus laevis oocytes (Koch, 1976).

Mitochondrial translation in rat liver (Kužela et al., 1980) and yeast (Esparza et al., 1981) gives rise to a polypeptide of molecular weight 10,000 and 9000, respectively, which is marked by high affinity for inorganic phosphate and presumably plays the part of phosphate carrier in H^+-ATPase.

On the whole there is substantial discord on the topography of synthesis of the H^+-ATPase membrane sector components. Different authors reported the number of polypeptides made intramitochondrially in yeast to be four (molecular weights 29,000; 22,000 - subunit 6; 12,000; and 7500 - subunit 9) (Tzagoloff et al., 1973), three (molecular weights 21,000, 10,000 and 8000) (Esparza et al., 1981), and even two (molecular weights 20,000 and 7800) (Somlo et al., 1982). In liver cells (Kužela et al., 1980) mitoribosomes make two polypeptides with molecular weights of 21,500 and 9000 (the P_i-binding polypeptide). In Neurospora crassa polypeptides of molecular weights 19,000 and 11,000 were claimed to be mitochondrial translation products (Jackl and Sebald, 1975) and in Xenopus laevis oocytes three polypeptides of molecular weights 29,000, 21,000 and 17,000 (Koch, 1976). Finally, in the fungus Botryodiplodia theobromae only one (molecular weight 20,000) out of 12 polypeptides of H^+-ATPase is synthesized on mitochondrial polysomes, according to Wenzler and Brambl (1981).

5.2.2.6. *Adenine Nucleotide Carrier*

It was at first vague which genome - nuclear or mitochondrial - codes for the protein components of the adenine nucleotide carrier. On the one hand, Haslam et al. (1973) obtained data from which it followed that Saccharomyces cerevisiae growth in the presence of erythromycin or loss of mitochondrial DNA as a result of mutation entail substantial alterations in the carrier: ATP uptake by mitochondria becomes insensitive to atractylate, centers with high affinity for adenine nucleotides are lost. On the other hand, Kolarov and Klingenberg (1974) found no substantial differences between wild-type yeast and a cytoplasmic mutant with respect to the main characteristics of the adenine nucleotide carrier (dissociation constant for carboxyatractyloside and specific ADP-transfer rate). A similar conclusion was arrived at by Lauquin and Vignais (1973) and Lauquin et al. (1976) who studied the effects of the inhibitors of mitochondrial protein synthesis and mutation in mitochondrial DNA on the properties o the adenine nucleotide carrier in the yeast Saccharomyces cerevisiae and Candida utilis. Mitochondria of yeast grown in the presence of erythromycin or ethidium bromide proved to be able to perform nucleotide exchange at a high rate and retained high affinity for ADP as well as for atractyloside and bongkrekic acid. Mitochondria of a cytoplasmic mutant were also normal in respect of ADP binding and interaction with specific inhibitors of the adenine nucleotide carrier. At the same time, some authors (see, e.g., Kolarov et al., 1972; Kolarov and Klingenberg, 1974; Lauquin et al., 1976) observed that a nuclear mutation (op 1) is accompanied by significant alterations in the mitochondrial adenine nucleotide carrier; its affinity for ADP (Kolarov et al., 1972), atractyloside (Kolarov and Klingenberg, 1974), and bongkrekic acid (Lauquin et al., 1976) decreases.

Finally, Hackenberg et al. (1978) showed that chloramphenicol added to Neurospora crassa does not affect the incorporation of labeled leucine into a polypeptide with a molecular weight of 33,000, which could be isolated from mitochondria as a complex with carboxyatractyloside; at the same time, labeling of this protein is prevented by cycloheximide.

Thus, most of the experimental data available indicate that the mitochondrial adenine nucleotide carrier is coded for by the nucleus and synthesized on cytoplasmic ribosomes. Neither inhibitors of mitochondrial protein synthesis nor mutations in the mitochondrial genome elicit any substantial changes in the protein.

5.2.2.7. *Transport System for Tricarboxylic Acid Cycle Substrates*

All components of the system responsible for the transport of malate, succinate, citrate, and α-ketoglutarate across the inner mitochondrial membrane are in all probability coded for by the nucleus and made on cytoplasmic ribosomes. As shown by Perkins et al. (1973), the transport of these substrates is not appreciably altered when yeast are grown in the presence of an inhibitor of mitochondrial protein synthesis (erythromycin) or when mitochondrial DNA is lost through mutation.

5.2.2.8. *Other Inner-Membrane Components*

Here one should mention the enzymes responsible for heme synthesis. Beattie and Stuchell (1970) demonstrated that administration of allylisopropyl acetamide to rats at first induces a rise in the activity of δ-aminolevulinic acid synthetase in the postmitochondrial supernatant of rat liver. With time the enzyme activity decreases in this fraction and increases in mitochondria. Cycloheximide blocks the induction of aminolevulinic-acid-synthetase activity in both subcellular fractions, while chloramphenicol partially suppresses its increase in mitochondria and has no effect on the postmitochondrial supernatant. On the basis of these data it was suggested that δ-aminolevulinic acid synthetase is formed on microsomes and then transferred to mitochondria.

According to Birkmayer and Bücher (1969), mitochondria isolated from Neurospora crassa wild type and its cytoplasmic mutant (mi-1) as well as from wild-type cells grown in the presence of chloramphenicol contain the same amounts of ferrochelatase, as judged by its enzymic activity. This may mean that ferrochelatase is coded for by the nuclear genome and synthesized on cytoplasmic ribosomes.

Thus, at least two enzymes participating in heme synthesis are of cytoplasmic origin.

5.2.3. **Mitochondrial Matrix**

5.2.3.1. *Tricarboxylic Acid Cycle and Linked Systems*

In a number of works it was shown that chloramphenicol and ethidium bromide do not hinder the development of malate dehydrogenase (Vary et al., 1969; Mahler and Perlman, 1971), fumarase (Vary et al., 1969), and isocitrate dehydrogenase (Mahler and Perlman, 1971) activities during yeast respiratory adaptation, whereas cycloheximide blocks it completely (Vary et al., 1969). These data allow one to suppose that the synthesis of the enzymes listed, which are components of the mitochondrial matrix, takes place on cytoplasmic polysomes. This is supported by the results of Roodyn et al. (1962), according to which isolated rat liver mitochondria do not incorporate labeled amino acids into malate dehydrogenase. On the other hand, Bingham and Campbell (1972) detected the specific mitochondrial malate dehydrogenase in microsomes, and showed that incubation of isolated microsomes with [^{3}H]leucine under conditions providing for protein synthesis leads to incorporation of the label into this enzyme. Evidence for the synthesis of glutamate dehydrogenase on rat liver microsomes and its subsequent transport to mitochondrial was presented by Godinot and Lardy (1973).

Genetic analysis revealed that the enzymes of the tricarboxylic acid cycle and linked systems are coded for by the nuclear genome. Such data are available for malate dehydrogenase (Munkres et al., 1965; Munkres, 1968), NADP-dependent isocitrate dehydrogenase (Tait, 1970), fumarase (Knowles and Tait, 1972), and glutamate-oxaloacetate aminotransferase (Davidson et al., 1970).

It can be added that yeast petites retain high levels of fumarase and aconitase (Schatz et al., 1963) as well as L-malate dehydrogenase, NADP-dependent isocitrate dehydrogenase, citrate synthetase, and isocitrate lyase (Perlman and Mahler, 1970).

5.2.3.2. *Mitochondrial Protein-Synthesizing System*

Quite a number of studies have been devoted to elucidating the topography of synthesis of mitochondrial ribosomal proteins. Küntzel

(1969) showed that cycloheximide inhibits to an equal extent [^{3}H]-leucine incorporation into the proteins of mitochondrial (by 97.2%) and cytoplasmic (by 96.9%) ribosomes in Neurospora crassa cells. At the same time chloramphenicol decreases the radioactivity level in ribosomes only by 2.2 and 2.8%, respectively. Hence it was suggested that most if not all mitochondrial ribosomal proteins are produced on cytoplasmic ribosomes. Similar results were obtained for Neurospora crassa by Neupert et al. (1969a) who also found (Neupert et al., 1969b) that isolated mitochondria cannot incorporate an appreciable amount of labeled amino acids into their ribosomes. The cytoplasmic origin of the bulk of mitochondrial ribosomal proteins was confirmed with other organisms (see, e.g., Brega and Baglioni, 1971).

More detailed research along this line was carried out by Lizardi and Luck (1972). These authors reported that labeled amino acids are not incorporated into any of the 53 mitochondrial ribosomal proteins in isolated Neurospora crassa mitochondria, and that the synthesis of each and every one of these proteins is suppressed by anisomycin (an inhibitor of cytoplasmic protein synthesis) but is insensitive to chloramphenicol. These data were checked by Lambowitz et al. (1976). It turned out that one of the proteins (S-4a, molecular weight 52,000) of the small ribosomal subunit is synthesized on mitochondrial ribosomes. This protein has a very small free pool and probably controls the rate of ribosome assembly. Chloramphenicol blocks the synthesis of the S-4a protein (S-5 by the new nomenclature) and prevents the "maturation" of the 30 S subunit (LaPolla and Lambowitz, 1977). Later these data were verified and supplemented (Lambowitz et al., 1979).

All this correlates to a certain extent with the results of other authors (Millis and Suyama, 1972; Schmidt, 1972) who observed partial inhibition of the synthesis of mitochondrial ribosomal proteins by chloramphenicol in yeast and Tetrahymena. Unfortunately, no controls were made in the work mentioned on the contamination of ribosomes with mitochondrial membrane components. However, later Groot (1974) showed that even thoroughly purified yeast mitochondrial ribosomes contain a polypeptide of molecular weight 35,000 whose synthesis is inhibited by chloramphenicol. The polypeptide is tightly associated with the 38 S subunit; it is not essential to the ribosomal function proper and may probably be regarded as an initiation factor or a connecting link between the ribosome and mitochondrial membrane.

A still more detailed study was made of this problem by Butow's group (Terpstra and Butow, 1979; Terpstra et al., 1979) who established that the 38 S ribosomal subunit contains a single polypeptide (var1) whose synthesis is insensitive to cycloheximide but is inhibited by chloramphenicol and still better by erythromycin. The polypeptide is present in the ribosome in amounts equimolar to other 38 S -subunit proteins and can be considered its integral component.

Other explanations were put forward for the effects of cycloheximide and chloramphenicol on the synthesis of mitochondrial ribosomal proteins. In particular, Neupert et al. (1971) argued that the experimental data available are compatible with the suggestion that all these proteins are of mitochondrial origin but their synthesis is induced by a special "trigger protein" supplied into mitochondria from the cytoplasm and suppressible by cycloheximide but not by chloramphenicol. This suggestion, however, was purely speculative.

It is interesting that, unlike the proteins of mitochondrial ribosomes, their ribonucleic acids are synthesized in mitochondria (Wintersberger, 1967; Suyama, 1967; Fukuhara, 1967; Wood and Luck, 1969). Kuriyama and Luck (1973a) showed that these RNAs are initially transcribed in Neurospora crassa as a 32 S precursor which is then cleaved into the 19 S and 25 S species. Lambowitz et al. (1976) found in mitochondria "precursor particles" containing this 32 S RNA and more than 15 polypeptides of the large and small ribosomal subunits. These data implied that ribosome assembly takes place in mitochondria, with all proteins of both subunits except the S-5 polypeptide being imported from the cytoplasm.

In conclusion, data should be presented on the synthesis of elongation factors specific for mitochondrial ribosomes. As demonstrated by Richter (1971), mitochondria of yeast mutants devoid of mitochondrial DNA contain T- and G-factors which in their functional, chemical, and immunological features are identical with those of the parental strain. Analogous results were obtained by other authors (Parisi and Cella, 1971; Scragg, 1971). From this it follows that mitochondrial elongation factors are in all probability coded for by nuclear DNA and synthesized on cytoplasmic ribosomes.

5.3. RELATIVE CONTRIBUTION OF MITOCHONDRIAL AND CYTOPLASMIC TRANSLATION TO MITOCHONDRIAL FORMATION

Summing up all the above, it can be concluded that the overwhelming majority of mitochondrial proteins having catalytic activity are synthesized in the cytoplasm. The work of many authors demonstrates that in yeast (Schweyen and Kaudewitz, 1970; Kellerman et al., 1971; Bandlow, 1972), Neurospora crassa (Hawley and Greenawalt, 1970), and mammalian cells (Galper and Darnell, 1971; Bartelink and De Kort, 1973; Satav et al., 1977) cycloheximide suppresses the incorporation of labeled amino acids into the total cell protein and mitochondrial outer-membrane protein by 97-99%, and into the total mitochondrial protein only by 85-90%, the residual incorporation in the latter case being wholly sensitive to chloramphenicol. According to calculations made on the basis of these and other data, only 5-15% of the total mitochondrial protein is made on mitochondrial ribosomes (Sebald et al., 1968; Hawley and Greenawalt, 1970; Schweyen and

Kaudewitz, 1970; Coote and Work, 1971; Kellerman et al., 1971; Bandlow, 1972; Bartelink and De Kort, 1973; Ebner et al., 1973b; Jeffreys and Craig, 1976). The label incorporated into mitochondria in vitro (Beattie et al., 1967b, 1970; Ibrahim et al., 1973a) or in vivo in the presence of cycloheximide (Henson et al., 1968a; Sebald et al., 1969b; Coote and Work, 1971; Kellerman et al., 1971; Kužela et al., 1975) is found almost exclusively in the fractions of mitochondrial membranes or insoluble membrane proteins. At the same time the synthesis of outer-membrane components appears to be cycloheximide-sensitive to the same degree as that of cytosolic proteins (see, e.g., Neupert and Ludwig, 1971; Bandlow, 1972). Hence the products of the mitochondrial protein-synthesizing machinery are integrated into the inner membrane of the organelle. According to the data of different authors (Neupert and Ludwig, 1971; Bandlow, 1972), they can constitute from 20 to 40% of the overall inner-membrane protein. The single exception is the var1 protein, which is a component of the small ribosomal subunit in yeast. Upon disruption of mitochondria up to 80% of var1 associated with the 38 S subunit is recovered in the postmitochondrial supernatant (Terpstra et al., 1979).

As follows from the preceding sections of this chapter, in the cells of lower eukaryotes mitochondria make three cytochrome c oxidase subunits, two (or three) subunits of H^+-ATPase (namely, sixth and ninth), apocytochrome *b*, and one protein of the small ribosomal subunit (S-5 or var1). With higher eukaryotes, translation on mitochondrial ribosomes has been demonstrated for three cytochrome *c* oxidase subunits, one or two H^+-ATPase subunits, and apocytochrome *b*.

However, the total set of mitochondrial translation products is not exhausted by these polypeptides. Experiments on the incorporation of labeled amino acids into animal cells in the presence of cytoplasmic-translation inhibitors and into isolated mitochondria revealed 8-14 polypeptides synthesized on mitochondrial ribosomes (Coote and Work, 1971; Constantino and Attardi, 1975; Yatscoff and Freeman, 1977; Gellerfors et al., 1979; Abraham et al., 1981; Gellerfors and Lindén, 1981).

The recently elucidated complete sequences of human (Anderson et al., 1981) and murine mitochondrial DNAs (Bibb et al., 1981) suggested that there are 13 readable frames transcribed into presumably functional mRNAs (Ojala et al., 1980, 1981) and coding for polypeptides in the molecular weight range 7800-66,000. Of these, five polypeptides are well known while the remaining eight have not yet been identified.

In recent years data have appeared suggesting that the experimentally determined number of mitochondrial translation products may substantially exceed the number of readable frames. With the use of a highly efficient system for bidimensional resolution of protein mixtures Bhat et al. (1982) detected 19-24 polypeptides synthesized in Ehrlich ascites cells and rat and mouse liver cells in the presence of cycloheximide, as well as in mitoplasts isolated from these cells. Ching and Attardi (1982) identified 26 polypeptides as mitochondrial translation products in HeLa cells. Numerous explanations have been put forward for the excess of the number of actually detected polypeptides over that of readable frames in the mitochondrial genome, from the putative heterogeneity of mitochondrial DNA molecules (Bhat et al., 1982) to such phenomena as overlapping of coding sequences for some polypeptides, occurrence of spliced genes overlapping other genes, abortive translation, ribosome slippage and progression (Ching and Attardi, 1982). Highly probable is the secondary modification of the primary translation products, such as covalent linking between polypeptides with the formation of high-molecular-weight products (see the next chapter).

With yeast, selective labeling of mitochondrial translation products yields up to 21 polypeptides (Douglas and Butow, 1976). In what measure this number corresponds to the structure of yeast mitochondrial genome it is thus far impossible to say, since the latter has not yet been completely elucidated. The genes known to date are for apocytochrome *b* (Tzagoloff et al., 1976), three larger cytochrome *c* oxidase subunits (Tzagoloff et al., 1980), the sixth (Macino and Tzagoloff, 1980) and ninth subunits of H^+- ATPase (Macino and Tzagoloff, 1979; Hensgens et al., 1979), varl (Butow et al., 1980), and a maturase of mRNA for apocytochrome *b* (Jacq et al., 1089).

In sum, this chapter presents the general information on the topography of the synthesis of proteins belonging to different mitochondrial compartments. Data are also provided on the relation of cytoplasmic and mitochondrial translation, from which it follows that despite the relatively small contribution to the mass (5-15% of the total) and number (about 20 out of 150 polypeptides) of the organelle protein components, the mitochondrial translation plays a crucial part in the formation of respiratory-competent mitochondria, since the intramitochondrially made polypeptides are indispensable constituents of the respiratory-chain enzymic complexes and H^+ -ATPase. Further details of the topography of the synthesis of mitochondrial proteins, including the arrangement in the cytoplasm and mitochondria of polysomes involved in the synthesis of mitochondrial components, are considered in the next chapter.

6
Mitochondrial Assembly

In elucidating the molecular bases of mitochondrial formation it is necessary to try to answer two main questions: (1) how mitochondrial membranes are assembled and (2) how definite sets of protein and nonprotein components are made up in the intermembrane space and matrix. These questions arise regardless of which concept of mitochondrial origin and development is assumed as the framework. Indeed, expansion of membranes and accumulation of "soluble" components must take place both when new mitochondrial structures appear owing to proliferation of preexisting respiratory-competent organelles (Luck, 1963, 1965) and when they evolve from prostructures, or "promitochondria" (Jayaraman et al., 1966; Plattner et al., 1970). Such processes are implicit irrespective of whether mitochondrial development is regarded as multiplication of spatially discrete entities or as growth of a single "giant mitochondrion" (Keddie and Barajas, 1969; Calvayrac et al., 1971; Arnold et al., 1972; Hoffman and Avers, 1973; Osafune, 1973; Grimes et al., 1974; Attardi et al., 1975; Davison and Garland, 1975, 1977; Stevens, 1977).

The data presented in the preceding chapter on the topography of the synthesis of mitochondrial proteins indicate that all the proteins of the intermembrane space must be selectively transported from the cytoplasm across the outer mitochondrial membrane while the pathway of the matrix proteins and some inner-membrane proteins lies across two membrane barriers. This chapter will present material on the mechanisms of such addressed transport. In passing, the term "topography of synthesis" will be refined since we will deal with the concrete location of cytoplasmic and mitochondrial polysomes producing the proteins of different mitochondrial compartments.

The outer mitochondrial membrane in its enzyme pattern (Parsons et al., 1967), phospholipid composition and role in phospholipid metabolism (Stoffel and Schiefer, 1968), and other features resembles the membrane of the endoplasmic reticulum. Experiments of Borgese and Meldolesi (1980) and Meldolesi et al. (1980) directly demonstrated that NADH cytochrome b_5 reductase is an authentic component of the Golgi complex, microsomal, and mitochondrial membranes. Shore and Tata (1977) and Pickett et al. (1980) pointed out that mitochondria are "intimately" or "physically" associated with a certain fraction of the rough endoplasmic reticulum (see also Bracker and Grove, 1971; Franke and Kartenbeck, 1971; Morré et al., 1971). Although no one knows the actual nature of this association, the outer mitochondrial membrane, similarly to the Golgi complex, peroxisomes, glyoxysomes, and plasma membrane, is held to originate in the endoplasmic reticulum (see reviews by Morré and Ovtracht, 1977; Lord, 1980; Waksman et al., 1980), and its formation is regarded as a special case of proliferation and differentiation of the latter (see a review by Svardal and Pryme, 1980).

As to the assembly of the inner mitochondrial membrane and formation of specific protein sets in the intermembrane space and matrix, these processes can also be comprehended only in the framework of known mechanisms of protein insertion in various intracellular membranes and their transmembrane transport.

6.1. SOME MODEL SYSTEMS USED IN STUDYING MITOCHONDRIAL ASSEMBLY

Before presenting the material on mitochondrial assembly *in vivo*, it is reasonable to briefly describe the experimental objects and experimental conditions used in this field.

6.1.1. Respiratory Adaptation

Yeast grown under strictly anaerobic conditions are marked by a number of peculiarities of their respiratory apparatus, mitochondrial energy-transducing machinery, and mitochondrial ultrastructure. Such yeast cells possess virtually no cyanide-sensitive respiration and display only very low, if any, activities in partial reactions of the mitochondrial respiratory chain, e.g., NADH-cytochrome *c* reductase, succinate-cytochrome *c* reductase, or cytochrome *c* oxidase (Chin, 1950; Ephrussi and Slonimski, 1950a; Slonimski, 1956; Lindenmayer and Estabrook, 1958; Chaix, 1961; Rouslin and Schatz, 1969). Absorption spectra of anaerobic yeast show no peaks characteristic of cytochromes *b*, *c*, c_1, or aa_3. As shown in the work of several investigators (Swift et al., 1967; Rabinowitz, 1968; Wallace et al., 1968; Damsky et al., 1969; Plattner and Schatz, 1969; Nagata et al., 1975), anaerobically grown yeast cells contain particles resembling mitochondria in their dimensions and basic structural features (the presence of outer and inner membranes, and cristae). A morphological analysis of such particles isolated from *Saccharomyces cerevisiae* was carried out by Watson et al. (1970). It was pointed out that even the mitochondrial structures occurring in anaerobic yeast grown under the most favorable conditions (in a medium containing ergosterol and a source of unsaturated fatty acids) differ from normal mitochondria in the shape and number of cristae. Under lipid deficiency in the growth medium, yeast cells contain mitochondrial structures practically devoid of cristae but still having an outer and an inner membrane enclosing a granular matrix (Figure 36).

Such mitochondrial structures, termed "promitochondria," were described in most detail by Schatz and co-workers (Criddle and Schatz, 1969; Paltauf and Schatz, 1969; Plattner and Schatz, 1969) and by Watson et al. (1970). Table 18 compares the data on the enzymic properties of yeast mitochondria and promitochondria. It can be seen that the oxidase activities of the latter are close to zero whereas NADH-ferricyanide reductase and succinate dehydrogenase activities may reach 30-50% of those observed with mature mitochondria. Promitochondria display an appreciable ATPase activity (about 25% of normal), which is suppressible by 90% with the ATPase protein inhibitor and by 83% with antibodies raised against F_1-ATPase. This indicates that the promitochondrial ATPase activity is due to the presence of factor F_1. As this activity is almost wholly sensitive to oligomycin and is retained in the cold, the F_1-ATPase can be thought to be bound to promitochondrial membranes similarly to that in mitochondria (Criddle and Schatz, 1969). Promitochondria show an ability to catalyze the ATP-$^{32}P_i$ exchange reaction and to perform ion transport (Groot et al., 1971), which is indicative of a functional energy transformation system, albeit much less efficient than in the mitochondria of aerobically grown yeast cells.

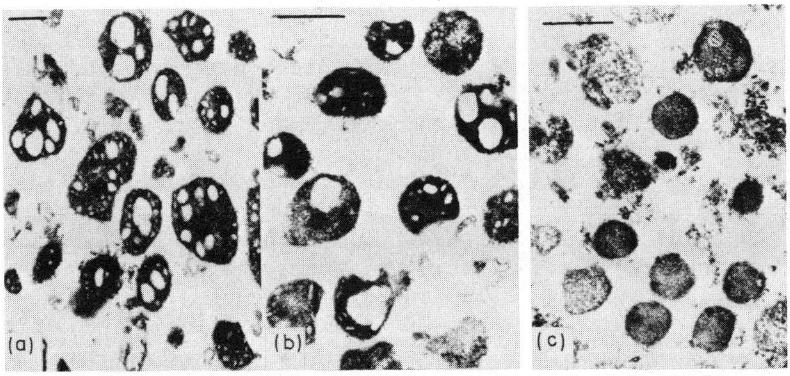

Fig. 36. Mitochondrial structures isolated from Saccharomyces cerevisiae grown anaerobically in media containing Tween-80, ergosterol, and galactose (a) or glucose (b), and in glucose medium without Tween and ergosterol (c). Bars, 0.5 μm. From Watson et al. (1970).

Table 18. Enzymic Composition of Saccharomyces cerevisiae Mitochondria and Promitochondria (from Criddle and Schatz, 1969).

Enzymic activity	Specific activity, µmoles substrate converted per min per mg protein	
	Mitochondria	Promitochondria
ATPase	3.94	0.95
Succinate dehydrogenase*	0.138	0.063
NADH-ferricyanide reductase	0.93	0.293
Succinate-ferricyanide reductase	0.144	0.0061
+ antimycin A	0.022	0.0060
Succinate-cytochrome *c* reductase	0.068	0.000
Succinate oxidase	0.100	0.000
NADH oxidase	0.75	0.002
+ KCN	0.003	0.003
Cytochrome *c* oxidase	1.10	0.002

* Succinate: 2-p-iodophenyl-3-p-nitrophenyl-5-phenyl-tetrazolium reductase.

Weighty evidence for the mitochondrial nature of promitochondria is the presence of specific mitochondrial DNA with a buoyant density of 1.682-1.685 g/cm^3 (Swift et al., 1967; Rabinowitz, 1968; Criddle and Schatz, 1969; Rabinowitz et al., 1969; Watson et al., 1970). According to Criddle and Schatz (1969), the content of this DNA is roughly equal in anaerobic promitochondria and aerobic mitochondria.

Discussing the question of the processes that must take place in promitochondrial structures in the course of their development into normal respiring mitochondria, one cannot neglect the resemblances in the lipid composition between them (Paltauf and Schatz, 1969). Phospholipids which, as pointed out in Chapter Three, play an exceptionally important part in practically all functions of the respiratory chain and energy-transducing machinery, are found in promitochondria in normal or even extra amounts depending on yeast growth conditions. However, the composition of the promitochondrial phospholipid fraction shows some peculiarities. The first to be noted among these are the lower phosphatidylethanolamine and cardiolipin contents and the higher phosphatidylinositol content compared with mitochondria. Regardless of the growth medium composition, promitochondria are almost devoid of ergosterol. Finally, the promitochondrial lipids (if yeast are cultivated in the absence of appropriate components from the medium) contain no unsaturated fatty acids, which usually constitute some 80% of total fatty acids and determine the plasticity of membrane structure.

Of basic importance appear to be the data on the comparative content of promitochondria in anaerobic yeast cells and of mitochondria in normal aerobically grown cells. Criddle and Schatz (1969) claimed that the total protein of each type of structures accounts for much the same portion of the protein of the corresponding cell homogenate. It must be commented, however, that the methods for isolating mitochondria and promitochondria are not identical, and therefore the comparison of their content in yeast cells is fairly arbitrary. Moreover, because promitochondria lack such distinct morphological features as mitochondria, it is very hard if at all possible to assess their homogeneity and purity.

According to Plattner et al. (1970), promitochondria are direct precursors of mitochondria. They convert into respiring mitochondria in the course of the respiratory adaptation of anaerobically grown yeast cells. These authors have shown, in particular, that the label ([^{2}H]leucine) selectively incorporated in vivo into promitochondria is recovered in the mitochondrial fraction after the completion of respiratory adaptation. As the adaptation proceeds, yeast cells develop cyanide-sensitive respiration and educe or markedly enhance the activities of respiratory-chain enzymes (Ephrussi and Slonimski, 1960a,b; Slonimski, 1956; Lindenmayer and Estabrook, 1958; Chen and Charalampous, 1969; Rouslin and Schatz, 1969; Vary et al., 1969, 1970; Cartledge and Lloyd, 1973); in parallel, cytochromes b, c_1, c, and aa_3 are formed (Chin, 1950; Ephrussi and Slonimski, 1950a,b; Lindenmayer and Estabrook, 1958; Chaix, 1961; Linnane, 1965; Chen and Charalampous, 1969; Cartledge et al., 1972) and the content of iron-sulfur proteins increases (Ohnishi et al., 1975). Since all the above-listed adaptational changes are suppressed by cycloheximide and by chloramphenicol (Chen and Charalampous, 1969; Rouslin and Schatz, 1969; Vary et al., 1969, 1970), they can be thought to involve protein synthesis both on cytoplasmic and mitochondrial ribosomes. Respiratory adaptation induces the synthesis of tricarboxylic-acid-cycle enzymes (Vary et al., 1969, 1970; Cartledge et al., 1972) metabolically linked with the mitochondrial respiratory chain. As follows from comparing other properties of promitochondria and mitochondria, aeration must also increase the content of F_1-ATPase and other energy-transducing components as well as alter the lipid composition of mitochondrial membranes.

6.1.2. Glucose Derepression

Another system widely used for studying mitochondrial formation is the facultatively anaerobic yeast grown in a medium with high glucose concentration. According to the data of several authors (Ephrussi et al., 1956; Schatz, 1963; Polakis et al., 1965; Jayaraman et al., 1966; Utter et al., 1967), such yeast have low respiratory activity and contain very low amounts of cutochromes b, c_1, c, and aa_3. The changes in mitochondria during yeast glucose repression were studied in detail by Jayaraman et al. (1966). Transfer of aerobically grown yeast into a repressive glucose-rich medium entails a rapid decrease in the cell respiration rate (repression); after exhaustion of glucose the cell respiration is restored (derepression). The repression is attended by a decline in NADH oxidase, succinate oxidase, lactate oxidase, and other mitochondrial activities, as well as in the content of cytochromes *b*, $c+c_1$, and aa_3. Mitochondria isolated from repressed cells display lowered respiratory control which can be considered one of the most sensitive criteria of the mitochondrial inner membrane integrity. Destructive alterations in mitochondria during repression (in particular, disappearance of cristae) are also demonstrable electron-microscopically. As the growth medium is exhausted of glucose (see, e.g., Tustanoff and Bartley, 1964; Jayaraman et al., 1966; Utter et al., 1967; Henson et al., 1968), all the above mitochondrial characteristics are gradually restored to normal. Jayaraman et al. (1966) regarded the mitochondrial alterations described as their partial dismantling and conversion into promitochondria at the repression state, with subsequent replenishment with lacking components and reassembly into normal mitochondria during derepression. Similar ideas were expressed by Schatz (1963).

Thus, in the cases of either respiratory adaptation of anaerobically grown yeast or glucose derepression we are dealing with the conversion of the more primitive promitochondrial structures into mature respiratory-competent mitochondria. This process requires synthesis of the components of the respiratory chain and energy-transducing machinery, their specific incorporation into preexisting membranes, changes in the membrane lipid composition, etc.

6.1.3. Fungal Spore Germination

Another model convenient for studying mitochondrial biogenesis is germinating fungal spores. Germination of Neurospora crassa spores is known to be accompanied by a pronounced increase in the cell respiration rate and in the contents of cytochromes *b*, c_1, *c*, and aa_3 (Stade and Brambl, 1981). The respiratory development is sensitive to inhibitors of cytoplasmic and mitochondrial translation and therefore involves polypeptide production by both protein-synthesizing systems (see, e.g., Hawley and Greenawalt, 1975; Stade and Brambl, 1981). The rise in cell respiration is due not to a simple increase in the content of mitochondrial material but to qualitative changes in mitochondria which in dormant spores are marked by a number of peculiarities. In particular, the mitochondria of Botryodiplodia theobromae spores were shown to contain three mitochondrially made subunits of cytochrome *c* oxidase but none of the cytoplasmically synthesized subunits of this complex (Brambl, 1980). These nonrespiring mitochondria also harbor two or three cytoplasmic subunits of H^+-ATPase which do not appear to be associated with each other (Wenzler and Brambl, 1981).

6.2 MECHANISMS OF PROTEIN TRANSPORT ACROSS BIOLOGICAL MEMBRANES

The mechanisms of protein transport across biological membranes have been extensively discussed in a number of reviews (see, e.g., Morré and Ovtracht, 1977; Parry, 1978; Chua and Schmidt, 1979; Wickner, 1979; Blobel, 1980; Svardal and Pryme, 1980; Waksman et al., 1980; Sabatini et al., 1982), which makes it possible to restrict this presentation to a brief account of some major issues. First of all it must be pointed out that polypeptide transport across the membrane lipid bilayer takes place either cotranslationally or post-translationally, and in the latter case it can proceed by several basically different pathways.

6.2.1. Cotranslational Protein Transport Across Membranes

The primary distinguishing feature of cotranslational transport is that the lipid bilayer is permeated by a growing polypeptide chain that has not yet folded upon itself. Such a chain has small cross-section and, consequently, high penetrating ability. Several mechanisms were proposed for the cotranslational transport of proteins across the endoplasmic reticular membrane in eukaryotes and the bacterial plasma membrane (see, e.g., Blobel and Dobberstein, 1975; Campbell and Blobel, 1976; Garnier et al., 1980; Sabatini et al., 1982). According to these concepts, the key part in the transport is played by a "signal" amino acid sequence at the NH_2-terminus of the growing polypeptide chain, which serves for polysome binding to a receptor site on the membrane (Campbell and Blobel, 1976; Sabatini et al., 1982) or for insertion of the chain into the lipid bilayer with subsequent acquisition of proper conformation (Wickner, 1979). To date, more than 20 such sequences have been deciphered. They all have amphipathic structure with a hydrophobic core of 9 to 24 amino acid residues and charged polar ends (Inouye and Halegoua, 1979; Garnier et al., 1980). The amino-terminal signal sequence may be a transient (Figure 37c) or permanent feature (Figure 37d) of the nascent polypeptide. Its cotranslational removal by a membrane-bound proteinase leads to the release of the polypeptide into the lumen of the endoplasmic reticulum (Figure 37e) unless vectorial discharge is interrupted by a "halt transfer" signal that prevents the passage of the carboxy-terminal portion of the polypeptide through the membrane (Figure 37f). In the absence of a halt-transfer signal, an uncleaved amino-terminal sequence may ensure permanent attachment of the protein to the membrane (Figure 37g). If such permanent insertion sequence is followed by a halt-transfer signal, both termini of the polypeptide chain remain on the cytoplasmic side of the membrane while the rest of the chain loops into the lumen. Examples illustrating all these variants are given by Sabatini et al. (1982).

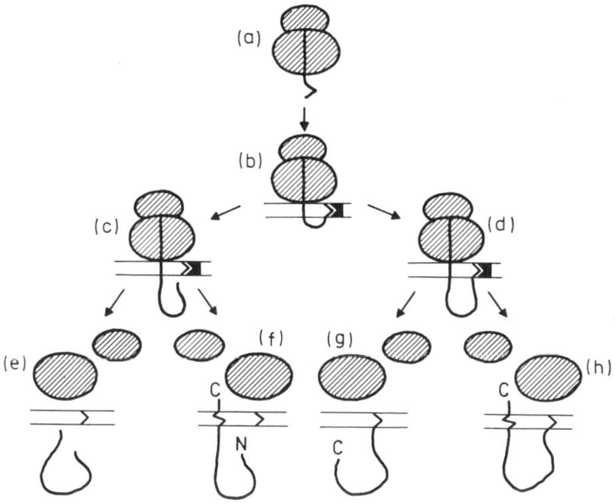

Fig. 37. Models describing the role of amino-terminal cotranslational insertion signals in the insertion of integral membrane proteins in the endoplasmic reticular membrane. In these models, membrane-protein biosynthesis is depicted as an interrupted vectorial discharge process. (>) An amino-terminal cotranslational insertion signal; (▶) receptor; (Ϛ) "halt transfer" signal. After Sabatini et al. (1982).

6.2.2. **Posttranslational Polypeptide Transport Across Membranes**

Unlike the preceding case where translation takes place on membrane-bound polysomes, some polypeptides are synthesized in the cytoplasm on free polysomes, which is readily demonstrable experimentally. Such polypeptides undergo folding prior to insertion into the membrane. Two mechanisms of posttranslational transmembrane protein transport are widely discussed at present. The first concerns the permeation of peptide toxins of microbial and plant origin. The other one pertains to the transport of membrane material from one intracellular structure to another and to the export of secreted material from the cell.

6.2.2.1. *Transmembrane Protein Transport Involving Subunit Separation and Reversible Unfolding*

Many peptide toxins, including diphtheria and cholera toxins, Pseudomonas aeruginosa exotoxin A, colicins E_2 and E_3, etc. (see references in Waksman et al., 1980), comprise two separate or separable components, A and B. The A-type component is responsible for the toxic effect. It penetrates the membrane and comes out on its cytoplasmic side or goes further into the cytoplasm. The structure of the A component is such as would permit reversible unfolding of the polypeptide chain to decrease its cross-sectional area and facilitate its passage through limited-size channels. The B-type component carries no toxic principle and only serves for toxin binding to the membrane surface.

For instance, diphtheria toxin (Pappenheimer, 1977) is synthesized as a single polypeptide chain of 62,000 daltons, and the A fragment (NH_2-terminal) is separable from the B fragment (COOH-terminal) upon cleavage of a peptide bond and one disulfide bridge. According to the model of Boquet and Pappenheimer (1976), the toxin molecule binds to the cell membrane by the hydrophilic surface of the B-portion COOH-terminus. The hydrophobic NH_2-region of this portion, either in association with the receptor molecule or by itself, forms a "channel" in the lipid bilayer. The A portion attached to this region is pulled through the channel, which is expedited by its capacity of reversible unfolding. This process is complete when the disulfide bridge and the site of the peptide bond between the two fragments emerge on the cytoplasmic side of the membrane. Here the disulfide bridge is broken and the A fragment goes into the cytoplasm of the host cell. It is as yet obscure whether the peptide bond is cleaved at the same moment or the molecule is nicked before.

This example may seem too abstract and not pertinent to mitochondrial biogenesis. However, something similar happens during the transport of the precursor to the small subunit of ribulose-1,5-biphosphate carboxylase, synthesized in the cytoplasm of *Chlamydomonas reinhardtii*, into the chloroplast stroma, and this is already quite close to the problem discussed in this chapter. As shown by Dobberstein et al. (1977), translation of poly(A) mRNA from *Chlamydomonas reinhardtii* in a cell-free system yields not the mature enzyme subunit with a molecular weight of 16,500 but a larger (20,000) polypeptide which is immunoprecipitated by an antiserum to the mature subunit. Upon addition of *Chlamydomonas reinhardtii* postribosomal supernatant this precursor is cleaved into two parts, S and F, the first one having the same size as the small subunit of ribulose-1,5-biphosphate carboxylase. This gave grounds for thinking that the precursor transport across the chloroplast envelope *in vivo* proceeds not cotranslationally but somehow otherwise. In particular, a mechanism was considered that was analogous to the one described above for the transfer of diphtheria toxin through the cell plasma membrane, with the smaller F sequence assigned the function performed by the toxin B fragment, i.e. binding to a receptor site on the chloroplast surface, and so on (see above). Such posttranslational transport mechanism coupled with the processing of the precursor of ribulose-1,5-biphosphate carboxylase was supported by the fact that no cytoplasmic polysomes bound to the chloroplast surface were found in *Chlamydomonas reinhardtii* cells. This implied that the synthesis of chloroplast proteins takes place on free rather than on membrane-bound polysomes and, consequently, that their transport through the chloroplast envelope is not cotranslational but posttranslational.

Later Chua and Schmidt (1978) confirmed the possibility of posttranslational transport of the precursor to the small subunit of ribulose-1,5-biphosphate carboxylase in a reconsituted system containing the pS precursor synthesized in a cell-free system, and isolated chloroplasts. The precursor was processed so as to attain the size and charge of the mature S subunit. Moreover, the labeled subunit was shown to be integrated into the holoenzyme, which testifies that it finds its way into the chloroplast stroma and interacts with the large L subunit of ribulose-1,5-biphosphate carboxylase localized there.

Recently evidence has been obtained that posttranslational transport into chloroplasts is characteristic not only of the ribulose-1,5-biphosphate carboxylase S subunit but of many other proteins of the organelle. Grossman et al. (1982) incubated chloroplasts with the translation products of a cell-free system programmed with the total pea poly(A) mRNA, and observed in the soluble fraction and thylakoid membranes more than 100 labeled polypeptides most of which were identical in electrophoretic mobility to the mature chloroplast proteins. Treatment of the translation mixture with antisera to plastocyanin, ferredoxin-NADP reductase, and fructose-1,6-biphosphatase showed the presence of larger precursors for the first two proteins. Their import into chloroplasts was accompanied by processing, as was the case with the S subunit of ribulose-1,5-biphosphate carboxylase.

6.2.2.2. *Transfer of Soluble and Membrane Proteins in Transport Vesicles*

Externalization and internalization of many proteins in the cell as well as biogenesis of cell membranes are at present held to be effected through a mechanism including the stage of transfer of membrane-bound and soluble proteins in transport vesicles (for review see Morré and Ovtracht, 1977; Parry, 1978; Morré et al., 1979; Waksman et al., 1980). Palade (1975) put forward a suggestion ("the membrane flow hypothesis") that the plasma membrane is built of membranous elements budding from the Golgi apparatus which in its turn is replenished by vesicles budding from smooth endoplasmic reticulum. Such a multistep pathway of the plasma-membrane elements is determined by the complex compartmentation of the cell and strict functional specialization of its individual compartments. In particular, it is known that attachment of some saccharide residues (e.g., galactose and fucose) to polypeptide chains, synthesis and sulfatation of carbohydrate chains of proteoglycans, and other processes of protein modification take place in the Golgi apparatus, while the initial polypeptides are synthesized in the rough endoplasmic reticulum. A graphic illustration of this mechanism of protein transport is the data on the G envelope protein of vesicular stomatitis virus. This protein was shown (Rothman and Fine, 1980) to be synthesized in the endoplasmic reticulum of infected cells and inserted cotranslationally into the reticular membrane, losing its NH_2-terminal insertion signal. Simultaneously, saccharide residues (mannose and N-acetyl glucosamine) are attached to the growing polypeptide chain. This intermediate (high-mannose) form of the protein, G_1, is found in the fraction of coated vesicles isolated from host cells shortly after [^{35}S]methionine incorporation in pulse-chase experiments. The final G_2 form, containing in addition galactose, fucose, and sialic acid and insensitive to endoglucosidase, appeared in the coated vesicles only after some time of the chase. Since attachment of these carbohydrate residues could take place only in the Golgi apparatus, it could be thought that there exist two types of vesicles (namely, G_1-containing and G_2-containing) performing the two stages of protein transport from the endoplasmic reticulum first to the Golgi apparatus and then to the plasma membrane.

It must be noted at once that an analogous transfer mechanism has been considered for two mitochondrial enzymes, glutamate dehydrogenase and malate dehydrogenase. Kawajiri et al. (1977a,b) supposed that these proteins, synthesized in the endoplasmic reticulum of liver cells, are transported into mitochondria in so-called "microparticles." However, this idea does not seem to hold true as evidenced, in particular, by the data of Mihara et al. (1982).

6.3 PROTEIN TRANSPORT ACROSS MITOCHONDRIAL MEMBRANES

The history of mitochondrial biogenesis is such that all the above-mentioned mechanisms of transmembrane protein transport have been, albeit with a certain lag, proposed with this organelle. In this case the problem has some "local color." Firstly, some mitochondrial proteins in order to reach their destination have to cross the lipid bilayers of two membrane systems, which throughout most of their expanse are separated by a distance exceeding the dimensions of protein molecules. Secondly, incorporation of some polypeptide chains into the mitochondrial membrane is associated with the attachment of prosthetic groups, i.e. with additional modification that is effected in definite organellar compartments. Thirdly and finally, in mitochondria there are two contrary fluxes of polypeptides: one from the cytoplasm into the intermembrane space, inner membrane, and matrix, and the other from the matrix to both sides of the inner membrane.

6.3.1. Cotranslational Discharge of Cytoplasmically Synthesized Polypeptides into Mitochondria

This mechanism was first applied to mitochondria by Butow's group. The premise in these studies was the discovery in growing

Saccharomyces cerevisiae cells of a peculiar class of cytoplasmic ribosomes bound in the form of polysomes to the outer mitochondrial membranes by nascent polypeptide chains (Kellems and Butow, 1972, 1974). These ribosomes were not detached upon mitochondrial incubation in high-ionic-strength solution, but could be stripped off by adding puromycin which terminates translation. Electron-microscopic investigations showed (Kellems et al., 1975) that the mitochondria-bound ribosomes are located mostly at the inner-outer membrane junctions, i.e. where the membranes are closely apposed (see also Keyhani, 1973).

A number of arguments were put forward in favor of the participation of the mitochondria-bound 80 S cytoplasmic ribosomes in the synthesis of mitochondrial proteins (Kellems and Butow, 1974; Ades and Butow, 1980a). Among other things, it was demonstrated that the mitochondria-bound polysomes produce in a yeast read-out system a different set of polypeptides than the free cytoplasmic polysomes. About 85% of polypeptides made by the mitochondrial 80 S polysomes are not cleaved by exogenous proteinase, but their susceptibility to digestion is enhanced by a detergent (deoxycholate). At the same time, more than 80% of the products of free polysomes after incubation with mitochondria remain proteinase-sensitive even without any detergent. Once stripped off the organelles, the mitochondrial 80 S polysomes displayed a far higher ability to synthesize F_1-ATPase subunits than free cytoplasmic polysomes. It is interesting to note that among the products of translation on mitochondria-bound 80 S polysomes only mature α, β, and γ ATPase subunits were observed, but not their precursors (see Section 6.3.3 of this chapter). This finding implied vectorial translation and cotranslational processing of the growing polypeptide chains. However, the results could not be interpreted unambiguously because of experimental difficulties.

Further evidence in support of the cotranslational transfer mechanism was obtained by comparing the kinetics of [^{35}S]methionine incorporation into the total mitochondrial and total cytosolic protein as well as into some individual proteins of these two subcellular fractions (Ades and Butow, 1980b). The labeling of mitochondrial proteins proved to be not in the least slower than that of cytoplasmic proteins. The only exception was when 80 S ribosomes in the yeast cells were first made to dissociate from mRNA molecules (e.g., by cooling to 0°C or by adding an inhibitor of translation initiation), and then translation initiated again. In such conditions the synthesis of cytoplasmic proteins began without any lag while label incorporation into mitochondrial proteins was somewhat delayed. This delay was supposed (Ades and Butow, 1980b) to be the time needed for the migration of initiated polysomes from the cytoplasm to mitochondria and their binding to the outer mitochondrial membrane via the nascent polypeptide chains, which are then vectorially discharged through the membrane.

On the whole, the results mentioned allowed Ades and Butow (1980b) to conclude that at least under certain yeast growth conditions there hardly are any extramitochondrial pools of polypeptides synthesized in the cytoplasm and awaiting their transport and integration in mitochondria.

Some facts, however, do not fit into the concept of vectorial translation of nuclear-coded mitochondrial proteins through mitochondrial membranes. In particular, no functional cytoplasmic ribosomes associated with mitochondria can be found in some cell types, such as *Neurospora crassa* (Harmey et al., 1977) or rat liver (Shore and Tata, 1977). It must also be noted that the studies of Butow and coworkers shed no light either on the portion of mitochondrial protein cotranslationally transferred into the organelles (i.e. not appearing in the cytosol) or on which particular proteins can be integrated in mitochondria by this mechanism. An example might have been the three larger ATPase subunits, were the data for them not debatable (see Section 6.3.3).

6.3.2. Posttranslational Transport of Cytoplasmically Synthesized Polypeptides into Mitochondria

In contrast to the data of Ades and Butow (1980a), experiments with *Neurospora crassa* cells and cell homogenates involving double labeling and immunoprecipitation (Hallermayer et al., 1977; Harmey et al., 1977) demonstrated that the appearance of label in the mitochondrial proteins is delayed compared to the total cell protein or microsomal and cytosolic proteins. Figure 39 depicts the results of an experiment in which *Neurospora crassa* cells were subjected to pulse and pulse-chase labeling. Microsomal, cytosolic, and mitochondrial fractions were obtained from the cells at times indicated and treated with antibodies to mitochondrial matrix proteins. It can be seen that the immunoreactive labeled polypeptides accumulate in significant amounts in the cytosol and only after a lag of 90 sec appear in mitochondria. It is noteworthy that, upon addition of excess cold leucine, labeling in the cytosol stops at once whereas in the mitochondria it proceeds to reach the level observed in the cytosolic protein at the start of the chase. A lag similar to that seen in Figure 38 was found for the accumulation in mitochondria of two individual proteins of the inner membrane, cytochrome *c* and the carboxyatractyloside-binding protein (Hallermayer et al., 1977). These data pointed to the existence of extramitochondrial pools of polypeptides which are transferred from the cytosol to mitochondria with different kinetics, in the time ranging from 1.5 to 10 min.

This issue was further supported by experiments with the translation inhibitors cycloheximide and puromycin (Hallermayer et al., 1977; Harmey et al., 1977). It turned out that these inhibitors at once block the labeling of cytosolic proteins, free ribosomes, and

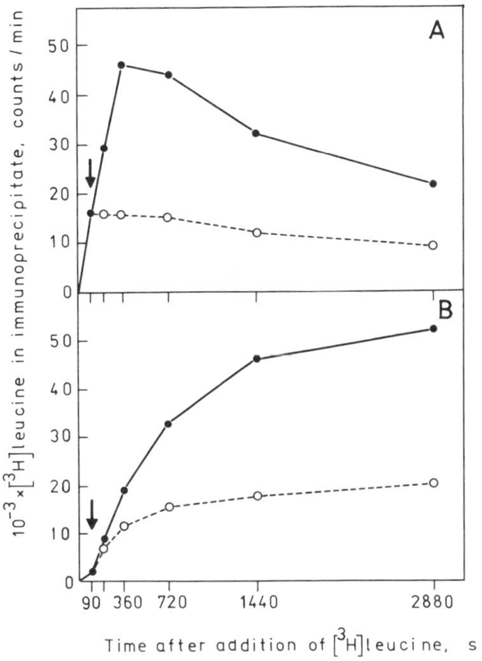

Fig. 38. Incorporation of $[^3H]$leucine into matrix proteins precipitable with antibodies to these proteins from the cytosol (A) and mitochondria (B) after adding the label to Neurospora crassa incubation medium. Dashed lines correspond to an experiment in which after 90 sec of incubation with the label it was chased with excess cold leucine. After Hallermayer et al. (1977).

microsomes, but have no immediate effect on the proteins of mitochondrial matrix or the carboxyatractyloside-binding protein of the inner mitochondrial membrane (see Figure 39). This is only possible if the cytoplasmically synthesized mitochondrial polypeptides initially accumulate in the cytoplasm. However, dissonant results were obtained later by Ades and Butow (1980b) for yeast mitochondria (cf. the preceding subsection).

Thus, the above data of Neupert and colleagues suggest that at least a considerable portion of the matrix proteins and some inner-membrane proteins are imported posttranslationally. A question arises about the actual mechanism of such transport.

According to Harmey et al. (1977), the overwhelming majority of prelabeled mature proteins obtained from the mitochondrial matrix do not enter the mitochondria in Neurospora crassa cell homogenates. This means that mitochondria can differentiate between the mature and

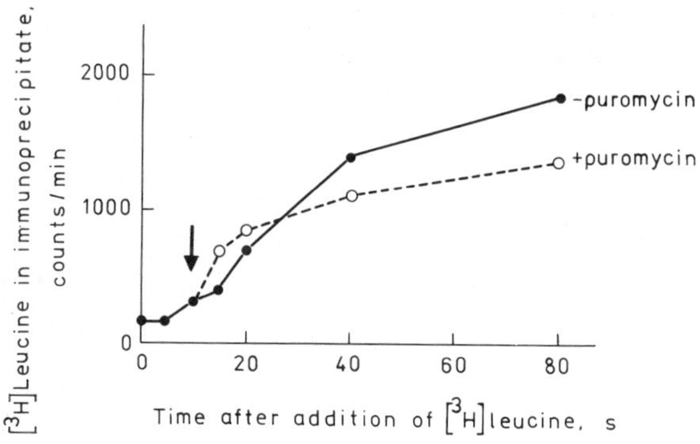

Fig. 39. Effect of puromycin on [^{3}H]leucine incorporation into mitochondrial matrix proteins in Neurospora crassa cell homogenates. The arrow indicates the moment of addition of puromycin. After Harmey et al. (1977).

the newly synthesized matrix proteins while antibodies make no such distinction. Further, mature proteins do not compete with the newly synthesized ones for the "transport sites" on the membrane. On the basis of these data it was suggested that cytoplasmically synthesized mitochondrial proteins initially carry a certain signal that enables them to find and negotiate mitochondrial membranes, and that during this passage the polypeptides are somehow modified so as to make the transfer irreversible (Harmey et al., 1977).

The concept of the posttranslational transport of cytoplasmically synthesized proteins to mitochondria, the basic points of which were formulated in a general form by Neupert and colleagues, has been developed in recent years. One of the working hypotheses (Harmey et al., 1977) was that polypeptide transport from the cytoplasm to mitochondria is similar to the transport of the precursor to the S subunit of ribulose-1,5-biphosphate carboxylase to chloroplast stroma in Chlamydomonas reinhardtii, the case considered above (see Section 6.2.2.1).

6.3.3. Precursors of Cytoplasmically Made Mitochondrial Proteins

In recent years evidence has accumulated for the existence of precursors to about thirty mitochondrial proteins (see Table 19)*. All these precursors were synthesized in heterologous or homologous cell-free systems and isolated by immunoprecipitation with antisera to corresponding mature proteins. It is easy to see from the Table that there are precursors to the proteins of all mitochondrial compartments, i.e. outer and inner membranes, intermembrane space, and matrix. Almost all these precursors have a somewhat lower electrophoretic mobility than the mature proteins, which is believed to

* See Addendum.

Table 19. Cytoplasmically Synthesized Precursors to Some Mitochondrial Proteins and Their Import into Mitochondria.

Polypeptide	Mitochondrial localization	Cell type	Appl. mol. wt.		Precursor found *in vivo*	Import shown *in vitro*	Reference
			Mature	Pre-			
35,000-dalton protein	Outer membrane	Rat liver	35,000	35,500	-	-	Shore et al. (1981)
Cytochrome *c* peroxidase	Intermembrane space	Yeast	33,500	39,500	+	+	Maccecchini et al. (1979b)
Cytochrome *b*$_2$			58,000	68,000, 65,000	-	+	Gasser et al. (1982)
Sulfite oxidase		Rat liver	55,000	59,000	-	+	Mihara et al. (1982); Ono and Ito (1982)
ADP/ATP carrier	Inner-membrane	Neurospora	32,000	32,000	-	+	Zimmerman and Neupert (1980)
Ubiquinol-cytochrome *c* reductase:							
subunit I (core protein)			50,000	51,500	-	+	Teintze et al. (1982)
subunit II (core protein)			45,000	47,500	-	-	
subunit IV (cytochrome *c*$_1$)			31,000	38,000	+	+	
subunit V (Fe-S protein)			25,000	28,500	-	+	
subunit VII			11,500	12,000	-	+	
subunit VIII			11,200	11,600	-	-	

Table 19. (continued)

Polypeptide	Mitochondrial localization	Cell type	Appl. mol. wt.		Precursor found *in vivo*	Import shown *in vitro*	Reference
			Mature	Pre-			
cytochrome *c*$_1$		Yeast	31,000	35,000, 34,000	-	-*	Gasser et al. (1982)
subunit V			25,000	27,000	+	-	Coté et al. (1979)
Cytochrome *c* oxidase:							
subunit IV	Inner-membrane	Rat liver	19,500	22,500	-	-	Schmelzer and Heinrich (1980)
subunit V		Yeast	13,000	15,000	+	-	Lewin et al. (1980)
subunit VI			12,500	20,000	+	-	
Cytochrome *c*		Neurospora	12,000	12,000	+	+	Korb and Neupert (1978)
F_1-ATPase:							
α subunit		Yeast	58,000	64,000	+	+	Maccecchini et al. (1979a)
β subunit			54,000	56,000	+	+	
γ subunit			34,000	40,000	+	+	
H^+-ATPase: proteolipid		Neurospora	8,000	12,000	-	-	Michel et al. (1979)
D-β-hydroxybutyrate dehydrogenase		Rat liver	32,000	37,000	+	-	Michara et al. (1982)
Cytochrome P-450$_{scc}$		Adrenal	49,000	54,500	-	-	Kramer et al. (1982)
Cytochrome P-450$_{11\beta}$		cortex	48,000	53,500	-	-	
δ-Aminolevylinic acid synthetase	Matrix	Chick embryo liver	49,000	70,000	+	-	Whiting (1976)

Table 19. (continued)

Polypeptide	Mito-chondrial localiz-ation	Cell type	Appl. mol. wt.		Pre-cursor found *in vivo*	Import shown *in vitro*	Reference
			Mature	Pre-			
Aspartate amino-transferase (subunit)		Chick heart	44,500	47,500	-	-	Sonderegger et al. (1982)
		Rat liver	45,000	47,000	-	+*	Sakakibara et al. (1980)
Mn-Superoxide dismutase (subunit)		Yeast	24,000	26,000	+	-	Autor (1982)
mRNA polymerase (subunit)			45,000	49,000	-	-	Lustig et al. (1982)
Malate dehydrogenase		Rat liver	37,000	38,000	-	-	Michara et al. (1982)
L-Glutamate dehydrogenase			54,000	60,000	+	-	Michara et al. (1982)
			47,600	50,200	-	+	Miralles et al. (1982)
Ornithine amino-transferase			43,000	49,000	-	-*	Mueckler et al. (1982)
Carbamyl phosphate synthetase (subunit)	Matrix	Rat liver	160,000	165,000	+	+	Shore et al. (1979); Raymond and Shore (1981); Mori et al. (1981); Campbell et al. (1982)
Ornithine carbamoyl-transferase (subunit)			36,000	39,600, 37,000	+	+	Mori et al. (1980, 1981); Conboy et al. (1982); Morita et al. (1982)
Adrenodoxin		Adrenal cortex	12,000	19,000	-	-	Kramer et al. (1982);
			12,000	20,000	-	-	Nabi and Omura (1982)
Adrenodoxin reductase			51,100	53,400	-	-	Kramer et al. (1982)

* Processing demonstrated *in vitro*.

reflect their larger molecular weight owing to the presence of an additional amino acid sequence. By the data of sodium dodecyl sulfate-polyacrylamide gel electrophoresis, precursors have molecular weights exceeding by 2000-6000 those of their mature counterparts. Table 19 also indicates the cases when precursors to mitochondrial proteins were detected in vivo together with the mature forms.

For a large number of mitochondrial proteins and their precursors, similar proteolytic fingerprints were obtained. Such data are available, in particular, for mature and precursor forms of α, β, and γ subunits of F_1-ATPase (Maccecchini et al., 1979a); cytochrome *c* peroxidase (Maccecchini et al., 1979b); subunit IV of cytochrome *c* oxidase (Schmelzer and Heinrich, 1980); subunit V of the cytochrome bc_1 complex (Côté et al., 1979); manganese superoxide dismutase (Autor, 1982); carbamyl phosphate synthetase (Shore et al., 1979); orithine aminotransferase (Mueckler et al., 1982); and RNA polymerase (Lustig et al., 1982).

The reactivity to the same antiserum and similarity of proteolytic fingerprints are obviously necessary but not sufficient proof that the polypeptide chain of the putative precursor is identical to the mature protein (of course apart from the amino acid extension). More reliable would have been data of a comparative analysis of their complete or at least NH_2-terminal and COOH-terminal amino acid sequences. Such data, however, are lacking at present*. In this connection it is timely to remind oneself that Poyton and McKemmie (1979a,b) reported a polyprotein precursor to all four cytoplasmic subunits of cytochrome *c* oxidase in Saccharomyces cerevisiae. This protein of molecular weight 55,000 could be synthesized in vitro and precipitated with antisera to subunit IV, subunit VI, and subunits V+VII. Comparative analysis of tryptic digests of these subunits and the precursor showed that the latter contained virtually all peptides of subunits IV to VII. Still, results obtained later (Lewin et al., 1980; Schmelzer and Heinrich, 1980) make it unlikely that this polypeptide is indeed the polyprotein precursor.

Several cytoplasmically made mitochondrial polypeptides are known to be synthesized in cell-free systems in a form indistinguishable by electrophoretic mobility from their mature counterpart. This pertains to apocytochrome *c* (Korb and Neupert, 1978; Zimmerman et al., 1979a), ADP/ATP carrier (Zimmerman et al., 1979b; Zimmerman and Neupert, 1980), and adenylate kinase (Watanabe and Kubo, 1982). For apocytochrome *c* the in vitro product was shown to have the same NH_2-terminal amino acid sequence as the functional protein in mitochondria (Zimmerman et al., 1979a). Nevertheless, the in vitro product is not precipitable with an antiserum to holocytochrome *c*, implying structural differences between the two forms (Korb and Neupert, 1978). In accordance with this, holocytochrome *c* does not compete with the in vitro-synthesized product for incorporation into mitochondria (Matsuura et al., 1981). These data were considered to

* See Addendum.

indicate the existence of a region in the apocytochrome *c* polypeptide chain acting as an "addressing signal" for mitochondria. Heme attachment entails shielding of this signal and deprives holocytochrome *c* of the ability to bind to a receptor site on the mitochondrial surface through which apocytochrome *c* is translocated. There are some grounds for thinking that the addressing signal is located somewhere in the region from the 66th residue to the carboxy terminus of the molecule.

Structural differences between the in vitro product and the membrane-integrated form were also shown for the ADP/ATP carrier protein. The in vitro product was found (Zimmerman and Neupert, 1980) to exist as a water-soluble oligomer that can form mixed micelles with a detergent. The extramitochondrial form of the carrier binds to hydroxylapaptite in the presence of Triton X-100 whereas the carrier isolated from mitochondria does not.

To understand the mechanism of the transport of cytoplasmically synthesized mitochondrial proteins, it is extremely important to know on which polysomes - free or membrane-bound - they are made. In virtually all the cases when this question was purposefully studied it turned out to be that free polysomes exclusively are translationally active. This was proved for the precursors of quite a number of mitochondrial enzymes: δ-aminolevulinic acid synthetase (Whiting, 1976); carbamyl phosphate synthetase (Raymond and Shore, 1979; Sonderegger et al., 1982); aspartate aminotransferase (Sakakibara et al., 1980); ornithine carbamoyltransferase (Morita et al., 1981); L-glutamate dehydrogenase, malate dehydrogenase, and D-β-hydroxybutyrate dehydrogenase (Mihara et al., 1982); and ornithine aminotransferase (Mueckler et al., 1982). Only in several cases were mitochondrial protein precursors synthesized both on free and membrane-bound polysomes; such examples are adrenodoxin (Nabi and Omura, 1980) and subunit IV of cytochrome *c* oxidase (Heinrich et al., 1981)*.

As to the above-mentioned data of Butow and co-workers (Kellems and Butow, 1974; Kellems et al., 1975; Ades and Butow, 1980a,b) on the cytoplasmic polysomes associated with yeast mitochondria, it has been argued (Neupert and Schatz, 1981) that the contribution of such polysomes to the production of mitochondrial proteins must be small, since more than 90% of the total mRNA for these proteins is bound in free polysomes which consequently must make the bulk of the polypeptides imported into the organelle. Such polypeptides can obviously be transported across mitochondrial membranes only posttranslationally.

The discovery of a precursor to a certain protein makes doubtful the expedience of studying the transport of its mature form in mitochondria in vitro with a view to elucidating what happens to this protein in vivo. This applies specifically to the work on the selective internalization of mitochondrial aspartate aminotransferase

---

* See Addendum.

(Marra et al., 1978, 1980; Hubert et al., 1979; Passarella et al., 1982), malate dehydrogenase (Passarella et al., 1980), and F_1-ATPase (Stratman et al., 1972) by isolated mitochondria and mitoplasts.

6.3.4. Maturation of Precursors to Cytoplasmically Made Mitochondrial Proteins In Vivo

Mori et al. (1981a,b) studied the synthesis, intracellular transport, and processing in isolated hepatocytes of two mitochondrial matrix proteins, carbamoyl phosphate synthetase and ornithine carbamoyltransferase. Their experiments included labeling the cells with [^{35}S]methionine for various periods, isolating the cytosolic and the mitochondrial fractions, subjecting them to immunoprecipitation with antisera against carbamoyl phosphate synthetase and ornithine carbamoyltransferase, and analyzing the immunoprecipitates by sodium dodecyl sulfate-polyacrylamide gel electrophoresis followed by fluorography. With this technique, only larger precursors were found in the cytosol whereas the mitochondria contained only mature enzymes. Figure 40 shows the kinetics of pulse label incorporation into the precursor and the mature ornithine carbamoyltransferase. Firstly, it should be noted that in the precursor the label appears almost without any delay while the mature enzyme is labeled with a more than 5-min lag. Secondly, the specific radioactivity of the precursor is far higher than that of its mature counterpart (see also Raymond and Shore, 1979). Important information on the fate of the precursor was furnished by pulse-chase experiments. As can be seen in Figure 41, in the chase following a 20-min pulse of [^{35}S]methionine the content of the labeled precursor of ornithine carbamoyltransferase rapidly decreases with a concomitant moderate increase in the label in the mature (i.e. mitochondrial) form. From these data the half-life of the precursor in the cytosol is estimated at only 2 min. Therefore, the low steady-state content of the labeled precursor in the cytosol as compared with the content of the mature enzyme in the mitochondria is not at all surprising. Similar data were obtained for carbamoyl phosphate synthetase.

The exceptionally short half-lives of the cytoplasmically made precursors of mitochondrial proteins considerably complicate their detection in vivo. One of the techniques used (see, e.g., Côté et al., 1979; Maccecchini et al., 1979a,b) is the rapid lysis of labeled cells with sodium dodecyl sulfate at 100°C in the presence of a mixture of proteinase inhibitors (p-aminobenzamidine, phenylmethyl sulfonyl fluoride, tosylphenylalanyl chloromethylketone, and tosyllysyl chloromethylketone) followed by treatment of the lysate with an antiserum to the protein under study. Accumulation of a precursor in vivo can be induced with uncouplers which, as will be considered below (see Section 6.3.6), interfere with the processing by hindering proper insertion of the precursor into the mitochondrial membrane. This was done for the presursors to subunit V of cytochrome *c* oxidase

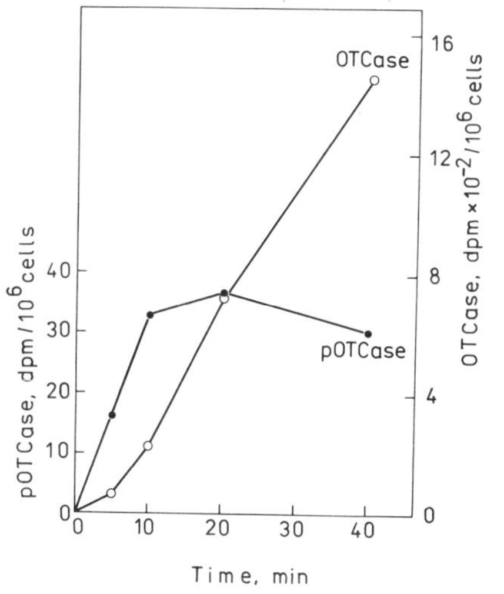

Fig. 40. Kinetics of accumulation of the precursor (pOTCase) and mature (OTCase) ornithine carbamoyltransferase (ornithine transcarbamylase) in isolated hepatocytes. After Mori et al. (1981).

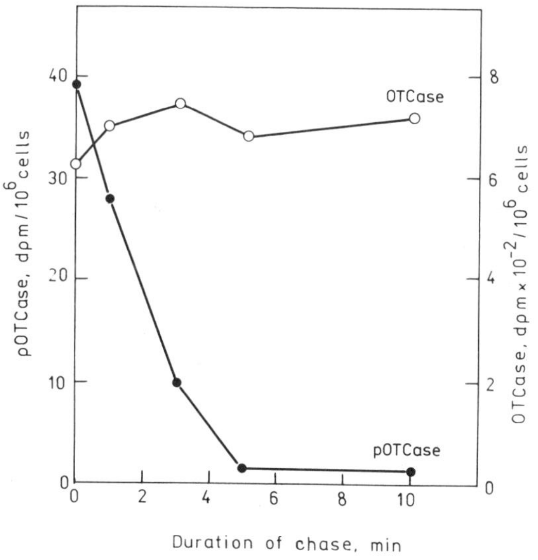

Fig. 41. Kinetics of the disappearance of ornithine transcarbamylase precursor from the cytosol in pulse-chase experiments. (●) Radioactivity of pOTCase in the cytosol; (○) radioactivity of OTCase in the mitochondrial fraction. After Mori et al. (1981).

(Lewin et al., 1980), cytochrome b_2 (Gasser et al., 1982) and manganese superoxide dismutase (Autor, 1982) in yeast and for the precursor to cytochrome c_1 in Neurospora crassa (Teintze et al., 1982). In some cases in order to observe precursors in vivo it is enough to make the labeling pulse short (Whiting, 1976; Maccecchini et al., 1979b; Raymond and Shore, 1979; Autor, 1982; Gasser et al., 1982; Mihara et al., 1982) or to lower the temperature during the pulse (Teintze et al., 1982).

Studying the processing and transport in mitochondria of the precursor of aspartate aminotransferase, Jaussi et al. (1982) found that these events take place on the background of the breakdown of the precursor in the cytosol of chicken embryo fibroblasts. Under normal conditions the half-life of the precursor in the cytosol is 30-60 sec. Addition of carbonylcyanide m-chlorophenylhydrazone to the cell culture results in accumulation of the precursor in the cytosol. As shown by pulse-chase experiments, under such conditions the protein is degraded with a half-life of about 5 min, which explains its only moderate steady-state level in the presence of the drug. These data point to an important role of intracellular proteolysis in the regulation of the content of precursors to cytoplasmically made mitochondrial proteins in the cell; precursors that for some reason have not been imported into mitochondria are rapidly degraded in the cytosol*.

6.3.5. Import and Processing of Protein Precursors in Isolated Mitochondria

More detailed information on the processing of the cytoplasmically made precursors to mitochondrial proteins came from in vitro experiments in which a cell-free translation system or its postribosomal fraction containing the labeled precursor was combined with nonlabeled isolated mitochondria, and accumulation of the mature protein in the latter was monitored by immunoprecipitation, sodium dodecyl sulfate-polyacrylamide gel electrophoresis of the immunoprecipitates, and fluorography of the gels. Such studies were carried out with the three larger subunits (α, β, and γ) of yeast F_1-ATPase (Maccecchini et al., 1979a), yeast cytochrome c peroxidase (Maccecchini et al., 1979b), ADP/ATP carrier protein (Zimmerman and Neupert, 1980), cytochrome c and H^+-ATPase ninth subunit (Zimmerman et al., 1981), subunit V of yeast cytochrome c oxidase (Lewin et al., 1981), rat liver ornithine carbamoyltransferase (Mori et al., 1980, 1981a,b; Morita et al., 1982), yeast cytochromes b_2 and c_1 (Gasser et al., 1982). In some cases to exclude the possibility of protein cotranslational insertion in the mitochondria the incubation medium contained cycloheximide, an inhibitor of cytoplasmic translation (Maccecchini et al., 1979a,b).

* See Addendum.

The criterion used to assess the incorporation of a certain protein into mitochondria was its sensitivity to exogenously added proteinases. As follows from Table 20, upon incubation of isolated yeast mitochondria with an in vitro-synthesized precursor to cytochrome b_2 50% of the labeled precursor is entrapped by mitochondria and converted to the mature protein whereas 42% of the label remains in the supernatant and is distributed between the precursor (31%) and the mature form (11%). After treatment with trypsin the label is almost wholly retained in the protein taken up by mitochondria, signifying that it is inaccessible to the proteinase; meanwhile the proteins remaining in the supernatant are completely broken down.

Similar results were obtained for the three larger F_1-ATPase subunits (Maccecchini et al., 1979a), cytochrome *c* peroxidase (Maccecchini et al., 1979b), subunit V of cytochrome *c* oxidase (Lewin et al., 1980), ornithine carbamoyltransferase (Mori et al., 1980; Morita et al., 1982), ADP/ATP carrier protein (Zimmerman and Neupert, 1980), and for subunits I, IV (cytochrome c_1), V (iron-sulfur protein), and VII of the ubiquinol-cytochrome *c* reductase complex (Teintze et al., 1982).

It is interesting to note that, unlike their precursors, the mature α, β, and γ subunits of F_1-ATPase are not incorporated into mitochondria. As shown by Maccecchini et al. (1979a), incubation of either assembled or dissociated F_1-ATPase with isolated yeast mitochondria does not protect its subunits from exogenously added proteinases.

Unfortunately it is yet too early to speak about exactly addressed incorporation of precursors to mitochondrial proteins into definite organellar compartments, to say nothing of definite oligoenzyme complexes. Mori et al. (1981b) found that ornithine carbamoyltransferase synthesized in vitro as a larger precursor is converted to the mature form by isolated liver mitochondria. The processed product was not extractable from mitochondria with 1 M KCl

Table 20. Import of Cytochrome b_2 Precursor into Isolated Mitochondria in the Presence of ATP (after Gasser et al., 1982).

	Mitochondria	Supernatant
Trypsin absent	50	42*
Trypsin present	47	1

* 31% precursor, 11% mature form; other values are the percentages constituted by the mature form, taking as 100% the total amount of the precursor precipitable from the reticulocyte lysate before incubation with mitochondria.

but could be released with digitonin. The digitonin concentration required for this was higher than that needed to liberate an intermembrane-space enzyme adenylate kinase but lower than that needed to extract the endogenous ornithine carbamoyltransferase from the mitochondrial matrix. From these data it was concluded that the precursor was imported deep into mitochondria but not ultimately into the matrix space.

On the other hand, Gasser et al. (1982), by fractionating yeast mitochondria into membranes, intermembrane contents, and matrix after completion of in vivo processing of the cytochrome b_2 precursor, found the mature product mostly in the intermembrane space which is its normal location. These authors reasoned that the import of cytochrome b_2 by isolated mitochondria is a "topologically specific process." However, the degree of such specificity is not yet clear since the experimental data available are of qualitative rather than quantitative nature*. Campbell et al. (1982) showed that the putative precursor of carbamoyl phosphate synthetase is imported into the matrix space during incubation with isolated rat fetal mitochondria. Still, up to two thirds of the added precursor remained bound to mitochondria but not imported into the matrix.

Another weak point in our notion of the import of cytoplasmically-made precursors of mitochondrial proteins into isolated mitochondria is that in none of the works was the imported and processed precursor shown either to be integrated into its "native" oligoprotein complex, as has been established for the precursor to subunit S of ribulose-1,5-biphosphate carboxylase in chloroplasts (Chua and Schmidt, 1978), or to be converted into an enzymically active protein. The only indication is the data of Campbell et al. (1982) that the import and processing of the carbamoyl phosphate synthetase precursor by isolated liver mitochondria is accompanied by a significant increase in the corresponding enzymic activity of mitochondria.

It should be noted that the import into isolated mitochondria has thus far been demonstrated only for a limited number of precursors (see Table 19). Mihara et al. (1982) failed altogether to observe translocation of precursors synthesized in a cell-free system to four mitochondrial proteins - malate dehydrogenase, L-glutamate dehydrogenase, D-β-hydroxybutyrate dehydrogenase, and sulfite oxidase - by isolated rat liver or yeast mitochondria. The first two enzymes are known to be components of the mitochondrial matrix, the third is localized in the inner membrane, and the fourth in the intermembrane space. Since all precursors are synthesized on free polysomes, they could be thought to be translocated into mitochondria posttranslationally. Probably the unsuccessful attempt at in vitro translocation was due to some flaws in the model system.

---

* See Addendum.

6.3.6. Energy Dependence of Precursor Transport into Mitochondria

The uptake of cytoplasmically made protein precursors by mitochondria appears to be an energy-dependent process. This is particularly well observable in experiments in vitro. It has already been mentioned in the preceding subsection that isolated yeast mitochondria can take up the cytochrome b_2 precursor from the postribosomal supernatant of a cell-free system programmed with total yeast mRNA and convert it to the mature form. This process only takes place in the presence of exogenously added ATP and is blocked when the medium is also supplemented with carboxyatractylate, an inhibitor of the ADP/ATP carrier. Addition of oligomycin, which blocks H^+-ATPase, or of valinomycin and K^+ ions, which collapse the membrane potential, also completely inhibits the import of the cytochrome b_2 precursor into isolated mitochondria. The polypeptide import can be supported by mitochondrial energization with oxidizable substrates, and in this case oligomycin proves ineffective. On the basis of the data listed Gasser et al. (1982) concluded that precursor import into mitochondria requires an electrochemical proton gradient across the inner mitochondrial membrane.

Analogous results were obtained in Neupert's laboratory. In particular, Zimmerman and Neupert (1980) showed that the transport of the in vitro-synthesized ADP/ATP carrier protein into isolated Neurospora crassa mitochondria, which renders it inaccessible for exogenous proteinases, is inhibited by an uncoupler, carbonylcyanide m-chlorophenylhydrazone. The same was found for the ninth subunit of H^+-ATPase (Zimmerman et al., 1981). With all this the uncoupler did not hinder precursor binding to mitochondria. Later Schleyer et al. (1982) obtained additional data from which it followed that it is the membrane potential and not intramitochondrial ATP that is necessary for the import and processing of the precursors to the ADP/ATP carrier protein and H^+-ATPase ninth subunit by isolated Neurospora crassa mitochondria. These authors demonstrated that import and processing do not take place when the potential is dissipated though the mitochondrial ATP content is high (in the presence of protonophores and valinomycin). Conversely, they proceed when the potential is sustained but the ATP level in mitochondria is low (in the presence of oligomycin and carboxyatractyloside). The same was observed (Teintze et al., 1982) with the posttranslational import of subunits I, IV, V, and VII of ubiquinol-cytochrome c reductase into isolated Neurospora crassa mitochondria. It is interesting to note that respiratory inhibitors rotenone and antimycin A only partially suppress precursor import, and cyanide has no effect at all. This is probably due to the existence of alternative electron-transfer pathways in Neurospora crassa mitochondria.

In accordance with these observations, Morita et al. (1982) found that uncouplers (carbonylcyanide m-chlorophenylhydrazone and 2,4-dinitrophenol) are inhibitory to the import-processing conversions in isolated liver mitochondria of the precursor to ornithine carbamoyltransferase synthesized in a cell-free system. However, neither antimycin A and cyanide nor oligomycin had any appreciable effect. Ono and Ito (1982) showed for another rat-liver enzyme, sulfite oxidase, that its larger precursor is neither imported into isolated organelles nor processed in the presence of carbonylcyanide m-chlorophenylhydrazone, oligomycin, or atractyloside.

As to in vivo experiments, uncouplers are known to promote the accumulation in yeast cells of precursors to cytochrome b_2 (Gasser et al., 1982), subunits V and VI of cytochrome c oxidase (Lewin et al., 1980), and manganese superoxide dismutase (Autor, 1982). Pursuing the question of what form of energy is required for the import of protein precursors into mitochondria in vivo, Nelson and Schatz (1979) found that processing of the precursor to F_1-ATPase β subunit in yeast cells is inhibited whenever oxidative phosphorylation and ATP transport into mitochondria are impaired. This was achieved by (1) rho^- mutation (which eliminates oxidative phosphorylation) plus inhibition of ATP transport into mitochondria by bongkrekic acid; (2) addition of antimycin A with bongkrekic acid into the medium; (3) op_1 mutation (which leads to the impairment of adenine nucleotide transport with oxidative phosphorylation remaining intact) plus cyanide. In the same report it was said that similar data were obtained for the precursors to α and γ subunits of F_1-ATPase and to apocytochrome c_1 and the 25,000-molecular-weight subunit of the cytochrome bc_1 complex. From these results the conclusion was derived that precursor import and processing require intramitochondrial ATP. This interpretation is in conflict with the already mentioned data of Neupert's group (see the discussion in Schleyer et al., 1982).

It must be noted that carbonylcyanide m-chlorophenylhydrazone and 2,4-dinitrophenol for some reason did not prevent the disappearance from the hepatocyte cytosol of the precursor of ornithine carbamoyltransferase (Morita et al., 1982). This illustrates the difficulties in interpreting the results of in vivo experiments with uncouplers.

According to Zimmerman et al. (1981), uncouplers interfere with the import and processing of precursors to those proteins that are integrated into the lipid bilayer of the inner mitochondrial membrane (such as the ADP/ATP carrier protein and H^+-ATPase ninth subunit), and do not affect the incorporation of cytochrome c from the outer face of this membrane. Still, the import of the precursor of another protein with similar destination, cytochrome c_1, and of the precursor of cytochrome b_2 situated in the intermembrane space of yeast mitochondria, is suppressed by uncouplers (Gasser et al., 1982). This was explained by the necessity for precursor transfer through the inner membrane into the matrix where the processor enzyme is localized (see the next subsection). In this context the situation is not quite clear with the precursor to sulfite oxidase, a rat-liver mitochondrial intermembrane-space enzyme. According to Ono and Ito

(1982), this precursor is processed on the C-side of the inner membrane of isolated mitochondria or mitoplasts, which apparently obviates the need of its transfer across this membrane. Nevertheless the processing is blocked by carbonylcyanide m-chlorophenylhydrazone and atractyloside plus cyanide.

Isolation of processor enzymes (maturases) accomplished in recent years from different sources, and their use to perform the processing of precursors to the three larger F_1-ATPase subunits and cytochrome *c* oxidase subunit V from yeast (Boehni et al., 1980), to yeast cytochromes b_2 and c_1 (Gasser et al., 1982), and to rat liver ornithine carbamoyltransferase (Conboy et al., 1982; Miura et al., 1982) has led to the conclusion that processing as such is not dependent on the transmembrane electrochemical potential. The latter consequently is needed for proper insertion and transfer of precursors through the inner mitochondrial membrane*.

From the above-said it follows that import of polypeptides into mitochondria would not depend on the transmembrane electrochemical potential if they are neither processed nor localized in the innermost mitochondrial compartment (i.e. on the M-aspect of the lipid bilayer of the inner membrane), or if the processing takes places in the intermembrane space or on the C-side of the inner mitochondrial membrane and the processed polypeptide is not further destined for the inner compartment. In addition to the already presented data for apocytochrome *c*, reference can be made to the 31,000-molecular-weight polypeptide that forms pores in the Neurospora crassa mitochondrial outer membrane for compounds of molecular weight not exceeding 6000. Freitag et al. (1982b) showed that this protein, synthesized in vivo on free cytoplasmic polysomes or in cell-free systems, has no larger precursor and is incorporated into the outer membrane of isolated mitochondria regardless of the transmembrane electrochemical potential.

Results similar to that for mitochondria were obtained in studying protein transport across the bacterial plasma membrane. In particular, the coat protein of phage M13 is known to be synthesized as a soluble precursor carrying and NH_2-terminal extension of 23 amino acid residues, which is processed and incorporated into the host-cell plasma membrane. As shown by Date et al. (1980), arsenate which exhausts the cellular pool of macroergic phosphate compounds does not affect the conversion of procoat to coat protein. At the same time, the uncoupler carbonylcyanide m-chlorophenylhydrazone blocks this conversion. The uncoupler effect is associated neither with inhibition of the so-called 'leader peptidase' (maturase) nor with impairment of nonspecific binding of the precursor to the bacterial cell membrane. From these data it was inferred that proper insertion of the procoat protein into the cell membrane to provide for its further processing implicates, at least indirectly, the electrochemical potential.

---

* See Addendum.

Later Daniels et al. (1981) studied the role of the transmembrane potential in the secretion of the leucine-binding protein into the periplasm of Escherichia coli cells and demonstrated that the processing of the precursor to this protein is abolished by carbonylcyanide m-chlorophenylhydrazone as well as by valinomycin. These authors proposed a model in which the transmembrane potential plays a paramount role in the correct orientation of the signal sequence in the membrane, which promotes the polypeptide processing and secretion (Figure 42).

The theoretical aspects of the influence of diffusion and pump potentials, as well as of "microscopic" membrane potential reflecting the surface-charge effects, on the orientation of proteins within the membrane have been considered by Weinstein et al. (1982).

6.3.7. Maturases: Properties, Multiplicity, and Localization in Mitochondria

Boehni et al. (1980) were the first to report that the processing activity for the precursors to the three larger F_1-ATPase subunits and to subunit V of cytochrome *c* oxidase from yeast could be found in a hypotonic (10 mM Tris-HCl pH 7.4) extract from yeast mitochondria. This activity was inhibited by o-phenanthroline and EDTA (but not EGTA) and persisted in the presence of phenylmethylsulfonyl fluoride, tosylphenylalanyl chloromethylketone, tosyllysyl chloromethylketone, diisopropyl fluorophosphate, and pepstatin. Similar processing activities were found in hypotonic extracts from mitochondria of rat heart and liver and maize seedlings. The extracts did not cleave bovine serum albumin, mouse immunoglobulin G, yeast hexokinase, or yeast tryptophan synthetase which has especially high sensitivity to proteinases. On the other hand, yeast proteinases A and B as well as aminopeptidase and preparations of disrupted yeast vacuoles digested the precursors to small fragments but could not convert them to mature forms. Hence it could be concluded that mitochondria contain an enzyme (or enzymes) having specific process-

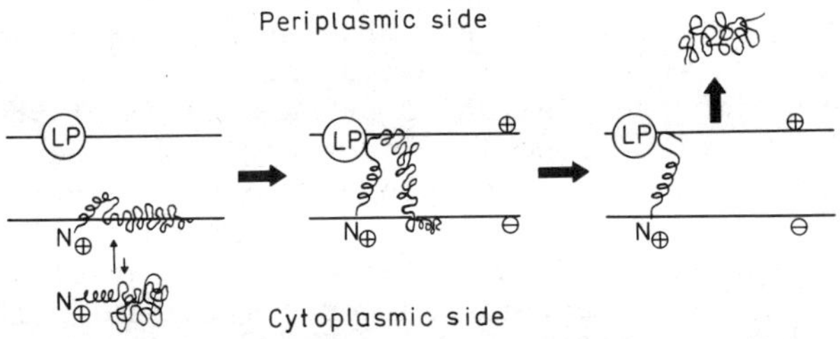

Fig. 42. A model of secretion of periplasmic proteins in Escherichia coli cells. (LP) Leader peptidase; (N+) the NH_2-terminus of the polypeptide chain. Redrawn after Daniels et al. (1981).

ing activity, i.e. ability to convert larger precursors to products having the same mobility in sodium dodecyl sulfate-polyacrylamide gel electrophoresis as the corresponding mature proteins.

A more detailed study was made (Miura et al., 1982) of the maturase processing the 39,400-molecular-weight precursor of ornithine carbamoyltransferase to a 37,000-molecular-weight polypeptide that was considered an intermediate maturation product. This protein was found along with the mature form (molecular weight 36,000) upon incubation of the _in vitro_-synthesized precursor with isolated mitochondria (Morita et al., 1982). Purification increased the specific maturase activity 140-fold. The enzyme had an apparent molecular weight of 108,000 and pI 5.5. It was not inhibited by a variety of microbial peptide proteinase inhibitors including antipain, chymostatin, pepstatin, elastatinal, bestatin, amastatin, phosphoramidon, and E-64. Inhibitors of serine proteinases were also ineffective. At the same time the maturase displayed high sensitivity to chelating reagents EDTA, o-phenanthroline, and zincon. The enzyme did not hydrolyze alkali-denatured casein and did not cleave the α-naphthyl ester derivatives of N_α-acetylglycyl-L-lysine, N_α-acetyl-L-tyrosine, L-prolyl-L-phenylalanyl-L-arginine, L-leucyl-L-phenylalanyl-L-tyrosine, and L-methionyl-L-leucyl-L-tyrosine; neither was it active with elastase substrates Congo red-elastin or succinyl-L-alanyl-L-alanyl-L-alanyl-p-nitroanilide*.

Thus, there is close resemblance between mitochondrial maturases from different sources (yeast and liver) in their sensitivity to proteinase inhibitors. The data listed allow them to be classed as metalloenzymes. These enzymes differ from the chloroplast maturase processing the precursor to subunit S of ribulose-1,5-biphosphate carboxylase (Dobberstein et al., 1977). Indeed, the chloroplast enzyme is not inhibited even with high concentrations of EDTA but is inactivated by reagents to sulfhydryl groups (Dobberstein et al., 1977), and was supposed to be a thiol proteinase.

The question arises, where in mitochondria are maturases localized? Having studied the distribution of the processing activity among rat-liver mitochondrial compartments, Boehni et al. (1980) arrived at a conclusion that the maturase resides in the matrix. Similar results were obtained by Miura et al. (1982) for the enzyme converting the ornithine carbamoyltransferase precursor to the 37,000-molecular-weight product. At the same time, Conboy et al. (1982) found this maturase not only in the matrix but also in the inner mitochondrial membrane; the other enzyme converting the precursor into the mature 36,000-molecular-weight protein had exclusively matrix location. Both processing activities displayed a dependence on Zn^{2+} ions, the second one requiring larger amounts of the activator.

---

* See Addendum.

As already mentioned in the preceding subsection, Ono and Ito (1982) studied the import and processing of another rat-liver mitochondrial enzyme sulfite oxidase. The mitochondrial processing activity proved to be concentrated in the mitoplast fraction, but neither matrix nor inner membranes could convert the sulfite oxidase precursor to the mature form. Since the inner-membrane fraction was prepared by sonicating the mitoplasts, the maturase could have been rendered inaccessible to the precursor in the inside-out submitochondrial vesicles resulting from such treatment. These data implied that the maturase is associated with the inner mitochondrial membrane and is localized on its C-side. It could not be ruled out, however, that the maturase simply had not survived mitoplast fragmentation.

Miura et al. (1982) presented data on the specificity of action of the maturase of ornithine carbamoyltransferase precursor. They found that the maturase does not process the precursor to another enzyme of rat-liver mitochondrial matrix, namely carbamoyl phosphate synthetase. An implication of these data might have been that each mitochondrial protein synthesized as a cytoplasmic precursor has its own particular maturase. However, further experiments demonstrated that this is not the case. Miura et al. (1982) observed that the precursor of ornithine carbamoyltransferase can be processed by mitochondria from tissues that do not contain this enzyme (Table 21). These results suggested that mitochondria may have transport-processing systems common for at least some groups of protein precursors. On the other hand, the processing activity in rat liver cells is associated only with mitochondria; it was detected neither in the cytosol, nor in the microsomal fraction, nor in proteinase-rich organelles accompanying mitochondria (Miura et al., 1982).

Table 21. Distribution of a Proteinase Processing the Precursor to Ornithine Carbamoyltransferase among Different Animal Tissues and Tumor Cells (taken from Miura et al., 1982).

Animal	Tissue or Tumor	Submitochondrial	Specific units/mg protein
Rat	Liver	Matrix	300
	Regenerating liver (48 h)	Matrix	350
	Kidney	Matrix	900
	Spleen	Sonicated mitoplast	270
	Heart	Sonicated mitoplast	250
	AH 13 (Yoshida ascites hepatoma cells)	Matrix	1500
	AH 130 (Yoshida ascites hepatoma cells)	Supernatant of sonicated mitochondria	1200
Mouse	Liver	Matrix	260
	Erlich ascites cells	Matrix	1600

Summarizing all that was said above, it can be concluded that the multiplicity of maturases in mitochondria remains an open question. Evidence is available on the one hand for the existence of universal maturases (Boehni et al., 1980; Miura et al., 1982), and on the other hand for the selectivity of maturase action (Miura et al., 1982). Though the comparison may seem only superficial, it can be mentioned that the Escherichia coli leader peptidase performing the transformation of phage-M13 procoat (cytoplasmic precursor) into the coat protein (integral transmembrane protein) can also cleave the precursors to Escherichia coli periplasmic and outer-membrane proteins (Date and Wickner, 1981). This is probably due to the universal structure of the signal sequences in the protein precursors (see, e.g., Inouye and Halegoua, 1980). On the other hand, Tokugana et al. (1982) demonstrated that the Escherichia coli signal peptides for the prolipoprotein is distinct from the M-13 procoat signal peptidase.

In all the works cited above the mitochondrial maturases were termed proteinases. The reason for this was that they converted precursors with lower mobility in sodium dodecyl sulfate-polyacrylamide gel electrophoresis into mature forms with higher mobility. With all this, in none of the cases the cleavage of amino acid extensions during processing was directly demonstrated. It was reported though (Lewin et al., 1980) that during maturation α and β subunits of F_1-ATPase lose the label incorporated into their NH_2-terminus in the form of N-formyl[^{35}S]methionine. This result was claimed to prove the removal of the signal NH_2-terminal sequence of 2000-6000 daltons. It is known, however, that both in prokaryotes and eukaryotes the N-formylmethionine residue can be selectively cleaved off by a special aminopeptidase but performs no signal function (Chapeville and Haenni, 1974).

Moreover, cases are known when the change (specifically, increase) in the protein electrophoretic mobility was not associated with the alteration of its molecular weight. An example of this is the conversions of Escherichia coli nitrogenase (MacGregor and McElhaney, 1981). Subunit B of this enzyme is synthesized as a B' precursor which has lower electrophoretic mobility than the mature counterpart. Immediately upon insertion into the cytoplasmic membrane the B' form is converted by posttranslational processes into the B form, and this conversion, albeit inhibited by some proteinase inhibitors (benzamidine and tosyllysyl chloromethylketone), is not proteolytic. The increase in electrophoretic mobility in this instance results rather from modification of amino acid residues.

Obviously, the proteolytic nature of the processing of the cytoplasmically synthesized precursors to mitochondrial proteins calls for more convincing proof.

6.3.8. Stages of Processing of Cytoplasmically Made Precursors to Mitochondrial Proteins

It was already pointed out in the preceding subsection that the conversion of the precursor of ornithine carbamoyltransferase into the mature 36,000-molecular-weight enzyme proceeds through an intermediate 37,000-molecular-weight product*, the two stages being probably performed by two distinct maturases, one of which is found in the inner membrane and in the matrix and the other only in the matrix.

According to Gasser et al. (1982), the chelator-sensitive maturase from yeast mitochondrial matrix likewise converts the cytochrome b_2 precursor of molecular weight 68,000 not into the 58,000-molecular weight mature form but into a 65,000-molecular-weight intermediate. Similarly, the precursor of cytochrome c_1 (molecular weight 35,000) is first processed by the maturase to a 34,000-molecular-weight product.

This step-wise behavior is not an experimental artifact, because in some circumstances all the three forms of cytochrome b_2 or cytochrome c_1 are observable in yeast cells. The intermediate form is the only species of the cytochrome c_1 polypeptide that is accumulated in a heme-deficient yeast mutant. From this it follows in particular that the second stage of maturation is a heme-requiring event.

The data mentioned indicate that maturation of the cytoplasmically made precursors to some mitochondrial proteins is likely to be complexly organized topographically. Gasser et al. (1982) supposed (see Figure 43) that in the first step the cytochrome b_2 and c_1 precursors are translocated in an energy-dependent manner through the inner mitochondrial membrane and, while being still bound to it, lose a certain amino acid sequence under the action of a chelator-sensitive matrix proteinase. With cytochrome b_2, its intermediate following the second maturation step is liberated into the intermembrane space irrespective of the availability of the heme. As to cytochrome c_1, the intermediate form first binds the heme and only then undergoes final processing so that the resulting product remains membrane-bound. It is not surprising therefore that maturation of the cytochrome c_1 precursor takes much more time than its translation. The pulse-chase experiments of Gasser et al. (1982) showed that the time of its complete conversion to the mature holocytochrome c_1 in yeast cells exceeds 20 min (Figure 44)**.

Another variant of the spatial transpositions of precursors in the course of their maturation in mitochondria is offered by ornithine aminotransferase. According to the data of Kolansky et al. (1982), its larger precursor is first converted into a form of intermediate size, probably on the C-side of the inner mitochondrial

---

*,** See Addendum.

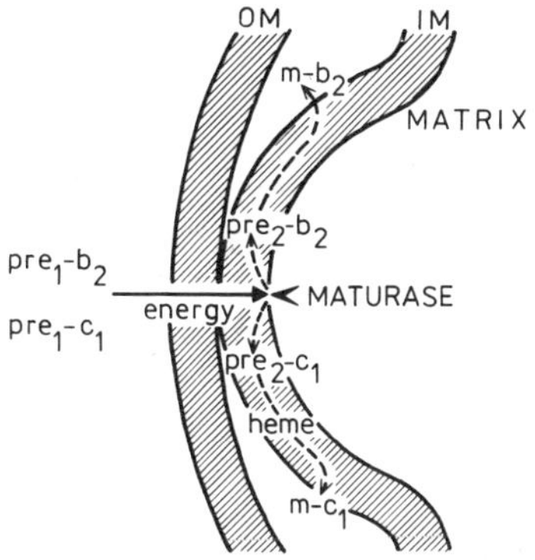

Fig. 43. Stages of import of the precursors to cytochrome b_2 and cytochrome c_1 into yeast mitochondria. (pre$_1$-b$_2$) The 68,000-molecular-weight precursor of cytochrome b_2; (pre$_2$-b$_2$) the 65,000-molecular-weight intermediate form of cytochrome b_2; (m-b$_2$) the 58,000-molecular-weight mature cytochrome b_2; (pre$_1$-c$_1$) the 35,000-molecular-weight precursor of cytochrome c_1; (pre$_2$-c$_1$) the 34,000-molecular-weight intermediate form of cytochrome c_1; (m-c$_1$) the 31,000-molecular-weight mature cytochrome c_1. (OM) Outer mitochondrial membrane; (IM) inner mitochondrial membrane. After Gasser et al. (1982).

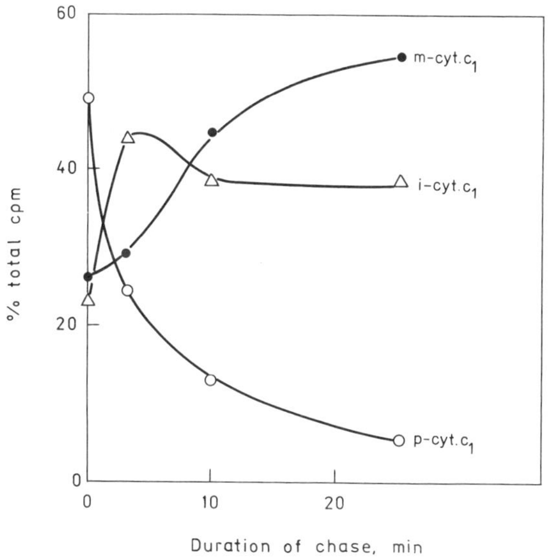

Fig. 44. Changes in the content of the precursor (○), intermediate (△), and mature (●) forms of cytochrome c_1 in wild-type yeast after pulse of [^{3}H]leucine. After Gasser et al. (1982).

membrane. This step is not suppressed by respiratory inhibitors or uncouplers. The intermediate is further processed to the mature form and transferred to the matrix. The second step is sensitive to compounds that collapse the membrane electrochemical potential; it is supposed to take place in the matrix (or at the inner membrane-matrix interface) and to be performed by a Zn^{2+}-dependent proteinase.

It is already obvious from the data presented that the role of heme in the binding by mitochondria of different cytoplasmically made polypeptides is not universal. The cytochrome b_2 polypeptide is imported and processed regardless of the heme, whereas the cytochrome c_1 polypeptide in the absence of the heme is only partially processed. As to apocytochrome *c*, its transfer across the outer mitochondrial membrane is somehow coupled with covalent heme attachment; the transfer is blocked if heme attachment is competitively inhibited with deuterohemin (Hennig and Neupert, 1981).

6.3.9. Receptor Sites Involved in the Transport of Cytoplasmically Made Proteins Across Mitochondrial Membranes

The addressed posttranslational transfer of cytoplasmically produced polypeptides into mitochondria implies the existence, on the one hand, of specific recognition sites in the polypeptide chains, and on the other hand, of corresponding receptor sites on the mitochondrial surface. Three variants of such transport are theoretically possible. First, the transport-processing system in the mitochondrial membranes may be universal for all imported polypeptides; then the signal or recognition regions of the polypeptide chains must have some universal features inherent only in the entities destined for mitochondria. Second, mitochondria may have transport-processing systems common for groups of polypeptides. Third, each imported polypeptide may have its own receptor site on the membrane.

As mentioned before, the ornithine carbamoyltransferase precursor can be processed in rat liver mitochondria by a system common to at least a group of polypeptides (Miura et al., 1982). On the other hand, Zimmerman et al. (1981) showed that the pathways of mitochondrial import are different for the ADP/ATP carrier protein or the ninth subunit of H^+-ATPase and for cytochrome *c*. According to these authors, a large (10^4-fold) excess of apocytochrome *c* does not interfere with the transport into Neurospora crassa mitochondria of in vitro-synthesized precursors to the ADP/ATP carrier protein or H^+-ATPase ninth subunit. Later Teintze et al. (1982) obtained analogous results while studying the influence of cytochrome *c* on the incorporation of the cytochrome c_1 precursor into isolated Neurospora crassa mitochondria. This was indicative of the existence of distinct receptors on the mitochondrial surface responsible for the specific recognition of the precursors at the first transport step.

The mechanism by which mitochondria recognize their cytoplasmically made polypeptides remains the most enigmatic issue in the scope of transport-processing events. The solution of this problem obviously awaits elaboration of techniques for isolating the receptor sites, as is already being done in the studies into protein transport across the membranes of the endoplasmic reticulum. Several years ago Walter and Blobel (1980) reported on the isolation from rough microsomal membranes of a protein complex comprising six polypeptides that is necessary for the translocation of secretory proteins. Further investigations revealed that this complex (signal recognition particle, SRP) also containing 7 SL RNA (Walter and Blobel, 1982) is capable of binding with *in vitro*-assembled polysomes synthesizing a secretory protein (Walter and Blobel, 1981a; Walter et al., 1981) and thereby imposing an arrest on chain elongation (Walter and Blobel, 1981b) which is released upon interaction of the SRP-polysome complex with the microsomal membrane. Later it was found (Gilmore et al., 1982a,b) that this is due to the presence in the membrane of a special receptor for SRP, which is a 72,000-molecular-weight polypeptide. These data imply that in the cytoplasm SRP interacts with the protein-producing polysome through the signal sequence of the growing polypeptide chain and stops its elongation. Then this complex binds with the SRP receptor in the endoplasmic reticular membrane, which is accompanied by the release of the arrest on elongation. The polypeptide chain resumes growing and traverses the membrane lipid bilayer while SRP and the receptor are freed to arrange the next step of transport.

6.3.10. Mixed Co-and-Posttranslational Mechanism for the Transfer of Cytoplasmic Polypeptides Across Mitochondrial Membranes

Ades and Butow (1980b) tried to reconcile their data with the work of other authors on the posttranslational transfer of cytoplasmically made polypeptides across mitochondrial membranes by suggesting that a mixed transfer mechanism may be operative, i.e. that both nascent and completed polypeptide chains may be transferred. The relation between these pathways would be determined, on the one hand, by the number of free binding sites for polypeptide chains on the mitochondrial membranes, and on the other hand, by the total number of initiated polysomes. If the latter are in large excess then the translation would take place mostly on free polysomes, which would increase the contribution of posttranslational transfer*.

Such a dual nature of transmembrane polypeptide transfer has been shown for bacteria. The *Escherichia coli* maltose-binding protein (molecular weight 38,500) localized in the cell periplasmic space is known to be initially synthesized in the cytoplasm as a larger precursor (molecular weight 41,000). Josefson and Randall (1981) established that the precursor processing *in vivo* occurs simultaneously both cotranslationally and posttranslationally, the

* See Addendum.

contribution of the former mechanism in their hands being only about 30%. It is interesting to note that processing begins only after the polypeptide reaches a critical size of 33,000 daltons.

According to Garwin and Beckwith (1982), the D-ribose-binding protein localized in Escherichia coli periplasm and synthesized in the cytoplasm as a precursor having an extension of 2000-3000 daltons can be processed to the mature form both cotranslationally and posttranslationally. The relation between these two processes is regulated by the rate of protein secretion which can be varied by special mutations.

6.4. INCORPORATION OF MITOCHONDRIAL TRANSLATION PRODUCTS INTO THE INNER MITOCHONDRIAL MEMBRANE

As indicated in the preceding chapter, up to 21 polypeptides in yeast mitochondria (Douglas and Butow, 1976) and up to 26 polypeptides in HeLa mitochondria (Ching and Attardi, 1982) are found to be synthesized by mitochondrial polysomes. Only one of these, the S-4a polypeptide (Lambowitz et al., 1976) or var1 (Terpstra et al., 1979) of the small mitochondrial ribosomal subunit of Neurospora crassa or Saccharomyces cerevisiae, respectively, is not a membrane protein. All the other entities are in all probability incorporated into the inner mitochondrial membrane. The intramembrane positions are so far known only for several polypeptides. It was established that subunit I of cytochrome *c* oxidase is deeply embedded into the lipid bilayer, subunit III of the same enzyme and cytochrome *b* span the membrane to be exposed on both its sides, subunit II of cytochrome *c* oxidase is on the C-side of the inner mitochondrial membrane (see Chapter One). Such positions of these polypeptides allow of their cotranslational transfer through the membrane lipid bilayer. On the other hand, it could be seen in the preceding sections of this chapter that polypeptide transfer across and insertion into mitochondrial membranes can occur posttranslationally (see, e.g., the data on the three larger F_1-ATPase subunits, apocytochrome c_1, and cytochrome *c* oxidase subunits V and VI). In this case the membrane protein is synthesized on free polysomes as a water-soluble precursor, enters the cytosolic pool, and only then is imported into the membrane, being concomitantly processed.

In order to find out how mitochondrial translation products are incorporated into the membrane it is necessary to ascertain on which polysomes - free or membrane-bound - their translation takes place, whether there are soluble precursors to these polypeptides, whether the mature forms retain their aminoterminal N-formylmethionine, and the like. Prior to considering these questions, let us briefly stop to review some properties of mitochondrial translation products.

6.4.1. Properties of Intramitochondrially Made Polypeptides

First of all it must be pointed out that the mitochondrial translation products are highly hydrophobic. They are extractable by 80-90% with a chloroform-methanol mixture from mitochondria of rat liver (Dianoux et al., 1976), Neurospora crassa (Rowe et al., 1974), and yeast (Tzagoloff and Akai, 1972, though cases are known when the extraction was not so thorough (Kadenbach and Hadváry, 1970; Burke and Beattie, 1974; Kužela et al., 1975). Amino acid analysis of many intramitochondrially made polypeptides (subunits I, II, and III of cytochrome *c* oxidase, ninth subunit of H^+-ATPase, and others; see Chapter One) reveals a high content of hydrophobic amino acids (Hadváry and Kadenbach, 1973; Dianoux et al., 1976).

The mitochondrial translation products show pronounced avidity for fatty acids, which can be removed only by prolonged treatment with 0.5 N NaOH in methanol (Dianoux et al., 1976), and for phosphatides (Kadenbach, 1971).

The high hydrophobicity of polypeptides associated with mitochondrial ribosomes is the cause of ribosome aggregation in vitro (Michel and Neupert, 1973). Ribosomes can only be disaggregated on addition of puromycin which releases the nascent polypeptide chains. The polypeptides themselves, either free or in the form of peptidyl-tRNA or peptidylpuromycin, can be maintained in solution only in the presence of detergents such as sodium dodecyl sulfate.

There are grounds for thinking that, being incorporated into the inner mitochondrial membrane, the intramitochondrially made polypeptides undergo modification manifesting itself specifically as a decrease in their electrophoretic mobility in polyacrylamide gels. As shown by Michel and Neupert (1973) in experiments with Neurospora crassa, [^{3}H]leucine pulse-incorporated into low-molecular-weight mitochondrial translation products is transferred during the chase into the fraction of higher-molecular-weight polypeptides. Analogous results were obtained by Kužela et al. (1975) who incubated in vitro-labeled low-molecular-weight mitochondrial polypeptides (extracted with chloroform-methanol) with nonlabeled rat liver mitochondria in the presence of a mitochondrial-translation inhibitor chloramphenicol. In this case the label content decreased in low-molecular- and increased in high-molecular-weight proteins of both the mitochondrial membrane fraction and postmitochondrial supernatant. The migration of the label was considered to result from the interaction of low-molecular-weight hydrophobic mitochondrial translation products with other mitochondrial proteins. It is interesting to note that the in vitro conversion of the primary mitochondrial translation products requires ATP (Kužela et al., 1975; Hofmann and Kadenbach, 1976). However, as long as the physiological significance of such conversions is not recognized, it cannot be ruled out that they are just artifacts of the experimental procedure.

6.4.2. Localization of Mitochondrial Ribosomes

The evidence on the location of mitochondrial ribosomes operative in translation is contradictory. Michel and Neupert (1973) reported that upon sonication of *Neurospora crassa* mitochondria or subjecting them to 15 cycles of freezing-thawing a considerable portion of the ribosomes (not less than 30%) is found in the free state. Assuming that these ribosomes participate in translation, and taking into account the high hydrophobicity of nascent mitochondrial polypeptides, Michel and Neupert suggested that completed polypeptide chains may be transported to the membrane in the form of peptidyl-tRNA associated with monomeric ribosomes. After cleavage of the ester bond the polypeptide is inserted into the membrane and the ribosome returns to the ribosomal pool.

In contrast to these authors, Kuriyama et al. (1973b) showed that the ribosomes engaged in protein synthesis are bound to the inner mitochondrial membrane. Their attachment depended on the presence of nascent polypeptide chains, since the ribosomes were liberated as subunits only upon treating submitochondrial vesicles with puromycin. At the same time, free ribosomes released on mitochondrial fragmentation were found to be inactive in protein synthesis.

Ribosomes tightly associated with mitochondrial membranes were observed by several researchers (Bunn et al., 1970; Green and Watson, 1972; Rogers and Küntzel, 1977). This was in indirect indication of protein synthesis taking place on the very surface of the inner organellar membrane. A more reliable demonstration of the issue was furnished by the studies of Marzuki et al. (1975b) on the temperature dependence of protein synthesis in isolated yeast mitochondra and the influence of unsaturated fatty acids on the assembly and function of the mitochondrial protein-synthesizing system. The Arrhenius plots for most membrane-associated enzymes are known to have several regions with different slopes. An inflection on such a plot is held to be indicative of a phase transition in the lipid bilayer: the phase transition may induce conformational changes in membrane enzymes which are accompanied by the changes in the activation energy of the reaction they perform.

As shown by Marzuki et al. (1975b), the temperature dependence of the protein synthesis in isolated yeast mitochondria containing 45% of unsaturated fatty acids in their phospholipids has the same pattern as those of oligomycin-sensitive ATP hydrolysis or NADH oxidation: in all cases the inflection points lie in a narrow range of 21-24°C. Moreover, the inflection points for all three processes shifted similarly with varying content of unsaturated fatty acids. Basing on this it was concluded that the ribosomes carrying out mitochondrial translation share the same lipid environment with ATPase and respiratory-chain components involved in the rate-limiting step in the NADH oxidase reaction. Thus, there were serious grounds for believing that mitochondrial protein synthesis proceeds on polysomes bound to the inner mitochondrial membrane.

Additional support to this is offered by the data on specific mutations in yeast mitochondrial DNA that enhance the resistance of mitochondrial protein synthesis in vivo to antibiotic spiramycin, carbomycin, and linkamycin (see, e.g., Spithill et al., 1978). Such mutations are not necessarily accompanied by the changes in the antibiotic sensitivity of mitochondrial ribosomes as such. Collation of these two facts leads to the idea of a "functional" linking of ribosomes with mitochondrial membranes.

Summing up, there is quite a body of experimental evidence for at least a portion of intramitochondrial polypeptides being made on the ribosomes bound to the inner mitochondrial membrane. This in its turn allows of cotranslational insertion of growing polypeptides into the membrane.

6.4.3. Intramitochondrially Made Precursors to Some Inner-Membrane Proteins

The existence of intramitochondrially made precursors to some proteins was initially inferred from the analysis of NH_2-terminal sequences of their mature counterparts. Machleidt and Werner (1979) found that subunit II of Neurospora crassa cytochrome *c* oxidase does not begin with N-formylmethionine but has an "open" sequence with an NH_2-terminal aspartic acid. This implied that subunit II is made as a precursor having an extension of at least a single N-formylmethionine residue and that this precursor undergoes proteolytic processing. Using their own technique for isolating free (i.e. not integrated into the complex) subunit II, Machleidt and Werner (1979) observed that it does not differ in molecular weight and position of some amino acids at the NH_2-terminus from the mature subunit II of the holoenzyme. From this it followed that processing takes place early in the subunit formation, probably cotranslationally. At any rate the precursor pool was so small that it could not be detected in that work.

This was later accomplished by Sevarino and Poyton (1980) with yeast mitochondria. It turned out that by suppressing mitochondrial translation in isolated organelles with aurintricarboxylic acid (a translation initiation inhibitor) some time after the addition of [^{35}S]methionine, a precursor to subunit II of cytochrome *c* oxidase can be accumulated. The kinetics of the mitochondrial polypeptide labeling with this amino acid in the presence of the inhibitor indicated that the label was incorporated into polypeptides that were already being made on the polysomes by the moment of inhibitor administration. The precursor was precipitable from mitochondrial lysates with an antiserum to the mature subunit, was larger by 1500 daltons by the data of sodium dodecyl sulfate-polyacrylamide gel electrophoresis, and showed proteolytic fingerprints rather similar to those of the mature protein. On the basis of some indirect observations Sevarino and Poyton interpreted precursor accumulation as the result of inhibition by aurintricarboxylic acid of a mitochondrial processing proteinase. Having studied the kinetics of precursor accumulation, they came to the conclusion that the precursor has an extra sequence on the NH_2-terminus and it is cleaved off cotranslationally. It is of interest that upon fragmentation of mitochondria treated with aurintricarboxylic acid the precursor is found only in the membrane fraction. This means that the unprocessed precursor is incorporated into and accumulated in the inner mitochondrial membrane. The complete absence of the precursor from soluble mitochondrial polypeptides was a further indication of its cotranslational insertion into the membrane.

The existence of a larger precursor to subunit II was confirmed by De Ronde et al. (1980) who carried out its synthesis in a heterologous system consisting of wheat-germ extract, partially purified mitochondrial mRNA for this protein, and the UGA-suppressor tRNA from Schizosaccharomyces pombe.

The most intriguing aspect of the work of Sevarino and Poyton (1980) was that aurintricarboxylic acid induced the appearance of only one additional polypeptide among the mitochondrial translation products. The authors offered two explanations for this fact. Firstly, it was admitted that aurintricarboxylic acid may selectively inhibit the processor proteinase specific for cytochrome *c* oxidase subunit II and not affect proteinases processing other protein precursors. Secondly, it was not excluded that other mitochondrially synthesized proteins have no precursors at all. Indeed, in some of these proteins the initiating amino acid, N-formylmethionine, was found at the NH_2-terminus. Such data are available for subunits I (Buse et al., 1980) and II (Steffens and Buse, 1979) of beef heart cytochrome *c* oxidase. Furthermore, according to Darley-Usmar and Fuller (1981), it is hardly probable that subunits I and II of beef heart cytochrome *c* oxidase could be synthesized as precursors with extensions at their NH_2- or COOH-termini that are subsequently cleaved off. Though in sodium dodecylsulfate-polyacrylamide gel electrophoresis the molecular weights for subunits I and III are estimated at 35,700 and 21,000, respectively, whereas the values calculated from their DNA sequences are 57,057 and 29,945, determination of the positions of characteristic amino acids at both termini shows exact correspondence between the coding sequences and the products. Thus there are reasons for thinking that in beef heart mitochondria subunits I and III of cytochrome *c* oxidase do not undergo proteolytic processing.

Quite another situation was observed with Neurospora crassa mitochondria. Van't Sant et al. (1981) found that among the mitochondrial translation products the 44,000-molecular-weight polypeptide (which most probably is the S-5 ribosomal protein), apocytochrome *b*, and a 8000-molecular-weight polypeptide retain their initial NH_2-terminus carrying the N-formylmethionyl residue. At the same time, subunits I and II of cytochrome *c* oxidase as well as several polypeptides of molecular weights 10,000-15,000 do not preserve the [^{14}C]formiate label and are consequently synthesized as larger precursors.

It proved possible to detect such precursors upon prolonged (120 min) labeling of the cells in the presence of cycloheximide or at temperature lowered to 9°C: under such conditions polypeptides with molecular weights of 45,000, 36,000, and 25,000 appeared in addition to the usual set. The first of them was identified as the precursor to cytochrome *c* oxidase subunit I. The other two were regarded as precursors to cytochrome *c* oxidase subunits II and III (or the 19,000-molecular-weight subunit of H^+-ATPase), respectively. The authors of the work cited supposed that the conditions they created slowed down the precursor processing. In particular, prolonged cell incubation with cycloheximide could have decreased the content of the processing enzymes if they had high turnover rates. In usual conditions the precursors could not be detected because of their exceptionally short half-lives. As the kinetics of precursor disappearance has not been studied in pulse-chase experiments, the exact half-lives are unknown; a rough estimate showed that they do not exceed 1 min.

Unexpected results were obtained when studying the localization of the precursor to subunit I of cytochrome *c* oxidase in Neurospora crassa cells. Since upon cell disruption the larger portion (50-80%) of the 45,000-molecular-weight polypeptide was recovered in the postmitochondrial supernatant, it was suggested that the precursor is either weakly or not at all bound to the membrane and is released from mitochondria during the experimental procedure. This also applied to the other two polypeptides (molecular weights of 36,000 and 25,000). Van't Sant et al. (1981) saw in such a behavior of intramitochondrially made proteins a resemblance to a number of cytoplasmically made precursors, which at first exist in a soluble state and then are converted to the membrane-bound form [see, e.g., the data for the precursors to the ADP/ATP carrier protein (Zimmerman and Neupert, 1980) and to subunit V of cytochrome *c* oxidase (Neupert and Schatz, 1981)]. The extra sequences carried by the 45,000-, 36,000-, and 25,000-molecular-weight polypeptides may serve as signals for their insertion into the inner mitochondrial membrane. It has to be admitted, however, that this assertion is in conflict with the data of Sevarino and Poyton (1980) according to which the precursor to subunit II of cytochrome *c* oxidase synthesized on mitochondrial ribosomes is found exclusively in the mitochondrial membrane fraction.

Another case is known when accumulation of the precursor to subunit I of cytochrome *c* oxidase could be observed. It is described by Werner and Bertrand (1979) who showed that this precursor of molecular weight 45,000 (versus 41,000 for mature subunit I) is accumulated in the cytoplasmic mi-3 mutant of Neurospora crassa without forming a complex with other subunits. When functional cytochrome *c* oxidase is induced in the mutant cells with antimycin A, the 45,000-molecular-weight polypeptide is converted to the 41,000-molecular-weight component which turns out to be integrated into the complex with other subunits.

Thus there seems to be no unified mechanism for the insertion of polypeptides synthesized on mitoribosomes into the inner mitochondrial membrane. Data are available on the existence of precursors to some polypeptides which can be processed and incorporated into the membrane cotranslationally (Sevarino and Poyton, 1980) as well as posttranslationally (Werner and Bertrand, 1979; Van't Sant et al., 1981). A number of polypeptides either do not have larger precursors (Darley-Usmar and Fuller, 1981) or do not carry NH_2-terminal extensions (Van't Sant et al., 1981). It is not yet known which polysomes, free or membrane-bound, make particular polypeptide chains. Virtually uninvestigated are the maturases, the topography of precursor processing and its energy requirements - questions already well studied for some cytoplasmically synthesized mitochondrial constituents.

6.5. ASSEMBLY OF ENZYMIC COMPLEXES IN THE INNER MITOCHONDRIAL MEMBRANE

Chapter One summarized the evidence for the existence in the inner mitochondrial membrane of a number of enzymic complexes comprising from 2 to 16 polypeptides. As could be seen from this and the preceding chapter, some of the complexes, namely respiratory-chain complexes III and IV and H^+-ATPase, are assembled from polypeptides produced by two spatially separated protein-synthesizing systems, mitochondrial and cytoplasmic. Others are built exclusively of cytoplasmically made polypeptides. Taking into account what is known about the mechanisms of protein transport, it can be supposed that in any event the complexes are assembled within the inner membrane itself.

The experimental data available at present demonstrate that after translation is completed most components of the complexes undergo various conversions, being subject to processing of probably proteolytic nature (see Table 19), binding prosthetic groups covalently (apocytochrome *c*, apocytochrome c_1) or noncovalently (apocytochrome *b*, subunits I and II of cytochrome *c* oxidase), or simply changing their conformation (ADP/ATP carrier protein). In view of all this, the early concepts of the assembly mechanisms, according to which mitochondrial translation products condensate with each other to form "templates" on which the ready-for-use cytoplasmically made proteins are arranged (Green and Hechter, 1965; Tzagoloff, 1972), look rather naïve today.

6.5.1. **Free Subunit Pools for the Complexes in the Inner Mitochondrial Membrane**

A study of the labeling kinetics of mitochondrial membranes and cytochrome *c* oxidase in Neurospora Crassa revealed a substantial disparity in the rates of these processes (Schwab et al., 1972). The labeling is half-maximal for the membrane protein already in 3 min while for cytochrome *c* oxidase only in about 20 min (Figure 45). This in principle may signify that the enzyme is assembled from pools of pre-accumulated polypeptides, though more explicit would have been data on the kinetics of the appearance of individual enzyme subunits in the membrane fraction. Figure 46 depicts the kinetics of incorporation of labeled cytochrome *c* oxidase subunits into the holoenzyme. As follows from these data, the intramitochondrially made 18,000-molecular-weight subunit enters the assembly with only a short delay after its appearance in the membrane, and its membrane pool is consequently very small. On the other hand, the cytoplasmically made 11,000- and 13,000-molecular-weight subunits as well as the 36,000-molecular-weight intramitochondrial subunit exist for a long time in the membrane in an unassembled form. Table 22 presents the half-lives of free individual subunits and the sizes of free subunit pools determined by Schwab and colleagues from these data.

Unfortunately, little is known to date about the rates of conversion of precursors to the mature cytochrome *c* oxidase subunits. According to the data available, these rates are so high that the

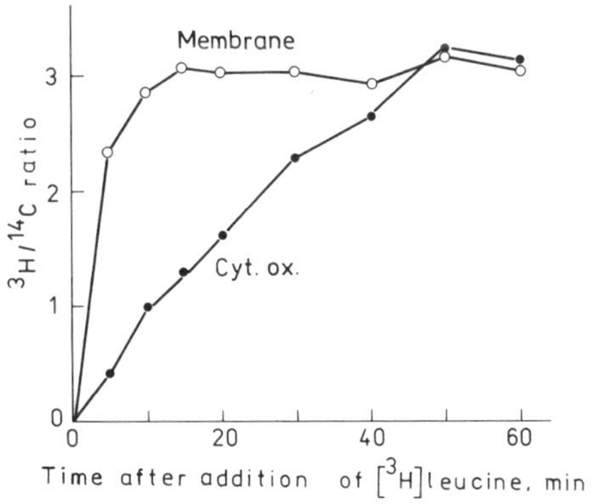

Fig. 45. Labeling kinetics of the mitochondrial inner-membrane proteins and cytochrome *c* oxidase in Neurospora crassa cells. After Schwab et al. (1972).

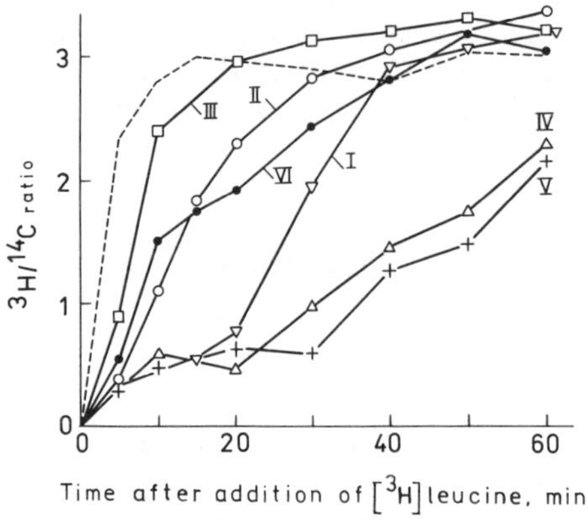

Fig. 46. Labeling kinetics of individual subunits of cytochrome *c* oxidase in Neurospora crassa. Numbering of subunits is as in Table 22. The dashed line corresponds to the labeling of the mitochondrial protein. After Schwab et al. (1972).

mitochondrially made precursors to subunits I, II, and III can be observed in Neurospora crassa cells only when their processing is artificially retarded (Van't Sant et al., 1981). This also applies to the precursor of subunit II (Sevarino and Poyton, 1980) and the cytoplasmically made precursors of subunits V and VI of yeast cytochrome *c* oxidase (Lewin et al., 1980). From this it is obvious that precursor processing and incorporation of subunits into the inner mitochondrial membrane are much faster than their assembly into the holoenzyme.

Table 22. Half-Lives and Sizes of Precursor Pools of Cytochrome *c* Oxidase Polypeptides (from Schwab et al., 1972). The pool sizes are given as a percentage of the amount of the corresponding polypeptide present in cytochrome *c* oxidase.

Apparent mol.wt.	Subunit No.*	Half-life of precursor,min	Pool size of precursor
8,000	VI	7	4
11,000	V	47	25
13,000	IV	37	20
18,000	III	3.5	2
28,000	II	10	5
36,000	I	23	10

* The notation is arbitrary and does not necessarily correspond to that in Table 1; it is introduced to correlate these data with the data of Fig. 46.

Similar results were obtained later by Hundt et al. (1980) who studied cytochrome *c* oxidase biosynthesis in isolated hepatocytes. The cytoplasmically made subunits, IV, V, VI, and VII were shown to be incorporated into cytochrome *c* oxidase somewhat more slowly than the intramitochondrially made subunits I, II, and III. These data were interpreted to indicate the existence of large pools of the cytoplasmic cytochrome *c* oxidase constituents.

The recent work of Wielburski et al. (1982) directly demonstrates that during a 1-h pulse of [^{35}]methionine the labeled subunit I of cytochrome *c* oxidase is accumulated in the inner mitochondrial membrane but is not integrated into the holoenzyme isolated then from mitochondria by different methods; it is transferred into the complex only in 2 h of the chase. Subunits II and III, on the contrary, are incorporated into the holoenzyme already within 10 min of the pulse.

With the use of inhibitors of cytoplasmic and mitochondrial translation, unassembled products of mitochondrial or cytoplasmic translation can be artificially accumulated in mitochondria. According to Sebald et al. (1972), when Neurospora crassa cells are incubated with labeled amino acids in the presence of chloramphenicol and cytochrome *c* oxidase is subsequently isolated, the label is found neither in its three larger subunits made on mitochondrial polysomes not in its four smaller subunits which are cytoplasmic products. Nevertheless, labeled subunits IV-VII are present in the mitochondrial membranes and when the label and chloramphenicol are removed from the medium they enter the cytochrome *c* oxidase fraction as the larger subunits are synthesized (Figure 47). Analogous results were obtained by selectively labeling subunits I-III of the enzyme in the presence of cycloheximide; appreciable quantities of these subunits were found in the isolated complex only if the labeling was preceded by treating the cells with chloramphenicol which favored accumulation of the cytoplasmically made enzyme subunits (Sebald et al., 1973).

A still more dramatic illustration of the existence of free cytoplasmic subunits of cytochrome *c* oxidase in mitochondria is the work of Poyton and Groot (1975) where subunits I, II, and III synthesized by isolated yeast mitochondria were shown to be bound with the preexisting subunits V and VII since they could be precipitated with antisera to the latter from mitochondrial lysates.

Several authors have demonstrated the assembly of enzymically active respiratory-chain components from preaccumulated cytoplasmic and mitochondrial translation products. Rouslin and Schatz (1969) showed that cycloheximide suppresses the development of respiration during the respiratory adaptation of anaerobically grown yeast and that respiration develops after the yeast are transferred to a medium free of cycloheximide but containing chloramphenicol (Figure 48). This may mean that yeast incubation with cycloheximide entails accumulation of mitochondrially synthesized polypeptides which are

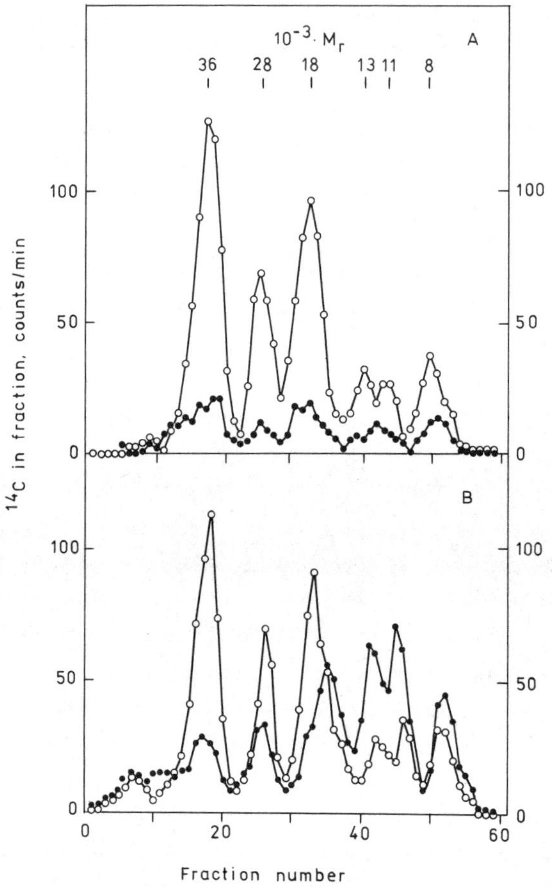

Fig. 47. Binding of the cytoplasmic subunits of cytochrome *c* oxidase with mitochondrial membranes under conditions of blocked mitochondrial protein synthetsis in Neurospora crassa. (o) Radioactivity of [^{14}C]amino acids in cytochrome oxidase after incubation without chloramphenicol (control); (•) radioactivity of [^{3}H]amino acids in cytochrome oxidase upon incubation in the presence of chloramphenicol (A) and in 1 h of culture growth after removal of chloramphenicol and label (B). After Sebald et al. (1972).

only integrated into respiratory assemblies after the removal of the cycloheximide block and the appearance of the cytoplasmic translation products required for such assembly.

The question of the pools of respiratory-chain precursors was examined in more detail by Chen and Charalampous (1969). These authors found that during the respiratory adaptation of anaerobically

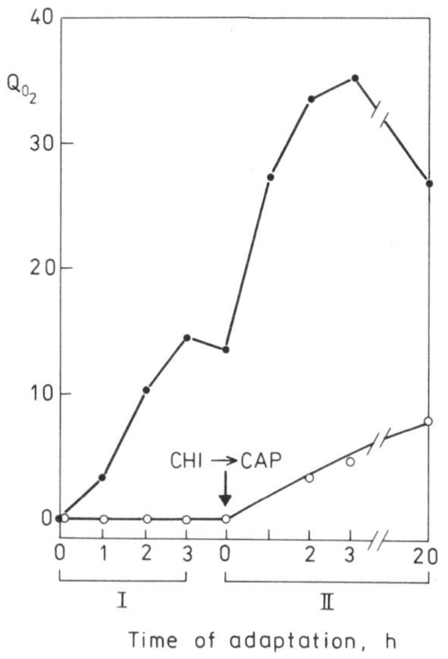

Fig. 48. Respiratory adaptation of anaerobically grown yeast cells during aeration in the absence (●) or in the presence (○) of the inhibitors of protein synthesis. The arrow indicates the moment of transfer from cycloheximide- to chloramphenicol-containing medium. After Rouslin and Schatz (1969).

grown yeast cytochrome *c* oxidase is formed even in the presence of chloramphenicol or cycloheximide, i.e. when the synthesis of either mitochondrial or cytoplasmic components of this complex is practically blocked. It must be noted, however, that the level of cytochrome *c* oxidase activity in cell homogenates did not exceed 20% and the cytochrome aa_3 content 33% of the control. Similar results were obtained by Mahler and Perlman (1971) and by Kim and Beattie (1973) who studied yeast glucose derepression. It was shown that neither chloramphenicol nor cycloheximide completely suppresses the development of cytochrome *c* oxidase, succinate-cytochrome *c* reductase, or NADH-cytochrome *c* reductase activities of yeast mitochondria, especially if the inhibitors are introduced at the later stages of derepression (Figure 49).

The data presented were considered to prove the existence of the pools of free products of mitochondrial and cytoplasmic translation, not integrated into complexes, and their ability to participate in the assembly of complexes after cessation of protein synthesis. Further information along this line can be gained from experiments on the transfer of yeast from chloramphenicol-containing to cycloheximide-containing medium and vice versa. A pertinent example is the data of Lin et al. (1974) for cytochrome *c* oxidase formation

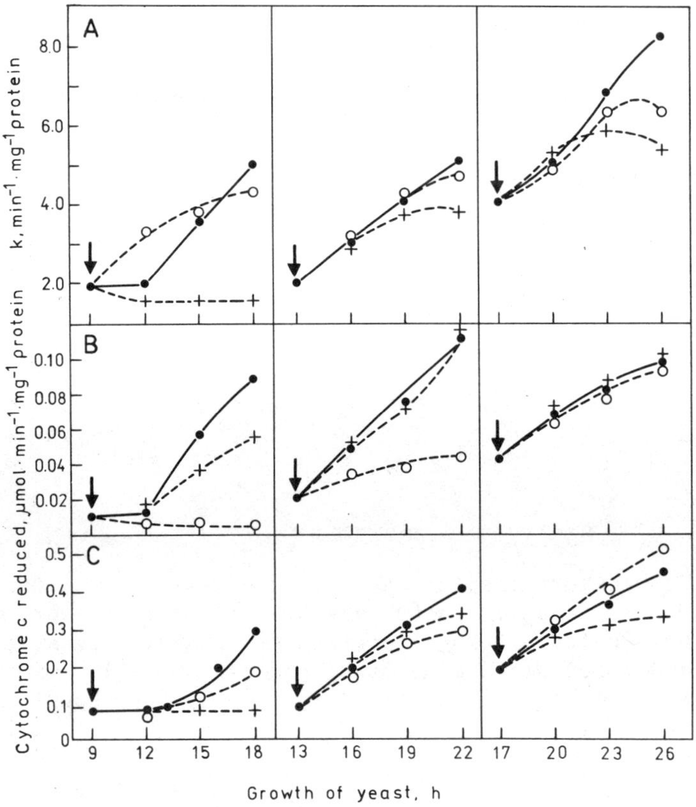

Fig. 49. Effects of cycloheximide (○) and chloramphenicol (+) on the development of cytochrome *c* oxidase (A), succinate-cytochrome *c* reductase (B), and NADH-cytochrome *c* reductase (C) activities of mitochondria upon addition of the inhibitors to yeast at different stages of growth on glucose (initial concentration 5%). (●) Control; arrows indicate the moments of inhibitor administration. After Kim and Beattie (1973).

during yeast glucose derepression. As follows from Figure 50, cycloheximide added to the growth medium at the beginning of derepression does not prevent the increase in the cytochrome *c* oxidase activity of yeast mitochondria. At this stage mitochondrial components of cytochrome *c* oxidase are synthesized and their prefabricated cytoplasmic partners are utilized for assembly. Upon yeast transfer to a medium devoid of cycloheximide but containing chloramphenicol there is a further increase in the activity, in all probability owing to the incorporation of preexisting cytochrome *c* oxidase mitochondrial subunits into complexes with newly formed cytoplasmic partners.

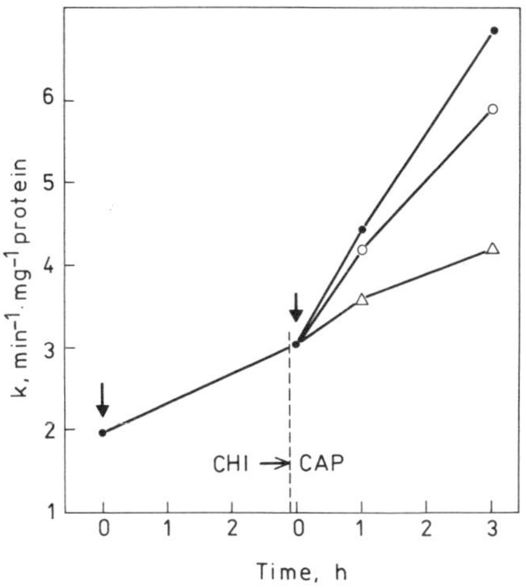

Fig. 50. Effect of consecutive administration of the inhibitors of cytoplasmic and mitochondrial protein synthesis on the development of cytochrome *c* oxidase activity during yeast glucose derepression. (●) Control (no inhibitors); (○) yeast incubated for 3 h with cycloheximide, then transferred into the inhibitor-free medium; (△) yeast incubated for 3 h with cycloheximide, then transferred to the medium with chloramphenicol. Cytochrome *c* oxidase activity was assayed in the mitochondrial fraction. After Lin et al. (1974).

Analogous experiments with oligomycin-sensitive ATPase and complex III also showed the accumulation of subunits made on cytoplasmic polysomes. However, mitochondrial translation products entering into the composition of these complexes were not accumulated in appreciable amounts. One of the conceivable explanations for this is that the steady-state concentration of free subunits of intramitochondrial origin is low because of their rapid incorporation into the complex. It is also possible that the half-life of some free mitochondrial translation products is short, and they are quickly broken down if the synthesis of the corresponding cytoplasmic partners is impaired. Besides, under interrupted cytoplasmic translation the mitochondrially produced polypeptides may be nonspecifically inserted into membranes and become inaccessible to their cytoplasmic partners in the complex.

By alternating cycloheximide and chloramphenicol in the growth medium of yeast undergoing respiratory adaptation, Brown and Beattie (1978) showed the accumulation of mitochondrially made components of cytochrome *c* oxidase and complex III of the respiratory chain. As to polypeptides synthesized on cytoplasmic polysomes, only cytochrome *c* oxidase subunits are accumulated under the conditions of blocked mitochondrial translation.

Experiments with two-stage respiratory adaptation, when mitochondrial protein synthesis is switched off at one stage and cytoplasmic at the other, allow one to estimate the pool sizes of free utilizable precursors of individual enzymic complexes of the inner mitochondrial membrane. In particular, Rouslin (1977) found that the functioning of only the mitochondrial translation system during yeast respiratory adaptation in a galactose medium can turn out such an amount of subunits as would provide for formation of up to 35% of the maximal content of complex III and up to 12% of cytochrome *c* oxidase in mitochondria. Another peculiarity of respiratory system formation in the two-stage adaptation is that the final attained activities of some respiratory-chain segments - e.g., cytochrome *c* oxidase, succinate-cytochrome *c* reductase, and NADH-cytochrome *c* reductase (Lin et al., 1974) - depend on the order in which inhibitors of mitochondrial and cytoplasmic translation are introduced into the growth medium. From this it follows that the simplest model for the formation of mitochondrial inner-membrane oligoenzyme complexes, according to which cytoplasmic and mitochondrial precursors are accumulated independently and precursor pool are completely utilized to form the complexes, can hardly hold true.

An interesting demonstration of the existence of the pools of unassembled subunits of mitochondrial inner-membrane enzymic complexes is the work of Brambl (1980) in which is is shown that in dormant spores of the fungus Botryodiplodia theobromae the mitochondria harbor three internally made subunits of cytochrome *c* oxidase whereas the postribosomal supernatant from the same spores contains four polypeptides precipitable with an antiserum to cytochrome *c* oxidase and slightly larger than the cytoplasmic subunits of the enzyme. These cytosolic polypeptides are imported into mitochondria at the onset of spore germination to unite with the preexisting mitochondrial subunits into a catalytically active enzyme. Experiments with the inhibitors of mitochondrial and cytoplasmic translation (Stade and Brambl, 1981) revealed that the free pool of mitochondrially made subunits I-III reaches 50% whereas that of cytoplasmic subunits does not exceed 15% of their content in germinated spores.

6.5.2. **Stepwise Assembly of Enzymic Complexes in the Inner Mitochondrial Membrane**

As follows from the preceding subsection, the pools of unassembled subunits of cytochrome *c* oxidase in Neurospora crassa substantially differ in their sizes. On assumption that the rates of precursor synthesis, import, and processing in mitochondria are approximately the same for different subunits, these data imply that the subunits are integrated into the complexes at rather different rates. In other words, the assembly of complexes proceeds stepwise, in a strictly defined sequence of incorporation of individual polypeptides.

The stepwise character of the assembly of separate mitochondrial inner-membrane fragments was indicated by the early observations of Chen and Charalampous (1973) that formation of cytochrome a_3 during yeast respiratory adaptation does not begin until the cytochrome a content reaches about two thirds of the final level.

Naturally the most complete and objective information on the sequence of events taking place in the course of mitochondrial membrane differentiation under various conditions can be obtained in experiments with synchronous cultures of microorganisms. Poole et al. (1974) measured the contents of cytochrome a and a_3 during the cell cycle of the yeast Schizosaccharomyces pombe 972 h^- and concluded that cytochrome a is probably a precursor to cytochrome a_3: its cell content increases while the a_3 content remains constant (phases 1 and 4, Figure 51) and then decreases with a concomitant rise in cytochrome a_3 (phases 3 and 5, Figure 51). Later data were reported (Lloyd and Edwards, 1976, 1977; Edwards and Lloyd, 1977) on the kinetics of the development of the yeast ATPase complex. These data, in particular, allowed one to judge of the sequence of appearance in the complex of sites responsible for binding of various inhibitors.

Studying the cytochrome c oxidase assembly in Saccharomyces cerevisiae synchronous cultures, Somasundaran and Jayaraman (1981a) found that the cytoplasmically made subunits are accumulated in the cytosol during the G_1 phase and early S phase, whereas large amounts

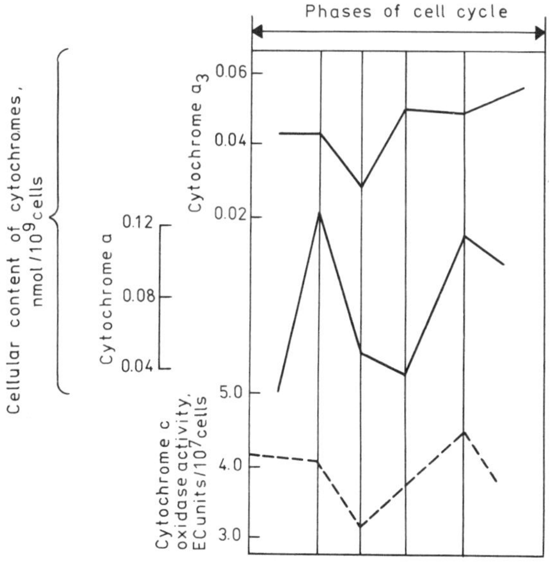

Fig. 51. Changes in the contents of cytochromes a and a_3 and in cytochrome c oxidase activity during the yeast cell cycle (after Poole et al., 1974).

of the mitochondrially made subunits are detected after the middle S phase. From this moment the cytoplasmic subunits start to be integrated into mitochondrial membranes. These authors also reported (Somasundaran and Jayaraman, 1981b) that F_1-ATPase is accumulated in the cytosol in the G_1 and early S phases. After the middle S phase, when the mitochondria have synthesized the membrane-sector components of the ATPase complex, F_1-ATPase is incorporated into the mitochondrial membrane. A drawback of these studies is that the experimental conditions allowed of the release of various soluble and weakly membrane-bound proteins from mitochondria into the medium during cell disruption and mitochondrial isolation; because of this it was impossible to assess the real localization of the subunits of either complex in the subcellular fractions.

One of the approaches to elucidating the order of assembly of mitochondrial enzymic complexes in lower eukaryotes would be the isolation and study of mutants in which assembly is interrupted at different stages, in analogy to the work undertaken by Cox et al. (1981) with various unc mutants of Escherichia coli to determine the sequence of individual polypeptide incorporation into the F_0-F_1-ATPase complex.

6.5.3. Mitochondrial Inner Membrane Formation Under Deficiency of Individual Respiratory Chain and Phosphorylation Components

Important information on the assembly of the inner mitochondrial membrane in vivo can be gained by studying this process under conditions of a deficiency of its individual protein or low-molecular-weight nonprotein components.

6.5.3.1. *Heme Deficiency*

There are several works in which mitochondrial composition and properties were studied in heme-deficient mutants of Saccharomyces cerevisiae. Saltzgaber-Müller and Schatz (1978) used such a mutant (GL1) lacking δ-aminolevulinic acid synthetase to investigate cytochrome *c* oxidase assembly. Mutant mitochondria displayed no cytochrome *c* oxidase activity. With this, they contained significant quantities of mitochondrially made subunits II and III and cytoplasmically made subunit VI. Other subunits were either not found at all (V and VII) or present in mitochondria in very small amounts (I and IV). Since an antiserum to subunit VI precipitated only this subunit itself from mitochondrial lysates, it could be thought that this polypeptide was not associated with any of the other subunits present in mitochondria (i.e. subunits I, II, or III).

Saltzgaber-Müller and Schatz explained the lack of certain cytochrome *c* oxidase subunits by assuming that the heme selectively controls the synthesis of individual polypeptides. Indeed, Kumar and Padmanaban (1980) showed in experiments with an iron-depleted Neurospora crassa culture that iron-ammonium citrate stimulates the synthesis of cytochrome *c* oxidase subunits I-III by enhancing δ-aminolevulinic acid dehydrogenase activity. However, according to Saltzgaber-Müller and Schatz (1978), the content of subunits II and III is not reduced in the heme-deficient GL1 yeast mutant.

Heme deficiency does not only influence the content of individual cytochrome *c* oxidase subunits and their binding with each other. Earlier Ross and Schatz (1976b) found that in the Saccharomyces cerevisiae GL1 mutant apocytochrome c_1 is synthesized and recovered in the cytosol but not incorporated into mitochondria. It was also reported (Saltzgaber-Müller and Schatz, 1978) that the cells of this mutant have a very low content of apocytochrome *b*. Later Clejan et al. (1980) showed that heme deficiency in the mutant does not affect the synthesis of the subunits of the cytochrome bc_1 complex. Incorporation of the label into the cytochrome *b* apoprotein in the mutant reached a high level though it was somewhat slower than in the parental wild-type cells. It should be noted that in heme-deficient yeast [^{35}S]methionine incorporation into mitochondrial translation products reaches 70% of that in control cells (Clejan et al., 1980).

The influence of heme on mitochondrial inner membrane assembly has a universal character. It is known, for instance, that the mitochondria of the already mentioned heme-deficient GL1 mutant have high oligomycin-sensitive ATPase activity but are incapable of performing the ATP-$^{32}P_i$ exchange reaction (Saltzagber-Müller and Schatz, 1978). Another mutant requiring δ-aminolevulinic acid for cytochrome synthesis was marked by decreased (down to 25% of control) ATPase and succinate dehydrogenase activities (Woods et al., 1975).

The universal character of the heme influence on mitochondrial inner membrane assembly may in principle arise from the fact that heme deficiency prevents the syntheses of unsaturated fatty acids, ergosterol, methionine, and other substances involving heme-containing enzymes (see, e.g., Astin et al., 1977). Such an explanation, however, would not apparently pass for the cases when the growth medium of heme-deficient mutants was supplemented with the lacking compounds (Woods et al., 1975; Saltzgaber-Müller and Schatz, 1978).

6.5.3.2. *Copper Deficiency*

Studying the influence of copper on cytochrome *c* oxidase assembly in Neurospora crassa mitochondria, Werner et al. (1974) found that Cu-depleted cells contain all cytochrome *c* oxidase polypeptides though in different relative amounts as compared to normal growth conditions. Since the anti-subunit III rabbit gammaglobulin with ovine anti-rabbit gammaglobulin precipitated only subunit III from a Triton lysate of Cu-depleted mitochondria, it could be thought that the subunits were not bound into an assembly. Hence, copper deficiency interrupts the assembly of the cytochrome *c* oxidase complex at a posttranslational level. It is of interest that the mitochondria of Cu-depleted cells contained no heme.

6.5.3.3. *Coenzyme Q Deficiency*

Tzagoloff et al. (1975) obtained a number of yeast nuclear mutants that displayed no NADH-cytochrome *c* reductase or succinate-cytochrome *c* reductase activities. Both activities could be restored on addition of coenzyme Q analogs. This observation indicated that the whole group of mutants have lesions affecting ubiquinone synthesis. In parallel, more or less pronounced alterations were observed in mitochondrial ubiquinol-cytochrome *c* reductase, cytochrome *c* oxidase, and ATPase activities. Out of 14 mutants studied, only in two these activities reached 50% of those in the parental wild type, and in a single mutant they were not diminished at all. In all other cases at least one of these activities was markedly decreased. Unfortunately, no other parameters of the mitochondria were studied, such as ATPase sensitivity to oligomycin and ability to perform the ATP-$^{32}P_i$ exchange. Nevertheless it is evident that ubiquinone deficiency as a rule leads to multiple defects in the mitochondrial inner membrane assembly manifesting themselves as impaired ability to carry out a number of reactions of the respiratory chain and the ATP-synthesizing system.

6.5.3.4. *Cytochrome b Deficiency*

Mutations in the BOX loci of yeast mitochondrial DNA (Slonimski et al., 1978) impair the normal synthesis of apocytochrome *b*, which results in the production of fragments of this polypeptide and in the changes in the cytochrome *b* spectrum. Some of such mutants (e.g. BOX3-1, BOX4-1) are also devoid of cytochrome aa_3 owing to the mosaicity of the BOX/COB region which appears to contain an insertion coding for subunit I of cytochrome *c* oxidase or somehow responsible for regulation of expression of the OXI3 gene (Haid et al., 1979; Hanson et al., 1979).

Recently it became known that the biochemical disturbances in the inner mitochondrial membrane of BOX/COB mutants affect not only the polypeptides mentioned. Capeillere-Blandin and Ohnishi (1982) showed that BOX2-1, BOX1-3, BOX4-1, and BOX3-1 mutants completely lack Rieske iron-sulfur centers, and in BOX6-1 and BOX4-3 mutants their content does not exceed 15% of that in mitochondria of the parental wild-type cells. This defect was most probably due to the inability of the iron-sulfur clusters to integrate in the absence of normal cytochrome *b* molecules from the inner mitochondrial membrane (it has been said in Chapter One that Rieske iron-sulfur clusters and cytochrome *b* together with cytochrome c_1 and several other polypeptides constitute complex III of the respiratory chain). The mutants were somewhat short of cytoplasmically made cytochromes *c* and c_1 (their content was 45-85% of normal). Even more pronounced was the effect of BOX mutations on the content of iron-sulfur centers S-1, S-2, and S-3 which fell to about 20% of that found in wild-type mitochondria. Since cytochrome *b* deficiency could directly affect neither synthesis nor assembly of these centers, which are encoded in the nuclear and not mitochondrial DNA, synthesized on cytoplasmic and not mitochondrial polysomes, and belong to complex II and not to complex III of the respiratory chain, the decrease in the content of S-1, S-2, and S-3 clusters was ascribed to a repression phenomenon whose mechanism has not been clarified to date. Thus, the inner mitochondrial membrane assembled under deficiency of cytochrome *b* is marked by multiple lesions affecting quite a number of respiratory chain links, and a part of these lesions can in all probability be regarded as defects of assembly.

6.5.3.5. *Deficiency of Cytochrome c Oxidase Polypeptides*

Ebner et al. (1973a) described several nuclear single-gene mutants of Saccharomyces cerevisiae lacking cytochrome *c* oxidase. Except for two of them (pet E11 and pet 494), such mutants displayed diminished succinate-cytochrome *c* reductase activity. Mitochondria of all these mutants showed decreased ATPase activity, lower ATPase sensitivity to rutamycin, and particularly decreases ability to perform ATP-$^{32}P_i$ exchange. Later Cabral and Schatz (1978) demonstrated that pet E11 is totally devoid of cytochrome *c* oxidase subunit II and pet 494 of subunit III. Both mutants contain the cytoplasmically made subunits IV-VII in amounts close to that characteristic of wild-type mitochondria. However, there is no tight association among these subunits, since antisera to individual subunits precipitate only their respective antigens whereas in extracts from wild-type mitochondria antisera to subunits V and VII precipitate all seven subunits of cytochrome *c* oxidase. Upon sonication of pet E11 and pet 494 mutant mitochondria, the cytoplasmically synthesized cytochrome *c* oxidase subunits are recovered exclusively in the high-speed supernatant. All this taken together means that in the absence of just one mitochondriallymade subunit all others do not tightly bind either with each other or with the inner mitochondrial membrane.

Quite similarly, Bertrand and Werner (1977) showed that in a Neurospora crassa cytoplasmic mi-3 mutant lacking subunit II of cytochrome *c* oxidase other subunits fail to form a definite complex among them.

Thus, a deficiency of a single subunit of the cytochrome *c* oxidase complex leads to dramatic alterations in the assembly of the complex itself as well as of the inner mitochondrial membrane as a whole. This can be further supported by the data of Capeillere-Blandin and Ohnishi (1982) according to which a mutation in OXI genes (OXI1, OXI2, and OXI3 mutants) of Saccharomyces cerevisiae mitochondrial DNA entails a substantial decrease in the content of Rieske iron-sulfur clusters and S-1, S-2, and S-3 clusters of succinate dehydrogenase. Finally, one should recollect the cytochrome *c* oxidase-lacking Saccharomyces cerevisiae nuclear mutants described by Tzagoloff et al. (1975), whose mitochondria had more or less pronouncedly lowered NADH-cytochrome *c* reductase and ATPase activities. Only in one (N9-142) out of 19 mutants studied were these activities not less than in wild-type mitochondria.

6.5.3.6. *Deficiency of ATP-Synthetase Components*

As shown by early studies of Tzagoloff et al. (1975), nuclear mutations in yeast leading to a 10-fold decrease in mitochondrial ATPase activity and complete loss of its rutamycin sensitivity are also attended by a substantial attenuation of NADH-cytochrome *c* reductase and cytochrome *c* oxidase activities.

One of the nuclear F_1-ATPase-depleted yeast mutants (pet 936) was subjected to a more detailed analysis by Ebner and Schatz (1973). Basing their argument on the data concerning the content of F_1-ATPase individual subunits, these authors suggested that the pet 936 mutation is a nonsense mutation in the structural gene for one of the F_1-ATPase larger subunits. The mutation is pleiotropic since it is accompanied by the loss of cell respiration, substantial decrease in cytochrome *bb*, c_1, and aa_3 contents, and impairment of mitochondrial protein synthesis. Ebner and Schatz (1973) admitted the possibility that F_1-ATPase plays the role of one of the organizing proteins in mitochondrial inner membrane assembly.

Coruzzi et al. (1978) found that in PHO2 mutants having an altered ninth subunit of H^+-ATPase the activities of NADH-cytochrome *c* reductase and cytochrome *c* oxidase do not exceed 30% and 10%, respectively, of that in wild-type mitochondria. The results of spectral measurements testified to a significant decrease in the cytochrome *b* and aa_3 contents in these mutants.

Mutation in the OLI2 gene of Saccharomyces cerevisiae mitochondrial DNA which alters another mitochondrially made (sixth) subunit of H^+-ATPase was also accompanied by a decline in NADH-cytochrome *c* reductase and cytochrome *c* oxidase mitochondrial activities by 15% and 50%, respectively. The mitochondria also almost completely lost the ability to carry out the ATP-$^{32}P_i$ exchange (Stephenson et al., 1981).

Finally, single-gene mutations in Schizosaccharomyces pombe resulting in altered α and β subunits of F_1-ATpase concomitantly lead to decreases in NADH-cytochrome *c* reductase and succinate-cytochrome *c* reductase activities and in the contents of cytochromes *b* and aa_3. Boutry and Coffeau (1982) suggested on the basis of these data that functional F_1-ATPase is needed for correct synthesis and/or assembly of cytochrome *c* oxidase and cytochrome bc_1 complex.

Summing up the results of this section, it is easy to note that a deficiency of any nonprotein (heme, copper, coenzyme Q) or protein component of the respiratory chain and ATP-synthesizing system as a rule results in the formation of a defective inner mitochondrial membrane carrying multiple lesions even in those components which are not directly affected by the initial defect. The few exceptions known to this rule are probably due to a not detailed enough analysis of mutant mitochondria.

Three explanations were offered for the multiplicity of defects caused by single-gene mutations in individual components of the oxidative phosphorylation system. According to one (see, e.g., Ebner and Schatz, 1973; Boutry and Coffeau, 1982; Capeillere-Blandin and Ohnishi, 1982), a lack of any type of polypeptides during mitochondrial inner membrane assembly leads to multiple alterations in the membrane because it is a cooperative system. This explanation cannot be applied at least to cases when membrane assembly is impaired because of coenzyme Q deficiency (Tzagaloff et al., 1975; Brown and Beattie, 1977) or lack of oxygen (Woodrow and Schatz, 1979).

According to another point of view (Capeillere-Blandin and Ohnishi, 1982), a shortage of any respiratory-chain component may bring about a phenomenon termed catabolite repression. Perlman and Mahler (1974) have earlier shown that antimycin A, which inhibits electron transfer through the respiratory chain, partially suppresses the increase in the activities of some dehydrogenases and cytochrome *c* oxidase and in the content of cytochromes *b*, c_1, *c*, and aa_3 during yeast glucose derepression. On the basis of these data they supposed that one of the phases of derepression, arbitrarily named "oxidative", requires the carrying out of mitochondrial functions. This is necessary not to make up for the cell energy expenditures but to regulate the derepression via a feedback mechanism (through heme synthesis, creation of templates for further mitochondrial membrane differentiation, and the like). The weak point of this hypothesis is that it is as yet speculative and has not been spelled out in detail.

Finally, the multiple defects in the inner mitochondrial membrane assembled under conditions of deficiency of an oxidative-phosphorylation component may be explained by its instability because of impaired energy transformation (see Chapters Three and Four). It must be borne in mind that, just as the half-lives of individual mitochondrial inner-membrane proteins and phospholipids are different, so the nonfunctioning regions of the system would most probably be degraded at different rates. To test and substantiate this concept, data are needed on the degradation rates of mitochondrial inner-membrane components in respiratory-deficient yeast mutants.

6.6. CONCLUDING REMARKS

As could be seen in this chapter, mitochondrial formation is an intricate multicomponent and multistage process. About 150 types of polypeptides, after their translation is completed, are selectively transported across preexisting mitochondrial (or promitochondrial) membranes, being incorporated into different organellar compartments. As this takes place, the polypeptides undergo proteolytic processing and other modifications including covalent heme attachment, noncovalent binding of prosthetic groups, etc. All these processes have a complex topographical arrangement; some of them are carried out in the intermembrane space, some in the inner mitochondrial membrane, some in the matrix. The proper incorporation of various polypeptides into mitochondrial compartments is thus conditioned not only by their individual properties but also by the availability and correctness of arrangement of the modifying enzymic systems. The existence of the pools of free unassembled polypeptides predetermines the necessity of their free migration in the mitochondrial inner membrane lipid bilayer in search of their partners in complexes; this requires the maintenance of certain fluidity of the bilayer which depends at least on the relative content of unsaturated fatty acids in phospholipid molecules and on ergosterol concentration in the membrane. Hindrance to the migration of a certain polypeptide within the membrane may result in its deficiency in some membrane regions and, as a consequence, in multiple disorders in the membrane composition and structure.

Let us assume that each mitochondrion is structurally arranged in a strictly defined manner, so that each component is present in a single compartment, each polypeptide in the membrane is orientated in a single possible way and integrated in a complex with a single possible set of partners, the mutual position of complexes obeys to stringent rules, and so on. Such rigorously determined principles of organization would imply a flawless performance of the receptor sites on the mitochondrial surface sorting out the cytoplasmically made polypeptides, as well as of enzymes processing and modifying protein precursors. It would require perfect matching of the syntheses of dozens of types of membrane polypeptides on cytoplasmic and mitochondrial polysomes to meet the rule of fixed stoichiometric relations between them, violation of which in any site brings about a wave of defects in the membrane. In other words, the more determined is the final mitochondrial structure the more exacting are the requirements applied to all biosynthetic processes related to the organelle formation. An alternative (or a supplement) to this may be the existence of controlling systems that would eliminate structures carrying any defects.

The next two chapters are devoted to what is known about the specificity of interaction between mitochondrial inner-membrane components in reconstitution experiments and about the *in vivo* control over mitochondrial assembly.

7
Reconstitution of the Inner Mitochondrial Membrane and its Fragments

Knowledge of how great is the affinity between mitochondrial inner-membrane components and what determines the orientation of certain components within the membrane can be partially gained from the works on reconstitution of the mitochondrial respiratory chain from isolated oligoenzyme complexes, reconstitution of individual complexes from their constituent elements, and reconstitution of membranous systems capable of energy transformation and other mitochondrial inner membrane functions. All data available are in this chapter conventionally divided into the following groups: (1) selective removal of separate protein or lipid components from mitochondria or submitochondrial vesicles, with subsequent reassociation of these entities with the membrane; (2) fragmentation of the inner membrane followed by separation of the fragments, and their spontaneous association into membranous polyenzyme systems; and (3) formation of membrane systems on the basis of artificial phospholipid vesicles and isolated protein complexes of the inner mitochondrial membrane.

7.1. SPECIFIC BINDING OF INDIVIDUAL PROTEIN COMPONENTS TO THE INNER MITOCHONDRIAL MEMBRANE

7.1.1. Binding of Cytochrome *c*

Cytochrome *c* is known (Jacobs and Sanadi, 1960; MacLennan et al., 1966) to be readily extracted from mitochondria with 0.15 M potassium chloride. As this takes place, the organelles appear to undergo no substantial destructive alterations. In particular, according to MacLennan et al. (1966), after extraction of about 85% of cytochrome *c* beef heart mitochondria retain their ability to carry out ATP-dependent transport of Ca^{2+} ions and transhydrogenase reaction; there is also no decrease in the efficiency of oxidative phosphorylation coupled to electron transfer between NADH and ferricyanide. Naturally the removal of cytochrome *c* from mitochondria almost completely interrupts respiration with such substrates as succinate, glutamate, α-ketoglutarate, citrate, or NADH. Addition of cytochrome *c* in amounts approximately equal to those preliminarily extracted with 0.15 M potassium chloride increases the rate of respiration coupled with ADP phosphorylation up to the control value (Jacobs and Sanadi, 1960; MacLennan et al., 1966). Cytochrome *c* binds to mitochondria so tightly as not to be removed by subsequent washing of the preparation with 0.3 M sucrose (Jacobs and Sanadi, 1960). On the basis of these data it was concluded that cytochrome *c* added to mitochondria "titrates" its binding sites. This evidences the exceptionally high selectivity of cytochrome *c* interaction with the inner mitochondrial membrane.

7.1.2. Binding of Soluble Succinate Dehydrogenase

Results basically similar to those presented above for cytochrome *c* were obtained in experiments on selective extraction of succinate dehydrogenase from submitochondrial vesicles. According to the data of King and colleagues (King, 1963; Lee et al., 1969, 1971), incubation of the vesicles at pH 9.5–10.0 in an anaerobic medium containing succinate leads to a reversible loss of succinate dehydrogenase. Since the vesicles retain therewith their ability to carry out a number of energy-dependent functions, it can be thought that the mild alkaline treatment induces no appreciable alterations in their membranes. Lowering of the pH of the medium from 10.0 to 7.6 results in a practically complete binding of the free succinate dehydrogenase. Indeed, vesicles sedimented by centrifugation from the medium subjected to the above-indicated pH shift contain not less than 90% of acid-nonextractable flavin and not less than 90% of the initial succinate dehydrogenase activity. On the other hand, vesicles sedimented from the alkaline medium carry not more than half of the initial amount of succinate dehydrogenase; the balance of the enzyme can be recovered in the supernatant. From this it follows that succinate dehydrogenase can be quantitatively bound by corresponding sites on the inner mitochondrial membrane.

7.1.3. Binding of F_1-ATPase

According to Kagawa and Racker (1966b) and Racker and Horstman (1967), F_1-ATPase accounts for about 10% of the total protein of inside-out submitochondrial vesicles. Complete removal of F_1 from the vesicles can be achieved by treating them with urea (Racker and Horstman, 1967) or silicotungstate (Racker et al., 1969). However, because the vesicles simultaneously lose other protein factors (e.g.,

F_2, F_3, and F_5), subsequent addition of F_1 does not restore their lost functions such as oxidative phosphorylation, ATP-$^{32}P_i$ exchange, ATP-dependent NAD^+ reduction by succinate, etc. Such lost properties are to a substantial extent reacquired upon binding the complete set of the protein factors (Racker and Horstman, 1967; Racker et al., 1969). Maximal restoration requires amounts of F_1 commensurate with its content in the initial vesicles. This indicates that F_1 interaction with the inner mitochondrial membrane is to a certain degree specific. However, the above-mentioned results of Racker and coworkers do not make it possible to assess the measure of such specificity, since even in the presence of the whole set of coupling factors the restoration is never complete. The latter circumstance may be due either to the fact that the specificity of the interaction between the extracted factors and the inner mitochondrial (or submitochondrial vesicle) membrane is not high enough or because the extraction is accompanied by irreversible destructive alterations in the membrane.

Somewhat more distinct in this respect appear to be the data of MacLennan et al. (1968) who showed that exposure of inside-out submitochondrial vesicles to 2 M sodium chloride at 0°C leads to selective extraction of F_1-ATPase subunits. As it does so, the vesicles lose much (75-80%) of their ATPase activity and ability to perform ATP-dependent transhydrogenation, but retain a high (65% of control) P/O ratio in ADP phosphorylation coupled to succinate oxidation, and near 100% of the rate of succinate-dependent transhydrogenase reaction. All the functions listed are virtually completely restored upon removal of the salt from the medium by dialysis of the vesicle suspension in the presence of ATP and magnesium chloride. Additional studies demonstrated that restoration of ATPase activity and other functions can be achieved by adding subunits of F_1-ATPase depolymerized in the presence of the salt to vesicles pretreated with 2 M sodium chloride. MacLennan and colleagues believed that the most plausible restoration mechanism is the reassociation of ATPase subunits with the vesicular membranes. Since in the absence of the vesicles these subunits could not reassociate to give an active enzyme, it had to be assumed that the membranes carried certain binding templates, which could be residues of F_1 molecules lacking some subunits. The high efficiency of restoration which proceeded without adding excess amounts of F_1 subunits suggested highly specific interactions with the membrane. It should be noted, however, that this specificity depended in no small measure on the presence of ATP and Mg^{2+} ions in the medium.

7.2. MITOCHONDRIAL INNER MEMBRANE FRAGMENTATION AND REASSOCIATION OF FRAGMENTS

7.2.1. Reassociation of Crude Fragments of the Inner Mitochondrial Membrane

According to Kopaczyk et al. (1968b), mitochondrial inner-membrane preparations are readily solubilized in the presence of potassium cholate, ammonium sulfate, and dithiothreitol. Electron microscopy of the solubilized preparation shows the absence of membrane structures (Figure 52b). Dilution of this preparation leads to formation of vesicles whose surface reveals the presence of mushroom-like particles analogous to those present in the initial material (Figure 52c). The reconstituted vesicles have rutamycin-sensitive ATPase activity equal to the corresponding specific activity of the solubized membrane preparation.

It must be emphasized that the membrane fragments obtained as indicated above give rise, upon lowering the concentration of the dispersing agent in the medium, not to random three-dimensional aggregates but specifically to vesicular membranes. Results similar to the above were obtained when reconstitution was carried out with preliminarily separated membrane fraction of the initial mitochondrial inner-membrane preparation and its soluble fraction containing oligomycin-sensitive ATPase. Dilution of the solubilized mixture of these two fractions also yielded vesicles covered with mushroom-like particles.

The ability to form vesicular membranes is found not only with crude fragments of the inner mitochondrial membrane but also with individual oligoenzyme complexes of the respiratory chain isolated therefrom. Usually the concentrated solutions of the complexes (with a concentration about 10 mg protein per ml) contain mostly lipoprotein particles and not membranous material. The particles are sustained in the dispersed state by detergents (specifically, bile acid salts or Triton X-100) adsorbed on their surface in the course of isolation. Dilution of such dispersions lowers the detergent concentration, which is accompanied by specific aggregation of the enzyme particles to form first unsealed membranes gradually transformed into vesicles. Such data were reported by Tzagoloff et al. (1967), Kopaczyk et al. (1968b), Sun et al. (1968), and others.

Isolated complexes show differential ability to interact with rutamycin-sensitive ATPase and F_1-ATPase. According to Kopaczyk et al. (1968b), the purified preparation of complex IV binds the former but not the latter. Such binding is achieved when the solubilized complex IV is combined with rutamycin-sensitive ATPase and the mixture is diluted to decrease the concentration of the solubilizing agent. The reconstituted membranes display high rutamycin-sensitive ATPase activity. Nevertheless, electron-microscopic observations

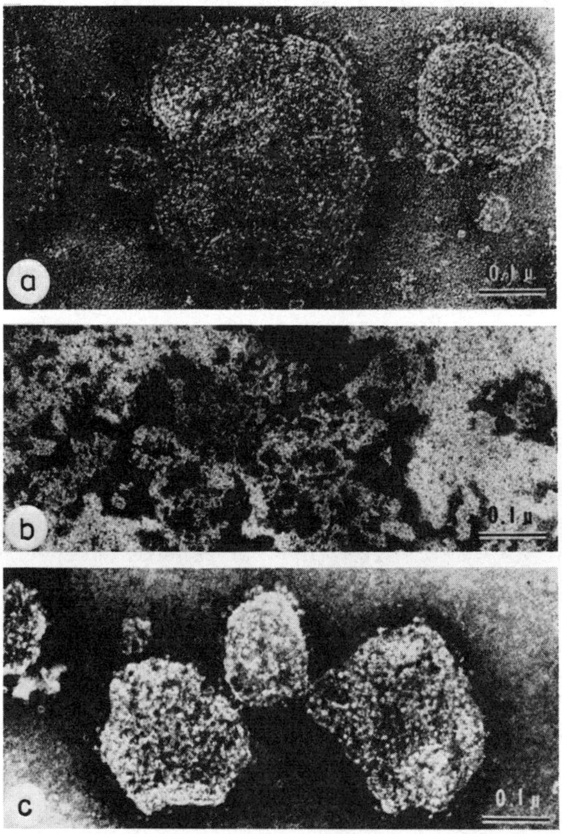

Fig. 52. Reconstitution of the mitochondrial inner membrane. (a) Preparation of inner mitochondrial membranes (inside-out submitochondrial vesicles; (b) the same preparation after treatment with cholate and removal of the "insoluble" material; (c) the reconstituted membrane fraction. Negative contrast. Bars, 0.1 µm.

testify to the absence of mushroom-like structures from their surface. On the other hand, the membrane system reconstituted from complex III and rutamycin-sensitive ATPase showed numerous mushroom-like particles in negatively contrasted preparations.

Thus, the results presented allow a conclusion that isolated complexes can spontaneously aggregate to form membranes resembling the inner mitochondrial membrane in their general morphology. Unfortunately, they tell nothing yet about the similarity between the native and reconstituted membranes in respect of the details of the structural arrangement (see Chapter One) and functional characteristics.

Of special interest is the question of the factors determining the ability of complexes to form membranes. Green et al. (1967) were of the opinion that such specific aggregation is due to the presence in the complexes of large surface regions covered with hydrophilic moieties ("heads") of phospholipid molecules. Such patches may prevent spontaneous aggregation leading to unordered three-dimensional conglomerates, providing thereby for specific interaction of complexes through the establishment of hydrophobic bonds between their certain protein regions. There is no doubt about the exceptionally important role of phospholipids in the formation of vesicular membranes from isolated complexes. In particular, one can refer to the data of Chuang et al. (1973) that cytochrome *c* oxidase preparations free of detergent and depleted of lipids (containing not more than 1.3-1.8 μg lipid phosphorus per mg protein) are represented by a mixture of globular or rodlike aggregates of the enzyme particles. Addition of phospholipids, starting from 4 μg phosphorus per mg protein, leads to formation of small unsealed membranes. With increasing phospholipid content in the medium the membranes grow in size and at a concentration of 28 μg phosphorus per mg protein form clearly discernible vesicles.

7.2.2. Respiratory Chain Reconstitution from Oligoenzyme Complexes

As established by Hatefi et al. (1962c), to reconstitute the whole respiratory chain or its fragments it is enough to mix a solution of the complexes at concentrations of 5-10 mg protein per ml and to dilute the mixture with a buffered sucrose solution to 0.05-0.10 mg protein per ml.

The crucial condition for the formation of a reconstituted system with high enzymic activity is to have a sufficiently high concentration of the complexes in initial solutions. Under such conditions it proved possible to reconstitute NADH-cytochrome *c* reductase from complexes I and III; succinate-cytochrome *c* reductase from complexes II and III; NADH- and succinate-cytochrome *c* reductase from complexes I, II, and III; NADH oxidase from complexes I, III, and IV and cytochrome *c*; succinate oxidase from complexes II, III, and IV and cytochrome *c*; and the complete respiratory chain from complexes I, II, III, and IV and cytochrome *c*.

Later Tzagoloff et al. (1967) obtained data explaining why the respiratory chain or its fragments are not reconstituted on mixing dilute solutions of the complexes. According to their electron-microscopic observations, concentrated preparations of the complexes (10-25 mg protein per ml), which are transparent solutions immediately upon isolation, are composed of dispersed particles. Potassium deoxycholate concentration in such solutions is high, amounting in the case of cytochrome *c* oxidase (complex IV) to 0.43 mg/mg protein. Upon a hundredfold dilution the complexes form vesicular membranes (see above) in which the deoxycholate content per unit protein is 10-15 times lower than in the initial preparations. From this it follows that in trying to reconstitute the respiratory chain from dilute preparations of the complexes, we simply mix two, three, or four suspensions of vesicles of different composition. In such conditions reconstitution does not take place because of the inability of the vesicles to fuse with one another and of the drastically restricted interaction between components belonging to different vesicles. On mixing concentrated solutions followed by dilution there is a possibility of formation of mixed membranes allowing for close interplay between individual components. The question arises how these interactions are carried out.

First of all it must be noted that the reconstituted respiratory chain appears to be in the functional sense identical to the uncoupled respiratory chain of mitochondria or submitochondrial vesicles, as evidenced by the results of inhibitor analysis (Hatefi et al., 1962c; Tzagoloff et al., 1967). Having varied the quantitative relations between the complexes during mixing their concentrated solutions, Tzagoloff et al. (1967) concluded that the same relations are retained in the reconstituted membranes. On this basis it was suggested that membrane formation is a statistical process. It was assumed that all the vesicles formed under fixed relations between complexes are of the same composition, and that complexes in the vesicular membranes are randomly scattered with respect to each other, forming no specific associates. In such systems the interaction between individual components can take place only through diffusion of mobile electron carriers (such as coenzyme Q and cytochrome *c*) and the overall catalytic activity is determined only by the content of the limiting component.

Later Ragan and Heron (1978) and Heron et al. (1978) presented evidence that in the reconstituted NADH-cytochrome *c* reductase the "molecules" of complexes I and III form a common supramolecular unit with a 1:1 stoichiometry, probably owing to protein-protein interactions. This unit also includes some amounts of coenzyme Q. Surplus molecules of complexes I and III do not enter the catalytically active assemblies and do not noticeably contribute to the overall rate of the NADH-cytochrome *c* reductase reaction. The formation of the complex I-complex III unit was considered to be a reversible event occurring faster than electron transfer in the reconstituted system.

On the whole, the existence of specific assemblies formed by the complexes in the reconstituted vesicular membrane remains an open question. The discussion of this issue appears as yet premature since the existence of associates of oligoenzyme complexes has not yet been reliably demonstrated even for intact mitochondrial membranes (see Chapter One).

7.3. FORMATION OF ENERGY-TRANSFORMING AND TRANSPORT SYSTEMS ON THE BASIS OF PHOSPHOLIPID VESICLES

7.3.1. Reconstitution of the Respiratory Chain

The paramount importance of phospholipids as an environment preventing spontaneous aggregation of respiratory-chain components and providing for functional contacts between them was demonstrated in experiments on reconstitution of succinate oxidase from its individual components. Yamashita and Racker (1969) showed that the necessary initial elements for this system are succinate dehydrogenase, cytochromes *b*, c_1, *c*, cytochrome *c* oxidase, a mixture of mitochondrial phospholipids, and coenzyme Q. A peculiarity of the reconstitution protocol was that the starting protein components (except cytochrome *c*) were mixed with large amounts of phospholipids (the weight ratio of protein to phospholipid not exceeding one) and coenzyme Q. The mixture was diluted, which was accompanied by vesicularization (see, e.g., Nishibayashi-Yamashita et al., 1972), and incubated for 1-5 h at 37°C.

Such conditions gave rise to a multienzyme system that could perform succinate oxidation by oxygen at a specific rate (related to the total protein) close to that observed with submitochondrial vesicles (Table 23). Succinate oxidase activity was inhibited by 97% by potassium cyanide and antimycin A, and also by 4,4,4-trifluoro-1-(2-thienyl)-1,3-butanedione (an inhibitor of succinate dehydrogenase). Addition of complex I to the reconstituted succinate oxidase resulted in the appearance of NADH oxidase activity sensitive to rotenone (an inhibitor of complex I) and antimycin A. It is easy to note that the phospholipid content in this reconstituted system exceeds by far that found in the membranes built of oligoenzyme complexes. Such a large amount of phospholipid was necessary in the first place to avoid spontaneous aggregation of the protein components; this applies particularly to cytochrome *b* which is poorly soluble in aqueous media (see, e.g., Yamashita and Racker, 1969).

The system proposed by Yamashita and Racker could be used for a graphic demonstration of highly specific interactions between two respiratory-chain proteins - cytochrome *b* and cytochrome c_1 - in the phospholipid bilayer. A measure of the efficiency of such interaction is the high rate of electron transfer through succinate oxidase or NADH oxidase chains. One of the drawbacks that complicate the interpretation of experimental data is that the preparations of cytochromes *b* and c_1 used for reconstitution were, in all probability, not individual proteins (see Berden and Slater, 1970; Nishibayashi-Yamashita et al., 1972). Further, the reconstituted system was not identical to the respiratory chain since it lacked the iron-sulfur protein of the bc_1 segment. The comparison of specific succinate oxidase activities in the reconstituted system and in submitochondrial vesicles allows one to make only tentative con-

Table 23. Components Necessary for Reconstituting Succinate Oxidase (after Yamashita and Racker, 1969).

Set of components	Succinate oxidase, ng-atoms O per min
Complete system	109 *
without cytochrome *b*	13.1
without cytochrome c_1	14.0
without succinate dehydrogenase	13.1
without cytochrome *c* oxidase	10.5
without phospholipids	29.7
without coenzyme Q	13.1
Complete system in the presence of:	
KCN (0.1 mM)	3.5
antimycin A (0.66 μg/mg protein)	3.5
rotenone (5.10^{-6} M)	109
4,4,4-trifluoro-1-(2-thienyl)-1,3-butandione (1 mM)	17.5

* Corresponds to a specific activity of 0.344 μg-atoms oxygen consumed per min per mg.

clusions. Indeed, the specific activities were calculated by Yamashita and Racker (1969) on the basis of the total protein although the two systems had entirely different protein compositions. It should also not be disregarded that the preparations of succinate dehydrogenase, cytochrome *c* oxidase, and cytochromes *b* and c_1 most probably contained some material that had been inactivated during isolation. This was firmly established, e.g., for cytochrome *b* (Yamashita and Racker, 1969; Berden and Slater, 1970). Finally, Berden and Slater (1970) do not exclude the possibility that the reconstitution involves fragments containing unresolved cytochromes *b* and c_1 and that the stimulatory action of exogenously added cytochrome c_1 is due to its influence on the structure of the system. This assumption is supported by the experimental fact that the reconstitutionally active cytochrome *b* preparations usually contain considerable amounts of cytochrome c_1.

Nishibayashi-Yamashita et al. (1972) obtained experimental data providing some notion of the conditions required for specific interaction between respiratory-chain components in the reconstituted system. According to their results, addition of cytochrome *b* to the already formed lipoprotein vesicles including succinate dehydrogenase, cytochrome c_1, and coenzyme Q-10 does not elicit any appreciable succinate-cytochrome *c* reductase activity. Likewise, the vesicles containing succinate dehydrogenase, coenzyme Q, and cytochrome *b* do not acquire this activity even 30 min after adding cytochrome c_1. Finally, when phospholipids and coenzyme Q are added to a mixture of all protein components mentioned, the resulting succinate-cytochrome *c* reductase activity amounts only to about 50% of the control value. These data give grounds for thinking that to interact specifically cytochromes *b* and c_1 have to be dispersed in the phospholipid milieu and incorporated into the same lipoprotein membrane. This condition is not obligatory for succinate dehydrogenase which is readily soluble in aqueous media, since its addition to phospholipid vesicles containing cytochromes *b* and c_1 and coenzyme Q already in 30 sec gives rise to an active system.

A very interesting feature of the succinate oxidase system reconstituted under conditions when the initial mixture before dilution comprised succinate dehydrogenase, cytochrome *b*, cytochrome c_1, cytochrome *c*, cytochrome *c* oxidase, coenzyme Q, and phospholipids, is that its activity is virtually insensitive to antibodies against cytochrome *c*. Suppression is observed if cytochrome *c* is added to the reconstituted system after dilution of the above mixture (excluding cytochrome *c*). Taking into account this fact together with that the activity of the reconstituted system does not depend on the moment of succinate dehydrogenase incorporation, it can be supposed that the spatial arrangement of the protein components in the reconstituted system was similar to that in the submitochondrial vesicles (succinate dehydrogenase on the outside and cytochrome *c* on the inside of the vesicular membrane). The reconstituted system, however, showed no respiratory control.

7.3.2. **Formation of Membrane Systems Capable of Energy Transformation and Energy-Dependent Functions**

7.3.2.1. *Reconstitution of an Energy-Transducing System from the Amorphous Membrane Fraction and Phospholipids*

Kagawa and Racker (1971) showed that a system capable of energy transformation can be reconstituted from phospholipid-depleted amorphous fragments of the inner mitochondrial membrane, phospholipids, and coupling factors: F_1-ATPase, oligomycin-sensitivity-conferring protein (OSCP), and factor F_6. The main stages of reconstitution are as follows: (1) preparation of a soybean-phospholipid solution by sonication followed by addition of sodium cholate to maintain the phospholipids in isotropic solution (Kagawa, 1972); (2) suspension of the amorphous membrane fraction (fraction CF_0 which may or may not contain some cytochromes *b* and c_1) in the phospholipids; (3) prolonged dialysis of the suspension thus obtained; and (4) addition of coupling factors.

Such a procedure gives rise to lipoprotein vesicles (Kagawa and Racker, 1971) which are demonstrable electron-microscopically. It is of interest that neither the amorphous membrane nor phospholipids taken separately can form such vesicles. Appearance of sealed phospholipid vesicles was confirmed in experiments with labeled inulin (molecular weight 5500) and ferritin. If either of them were added to the suspension of membrane fragments in the phospholipid - cholate mixture prior to dialysis, it could not be removed by extensive washing of the vesicles obtained or by passing them through a Sepharose column. On the other hand, inulin and ferritin could be practically completely removed from the suspension if added after dialysis, i.e. after the completion of vesicularization.

The lipoprotein vesicles obtained under the conditions indicated can catalyze the ATP-$^{32}P_i$ exchange reaction. In their presence ATP enhances 1-anilinonaphthalene-8-sulfonate fluorescence. Both these functions are suppressed by rutamycin and the uncoupler carbonylcyanide p-trifluoromethoxyphenylhydrazone (Kagawa and Racker, 1971). This points to a certain resemblance between the reconstituted vesicles and coupled submitochondrial particles.

As follows from a number of indirect data, the membranes in reconstituted vesicles are orientated as in inside-out submitochondrial particles. Indeed, the ATP-$^{32}P_i$ exchange catalyzed by the reconstituted vesicles is inhibited by a lipophilic anion tetraphenylboron but not by a lipophilic cation tetraphenylarsonium (Kagawa and Racker, 1971; see also Jasaitis et al., 1972). Both ions show quite the same action on submitochondrial particles and quite the opposite on mitochondria: the latter are sensitive to the lipophilic cation but not to the anion. The ATP-$^{32}P_i$ exchange catalyzed by the reconstituted vesicles is suppressed by combined action of valinomycin and nigericin in the presence of K^+ ions (Kagawa and Racker, 1971), which is also characteristic of submitochondrial particles. Unlike mitochondria and similarly to submitochondrial particles, the reconstituted vesicles enhance 1-anilinonaphthalene-8-sulfonate fluorescence in the presence of ATP. Finally, the reconstituted vesicles, like submitochondrial particles, accumulate H^+ ions on addition of ATP (Kagawa et al., 1973) whereas in mitochondria the H^+ flux is opposite. All these data, summarized in Table 24, indicate that the reconstituted membrane is anisotropic and orientated similarly to that of inside-out submitochondrial particles. The anisotropy of the reconstituted membrane may be due to the fact that the coupling factors F_1 and OSCP were added to the suspension only after the vesicles had already been formed, and could therefore be localized only on their surface. Indeed, antibodies to F_1 inhibited the ATP-$^{32}P_i$ exchange preformed by reconstituted vesicles, although to a lesser extent than with submitochondrial particles. It should be mentioned that even if some F_1 molecules were present in the vesicle interior regions (e.g., as part of the CF_0 fraction) they were inaccessible to adenine nucleotides because of the lack of the corresponding transport system. In other words, it may well be that the anisotropy of the reconstituted membrane is in great measure determined by its low permeability to water-soluble substances, and the similarity of some elements of its structure to those of an intact mitochondrial membrane is only an apparent one.

Table 24. Anisotropy of Reconstituted Lipoprotein Vesicles Comprising F_1-ATPase and the Hydrophobic Protein Fraction from Mitochondrial Membranes (after Kagawa, 1972).

Characteristic	Mitochondria	Inside-out submitochondrial vesicles	Reconstituted lipoprotein vesicles
Energized state:			
movement of H^+ ions	Outside	Inside	Inside
ANS* fluorescence	Decrease	Increase	Increase
Exchange ^{32}P -ATP or the P/O ratio:			
lipophilic anion	No effect	Inhibitory	Inhibitory
lipophilic cation	Inhibitory	No effect	No effect
Effective antibodies	Anti-cyt. *c*	Anti-ATPase	-
Subject to chemical modification	Cytochrome cI	ATPase	-

* ANS, 1-Anilinonaphthalene-8-sulfonate.

7.3.2.2. *Reconstitution of Coupling Sites I and III from the Amorphous Membrane Fraction and Phospholipids*

The structural differences between the inner mitochondrial membrane and the vesicles reconstituted on the basis of the amorphous membrane fraction (the fraction of hydrophobic proteins) and phospholipids are especially manifest in the cases when the vesicles include various respiratory-chain oligoenzyme complexes.

As indicated in Chapter One, cytochrome *c* oxidase in the inner mitochondrial membrane is orientated so that cytochrome *a* faces its outer side where the cytochrome *c*-binding sites are, while cytochrome a_3 is on the inner side in contact with the matrix. Racker and Kandrach (1971, 1973) showed that reconstitution of coupling site III under the conditions specified in the preceding subsection gives rise to lipoprotein vesicles in which cytochrome *c* oxidase is positioned symmetrically with respect to the outer and inner faces of the membrane. In these experiments the hydrophobic membrane proteins were mixed with cytochrome *c*, cytochrome *c* oxidase, and phospholipids in the presence of cholate, the mixture subjected to dialysis, and coupling factors (F_1 and OSCP) added to the resulting vesicles. The vesicles displayed an ability to perform ADP phosphorylation coupled to N-methylphenazinium sulfate oxidation with a fair P/O ratio (0.25–0.5). An indispensable condition for phosphorylation was the thorough removal of cytochrome *c* from the medium and vesicle surface. In other words, phosphorylation took place only when cytochrome *c* contained within the vesicles was oxidized and proton transport was directed only from the medium into the vesicles. Addition of cytochrome *c* to such vesicles increased the rate of oxygen consumption and lowered the phosphorylation efficiency. Contrariwise, administration of polylysine known to compete with cytochrome *c* for the membrane binding sites attenuated oxygen uptake and markedly enhanced the phosphorylation efficiency (Racker and Kandrach, 1973). From these data it was concluded that during formation of artificial lipoprotein vesicles the cytochrome *c* oxidase molecules are positioned so that some of them can oxidize cytochrome *c* enclosed within the vesicles and some from the surrounding medium.

Results analogous to those described above were obtained by Ragan and Racker (1973) while reconstituting coupling site I. In their experiments the reconstituted vesicles included the fraction of hydrophobic proteins, isolated complex I (NADH-ubiquinone reductase), phospholipids, and necessary coupling factors. Though the reconstituted vesicles showed highly efficient oxidative phosphorylation (up to 0.5 mole ATP formed for each mole of NADH oxidized), the molecules of complex I there were arranged differently from the inner mitochondrial membrane. It has been indicated in Chapter One that the NADH dehydrogenase belonging to complex I is situated on the M-side of the membrane. In intact mitochondria it cannot react with exogenous NADH while in inside-out submitochondrial vesicles it is completely reducible by the substrate. According to Ragan and Racker (1973), in reconstituted vesicles complex I is orientated so that about 50% of its molecules react with exogenous NADH. This manifests itself as incomplete reduction of flavin and nonheme iron in NADH dehydrogenase. The components mentioned become wholly assessible to NADH in the presence of 0.5% deoxycholate which makes the vesicular membranes permeable to the substrate. It is thus quite possible that in the reconstituted vesicles some molecules of complex I are turned to the surrounding medium and some to the interior. Reasoning from the concepts of Mitchell (1966, 1972, 1976), the phosphorylating activity of such vesicles, which requires directed proton transfer across vesicular membranes, can only be explained by membrane impermeability to NADH; the latter is oxidized by ubiquinone exclusively on the vesicle surface with the protons being pumped into the vesicles (see, e.g., Ragan and Hinkle, 1975).

7.3.2.3. *Formation of Asymmetric Lipoprotein Vesicles*

In recent years methods have been developed in Racker's laboratory for constructing lipoprotein vesicles with asymmetric arrangement of complexes I, III, and IV in the membrane. One of the techniques consists in dispersing an appropriate phospholipid mixture with potassium cholate, adding the enzymic complex needed to the solution, and diluting or dialyzing the mixture.

In this case the reconstitution proceeds without the hydrophobic protein fraction. According to Ragan and Hinkle (1975), the molecules of complex I are incorporated into the forming vesicles in such a way that only 13% of the total complex content are not reduced on addition of NADH, while no less than 85% are accessible to the exogenous substrate. Oxidation of NADH by ubiquinone mediated by the vesicles is coupled with H^+-ion transfer to the interior, just as with submitochondrial vesicles. Another feature attesting to the orientation of complex I in the reconstituted vesicles is their interaction with the lipophilic anion tetraphenylboron. The anion has been shown to be transferred into the vesicles on addition of NADH and ubiquinone. The same migration of tetraphenylboron coupled with electron transfer is known to occur in submitochondrial vesicles. Thus there are grounds for thinking that the orientation of complex I in the reconstituted vesicles is the same as in submitochondrial vesicles.

By means of the above-mentioned procedure Leung and Hinkle (1975) prepared lipoprotein vesicles containing complex III. Ubiquinol oxidation by cytochrome *c* in their presence was accompanied by expulsion of H^+ ions from the vesicles into the medium and uptake of K^+ ions from the medium in the presence of valinomycin. Such ion fluxes are characteristic of intact mitochondria performing the same redox reaction. However, hardly any conclusions can yet be made from the abovesaid about the manner in which complex III is arranged in the vesicular membranes. Indeed, the functional asymmetry of the latter could be due simply to their impermeability to cytochrome *c*. Nevertheless Leung and Hinkle believe that complex III in the reconstituted vesicles is orientated similarly to that in the inner mitochondrial membrane, because the maximal electron-transfer rate in the vesicles (in the presence of uncoupler) is close to that for free complex III not inserted into a phospholipid membrane. However, there are no direct data confirming this assertion, while the comparison of the rates of the ubiquinol-cytochrome *c* reductase reaction in the two systems encounters certain difficulties (Hinkle et al., 1972).

Another means of obtaining lipoprotein vesicles with asymmetric protein disposition (Eytan et al., 1976) is to bind various enzymic complexes with performed vesicles containing 20-30% acidic phospholipids (cardiolipin, phosphatidylserine, or phosphatidylinositol). Cytochrome *c* oxidase incorporated into such liposomes displays high respiratory control (ratio of the exogenous cytochrome *c* oxidation rates in the presence and in the absence of uncoupler). Since the maximal cytochrome *c* oxidase activity of the vesicles observed on addition of uncouplers or detergents (e.g., Tween 80) is comparable to that of the isolated enzyme, it may be thought that cytochrome *c* oxidase in the vesicles is orientated so that all cytochrome *c*-binding sites face the surrounding medium and are completely accessible for the substrate. The same orientation of cytochrome *c* oxidase was established for intact mitochondria (see Chapter One). Similar results were obtained with complex III (ubiquinol-cytochrome *c* reductase).

It must be pointed out that the criterion employed (Eytan et al., 1976) for assessing the orientation of complex III and IV molecules in reconstituted vesicles is indirect and not quite reliable. Unfortunately, the structure of lipoprotein vesicles (proteoliposomes) which include the respiratory-chain complexes has not yet been studied in as much detail as that of mitochondrial membranes (see Chapter One). Direct experimental data on proteoliposome structure were presented by Ruben et al. (1976) who showed by freeze-fracture that the cytochrome *c* oxidase particles traverse the phospholipid membrane. This technique, however, provides no information on how the enzyme is orientated with respect to the exterior surface and the internal volume of the vesicles. As reported by Eytan and Broza (1978), subunits II and III of cytochrome *c* oxidase are preferentially accessible to p-diazobenzene sulfonate in proteoliposomes and are hence on their surface. Exactly the same orientation of cytochrome *c* oxidase has earlier been reported for mitochondria (Eytan et al., 1975; Chan and Tracy, 1978).

A strictly uniform binding to phospholipid vesicles containing certain amounts of acidic phospholipids was demonstrated for oligomycin-sensitive ATPase. As indicated above, phospholipid vesicles are impermeable to ATP. Consequently, ATPase molecules which are bound so that their active sites are screened from the medium would not catalyze substrate hydrolysis. Experiments showed that the ATPase activity of lipoprotein vesicles containing acidic phospholipids (Eytan et al., 1976) is not enhanced by the detergent Emasol 1130, which at 3% renders the vesicles permeable for ATP but does not inhibit the enzyme. Thus the ATPase molecules in the reconstituted vesicles are indeed arranged in a strictly definite manner matching that in submitochondrial vesicles.

7.3.2.4. *Factors Determining the Incorporation of Enzymic Complexes into Phospholipid Vesicles*

The efficiency of incorporation of enzymic complexes into phospholipid vesicles depends on experimental conditions (Eytan and Racker, 1975). The first to be considered is the vesicle composition. Thus cytochrome *c* oxidase added to a mixture of vesicles (liposomes) containing either phosphatidylethanolamine alone or a combination of three phospholipids (phosphatidylethanolamine, phosphatidylcholine, and phosphatidylserine in a 3:1:2 ratio) binds exclusively to liposomes that have phosphatidylserine in their composition. A similar result was obtained with a mixture of liposomes made of phosphatidylcholine alone or the above phospholipid cocktail. On the basis of these data it was concluded that cytochrome *c* oxidase has an especially high affinity for liposomes containing acidic phospholipids. Since acidic amino acids prevail in the enzyme itself, the role of phospholipids in its insertion into the liposomes appears to be more complicated than with cytochrome *c*, which is a basic protein and can bind with phosphatidylserine by electrostatic interactions.

Another circumstance determining the ability of a certain protein to bind with liposomes is the presence or absence of other proteins. It was found that incorporation of cytochrome *c* oxidase into liposomes favors subsequent binding of the hydrophobic proteins of oligomycin-sensitive ATPase. Proteoliposomes containing hydrophobic proteins bind cytochrome *c* oxidase worse than those devoid of such proteins. Unlike cytochrome *c* oxidase, ubiquinol-cytochrome *c* reductase (complex III) incorporates with equal effectiveness into liposomes and into proteoliposomes carrying the hydrophobic proteins of the ATPase complex. Unfortunately, all the above notions on the factors governing the interaction of enzymic complexes with phospholipid vesicles were inferred from indirect data. Indeed, the only criterion used to assess the binding of complexes was the value of the respiratory control.

7.3.2.5. *Artificial Proteoliposomes Containing Transhydrogenase, and Adenine Nucleotide and Phosphate Carriers*

The techniques considered above made it possible to reconstitute lipoprotein vesicles which included many protein complexes of the inner mitochondrial membrane.

Shertzer and Racker (1976) showed that by ultrasonically dispersing a mixture of phospholipids (phosphatidylethanolamine and phosphatidylcholine) and a protein fraction obtained by fractionating the submitochondrial-particle cholate extract with ammonium sulfate (40-45% saturation), vesicles can be prepared that are capable of ADP and ATP transport between their internal contents and the surrounding medium. The transport was suppressed by usual inhibitors (atractyloside, bongkrekic acid, and palmitoyl-coenzyme A) in low concentrations. Special experiments revealed that the ADP/ATP exchange is electrogenic.

Several works were aimed at reconstituting lipoprotein vesicles carrying pyridine nucleotide transhydrogenase. Initially Rydström et al. (1975) used for the reconstitution a crude transhydrogenase preparation with specific activity only 7-fold higher than in submitochondrial particles. Lipoprotein vesicles prepared by dissolving transhydrogenase and appropriate phospholipids in 2% cholate with subsequent dialysis could accumulate the lipophilic anion tetraphenylboron on addition of NAD^+ and NADPH to the medium. This process was uncoupler-sensitive and testified to the generation of a transmembrane potential in the system. An analogous study was later carried out by Höjeberg and Rydström (1977) with a highly purified transhydrogenase obtained as an individual protein of molecular weight 97,000 (see also Earle et al., 1978). The transhydrogenase incorporated into lecithin liposomes had properties similar to those of the enzyme in submitochondrial particles. As shown in experiments with fluorescent pH probes, both systems are proton pumps generating an uncoupler-sensitive membrane potential (positive inside with NAD^+ reduction by NADPH and outside with $NADP^+$ reduction by NADH). In the reconstituted system, just as in submitochondrial vesicles, the pyridine nucleotide-binding sites face the surrounding medium, as judged by the lack of additional activity of uncoupled transhydrogenase upon membrane disruption with detergents (see also Rydström, 1979). Proton transfer coupled to transhydrogenation in reconstituted vesicles was directly demonstrated by Earle and Fisher (1980).

According to Ragan and Widger (1975), the transhydrogenase activity of complex I is severalfold higher than that of beef heart submitochondrial particles. Lipoprotein vesicles built of phospholipids, mitochondrial hydrophobic protein fraction, F_1-ATPase, and complex I are able to catalyze the ATP-dependent reaction: $NADP^+ + NADH \rightarrow NADPH + NAD^+$. The reaction is inhibited by oligomycin, uncoupling agents, and small amounts of the detergent Triton X-100.

Similar data were obtained by Rydström (1979) who showed that incorporation of homogeneous transhydrogenase into liposomes together with oligomycin-sensitive ATPase yields a vesicular system catalyzing an oligomycin-sensitive ATP-driven transhydrogenase reaction.

Proteoliposomes capable of phosphate transport between their contents and environment were reconstituted in Racker's laboratory (Banerjee et al., 1977). Reconstitution was accomplished by sonicating a suspension of soybean phospholipids and purified phosphate carrier. The carrier orientation in the reconstituted vesicles was apparently the same as in intact mitochondria since the transport was inhibited both by hydrophilic N-ethyl maleimide and by hydrophobic N-benzyl maleimide. As shown earlier by Rhodin and Racker (1974), with submitochondrial vesicles the inhibition is observed only from N-benzyl maleimide which can penetrate the inner mitochondrial membrane.

Shertzer et al. (1977) demonstrated coupled operation of two transport systems incorporated into phospholipid vesicles. By their data, in proteoliposomes containing purified preparations of adenine nucleotide carrier and phosphate carrier the exchange of endogenous ATP for exogenous ADP is stimulated by inorganic phosphate, the stimulation being abolished by mersalyl or other sulfhydryl reagents known as inhibitors of the phosphate carrier.

7.4. RECONSTITUTION OF ENZYMIC COMPLEXES OF THE INNER MITOCHONDRIAL MEMBRANE

As follows from the above presentation, studies on the reconstitution of various mitochondrial inner-membrane functions have to date achieved substantial progress. It should be easily noted, however, that the reconstitution was usually carried out with isolated protein complexes each of which can be regarded as a more or less large membrane fragment. Quite another picture is formed on acquaintance with work devoted to reconstituting the oligoenzyme complexes from their subfragments or individual proteins.

7.4.1. NADH-Ubiquinone Reductase (Complex I)

This complex is known (see Chapter One) to comprise at least 16 different polypeptides; it also contains some ubiquinone and phospholipids. The task of reconstituting complex I could be considered solved if it could be made from completely separated components. The feat is, however, far from accomplishment. Not even methods have been developed for isolating all individual polypeptides, as is the case with, e.g., complex IV. More simple appears to be work on resolving complex I into large subfragments, followed by its reconstitution from such blocks. As indicated in Chapter One, Hatefi and

Stempel (1967, 1969) performed complex I fragmentation with 2.5 M urea and obtained three fractions, one of which was an NADH dehydrogenase of molecular weight 70,000, another contained nonheme iron and was completely devoid of flavin, and the third was considered to be a mixture of "structural proteins." Later similar investigations were carried out by Ragan (1976a) and Dooijewaard et al. (1978). The latter authors showed that the flavin-rich fraction and the flavin-lacking fraction interact upon removal of the dispersing (chaotropic) agent from the medium or lowering of its concentration, which results in increasing NADH-ferricyanide reductase activity and declining NADH-cytochrome *c* reductase activity of the system (see Chapter One).

7.4.2. **Succinate-Ubiquinone Reductase (Complex II)**

It is known that complex II, besides flavin not extractable in acidic media and nonheme iron, contains cytochrome *b* in an amount equimolar to flavin. Bruni and Racker (1968) showed that succinate dehydrogenase upon addition of mitochondrial phospholipid mixture and beef heart cytochrome *b* acquires succinate-ubiquinone reductase activity sensitive, similarly to the nonfragmented respiratory chain, to 4,4,4-trifluoro-1-(2-thienyl)-1,3-butanedione. A peculiarity of the reconstituted system is the large excess of cytochrome *b* compared with the intact respiratory chain (the heme to flavin molar ratio up to 28:1). The activity of the system is markedly enhanced if it also includes coupling factor F_4. Bruni and Racker supposed that the latter can perform organizing functions by establishing correct contacts between succinate dehydrogenase, phospholipids, and cytochrome *b*. It is of interest that cytochrome *b*, though being an indispensable component of the reconstituted system and not replaceable by cytochrome c_1, cytochrome *c* oxidase, or cytochrome *b* from other sources (e.g., from red blood cells or Escherichia coli), undergoes no detectable redox conversions during succinate oxidation by ubiquinone. This allows structural functions to be ascribed to cytochrome *b* in reconstitution. The possible role of cytochrome *b* in the functioning of the succinate-ubiquinone respiratory-chain segment is considered more closely by Hanstein et al. (1971).

The experimental data mentioned by no means exhaust the problem of reconstituting complex II. Indeed, the experiments involved succinate dehydrogenase, which itself is a complex system. Davis and Hatefi (1971) and Righetti and Cerletti (1971) showed that succinate dehydrogenase can be resolved into a ferroflavoprotein of molecular weight about 70,000 containing flavine, nonheme iron, and labile sulfides in a molecular ratio of 1:4:4; and an iron-sulfur protein of molecular weight about 27,000. So far there are no data on reconstitution of succinate dehydrogenase from such subfragments.

Finally, it must be pointed out that polyacrylamide gel electrophoresis of the fully dissociated complex II reveals 5 to 10 polypeptides (Davis and Hatefi, 1971; Capaldi, 1974; Capaldi et al., 1977). One of them (molecular weight 70,000) corresponds to the ferroflavoprotein, another (24,000-27,000) to the iron-sulfur protein, a third one (13,500) probably to the apo-Q-protein (see Chapter One). The identity and function of the other polypeptides have not been established. It was supposed (Yu et al., 1977a) that attachment of the apo-Q-protein to soluble low-molecular-weight succinate dehydrogenase suffices for restoration of succinate-ubiquinone reductase activity. In other words, three polypeptides (70,000, 27,000, and 13,500 molecular weight) are enough to carry out the reaction between succinate and ubiquinone.

In more detail the reconstitution of such three-component complex was studied by Vinogradov et al. (1980) and by Yu and Yu (1980c). Reconstitution can be achieved by mixing PQ_S and soluble succinate dehydrogenase in a Triton X-100-containing medium where PQ_S is in a dispersed state. The system reconstituted under such conditions displays several features of the native succinate-ubiquinone reductase region of the respiratory chain: its activity does not depend on dilution, it is stable in aerobic conditions (unlike succinate dehydrogenase), and sensitive to 4,4,4-trifluro-1-(2-thienyl)-1,3-butanedione and carboxin (Vinogradov et al., 1980). In media not containing detergents PQ_S exists in an aggregated form; on addition of succinate dehydrogenase PQ_S disaggregates and binds to it. The reconstituted complex is not broken down in 0.2% Triton X-100 and has a minimal molecular weight of 120,000 by gel filtration (Yu and Yu, 1980c).

Electrophoretic analysis of the reconstituted complex showed that the Coomassie-staining intensities of the bands corresponding to 70,000-, 27,000-, and 15,000-molecular-weight polypeptides are directly proportional to their molecular weights (Yu and Yu, 1980c). However, because of the lack of reliable data on the polypeptide stoichiometry in complex II it is yet impossible to judge the similarity of the intact and reconstituted systems in this respect. It remains not clear therefore how "precisely" the assembly of the minimal tripartite complex is performed. At any rate it would now be risky to claim that such succinate dehydrogenase molecule binding with a PQ_S molecule gives rise to an enzymically active succinate-ubiquinone reductase unit and that this unit exactly corresponds in its composition and structure to the native complex in the membrane.

This aspect of the assembly of the respiratory-chain succinate-ubiquinone reductase segment was to some extent considered by Ackrell et al. (1980) who started the reconstitution with soluble dehydrogenase and a fraction of complex II extracted with perchlorate under special conditions and containing mostly two polypeptides of molecular weights 13,500 and 7000. It turned out that maximal utilization of succinate dehydrogenase from the reconstitution mixture requires a 6-7-fold excess of the polypeptides over their amount in complex II. Ackrell et al. (1980) explained this by the presence of inactive or denatured polypeptides in the initial preparations. It was also suggested that since the 13,500- and 7000-molecular-weight polypeptides are always to some extent aggregated, not all of them are accessible to the succinate dehydrogenase molecules under the reconstituting conditions.

7.4.3. Ubiquinol-Cytochrome *c* Reductase (Complex III)

The classical studies on the fragmentation of complex III carried out by Rieske et al. (1967) and Silman et al. (1967) showed that the main components of the complex (cytochrome *b*, cytochrome c_1, core protein, and a specific iron-sulfur protein) can be resolved with the use of taurocholate or guanidine salts. The fragmentation, however, proved irreversible. As indicated before, to restore the respiratory chain activity after mitochondrial inner membrane fragmentation into oligoenzyme complexes with bile acid salts it is enough to lower somehow the detergent concentration in the medium. Fragmentation of complex III, according to Rieske et al. (1967), cannot be reversed even at the early stages by removing taurocholate or guanidine salts.

The data of Yamashita and Racker (1969) on the assembly of succinate oxidase from succinate dehydrogenase, cytochromes *b* and c_1, cytochrome *c* oxidase, ubiquinone, and phospholipids might have been regarded as evidence for the reconstitution of complex III, were it not for the serious difficulties in their interpretation pointed out above (Section 7.3.1.).

There are yet no data on the isolation of the 7-12 polypeptides constituting complex III (Das Gupta and Rieske, 1973; Gellerfors and Nelson, 1975; Bell and Capaldi, 1976; Katan et al., 1976; Marres and Slater, 1977; Smith and Capaldi, 1977; Siedow et al., 1978; Sidhu and Beattie, 1982) in preparative quantities and reconstitution of the complex from individual components.

To date only one polypeptide has been successfully extracted and then reinserted into the depleted complex III. Trumpower et al. (1980) showed that treatment of the succinate-cytochrome *c* reductase fragment of the respiratory chain with guanidine chloride in the presence of cholate and dithionite leads to extraction of a 24,500-molecular-weight polypeptide, with concomitant disappearance of the g=1.90 epr signal and loss of ubiquinol-cytochrome *c* reductase and succinate-cytochrome *c* reductase activities. When the purified iron-sulfur protein is incubated with the depleted complex both activities are restored; simultaneously the complex acquires the g=1.90 epr signal and the 24,500-molecular-weight polypeptide. The maximal ubiquinol-cytochrome *c* reductase activity of the reconstituted system reaches about one third of that of the initial complex prior to extraction of the iron-sulfur protein. This was supposed to result from the lack of an unknown component extracted together with the iron-sulfur protein. Another cause of the low activity of the reconstituted system could be the partial inactivation of the iron-sulfur protein, which is especially pronounced in dilute solutions (Trumpower et al., 1980).

7.4.4. Cytochrome *c* Oxidase (Complex IV)

Complex IV is the only respiratory-chain component for which methods of preparative resolution of all constituent polypeptides have been devised (Sebald et al., 1973; Poyton and Schatz, 1975a; Steffens and Buse, 1976; Werner, 1977). Reconstitution of the complex from individual polypeptides has not been carried out.

7.4.5. Oligomycin-Sensitive ATPase (Complex V)

As mentioned in Chapter One, high-resolution polyacrylamide gel electrophoresis reveals up to 17 types of polypeptides in the preparations of oligomycin-sensitive ATPase. Glaser et al. (1980) showed that treatment of this complex (isolated from beef heart mitochondria) with 3.5 M NaBr entails detachment of F_1-ATPase. The residue, termed F_0, contains seven components according to the data of sodium dodecyl sulfate-polyacrylamide gel electrophoresis. Five polypeptides are stained in the gel with Coomassie Blue; two of these correspond to OSCP (molecular weight 21,000) and coupling factor F_6 (molecular weight 9400), three others (molecular weights 23,000, 10,500, and 8600) have not been identified. Finally, two polypeptides from F_0 were detected after incubating F_0 with [^{14}C]dicyclohexylcarbodiimide and were in all probability the monomeric and oligomeric DCCD-binding protein. Addition of F_1 to F_0 leads to a decrease in ATPase activity and an increase in its sensitivity to oligomycin. The oligomycin sensitivity reached a level (about 85% inhibition) close to that observed with the initial oligomycin-sensitive ATPase, on condition that a 4-5-fold weight excess of F_0 was taken for reconstitution. The reconstituted F_0-F_1 sedimented from the reconstitution mixture of the specified composition (i.e. F_0/F_1 about 5) showed approximately the same relation of polypeptides as the intact complex (Glaser et al., 1977b). The necessity for a large excess of F_0 for proper reconstitution is apparently explained by its being present in an aggregated form and only partly interacting with soluble F_1-ATPase.

Somewhat different results were obtained by Galante et al. (1981) for the fragmentation and subsequent reconstitution of the ATP-synthetase complex (complex V). Extraction of the latter with NaBr resulted in the removal of almost all β, δ, and ε subunits of F_1-ATPase and about 50% of the γ subunit while the α subunit was almost completely retained in the insoluble residue. On addition of F_1 to the depleted system (NaBr-F_0) it again acquired ATPase activity reaching 30-50% of that of the intact complex and inhibited by oligomycin by no more than 70-75% (compared with the more than 90% inhibition in complex V).

More close in its features to the intact ATP-synthetase complex is the system reconstituted from the insoluble residue after treating the complex with urea (urea-F_0), OSCP, and F_1-ATPase. The residue was almost completely devoid of δ and ε subunits of F_1-ATPase, OSCP, and ATPase protein inhibitor, and of most of α, β, and γ F_1-ATPase subunits. However, even in the most favorable case restoration of a more than 90% sensitivity to oligomycin requires at least a twofold weight excess of urea-F_0 over F_1 in reconstitution, and 94% of inhibition is attained in a system reconstituted with a sixfold excess of urea-F_0.

SUMMARY

In the beginning of this chapter a hope was expressed that the information gained in studying the fragmentation and reconstitution of the inner mitochondrial membrane may cast light on the specialities of interactions among its component parts. If the membrane could be reconstituted *in vitro* from isolated individual polypeptides, phospholipids, and other components, then it could be supposed that the sole principle governing the assembly is the mutual recognition of the constituent elements based on their exceptionally high affinity for each other. This would imply that each polypeptide must bind in the membrane with a strictly defined set of partners, regardless of the place in which it has been synthesized and inserted into the membrane.

What are the actual achievements of the reconstitution experiments? First, mild membrane fragmentation permits isolation of lipoprotein complexes still carrying partial catalytic functions of the respiratory chain. Such complexes without any additions can form a membranous system having the properties of an uncoupled respiratory chain. Second, respiratory complexes, pyridine nucleotide transhydrogenase, components of oligomycin-sensitive ATPase, and adenine nucleotide and phosphate carriers can be incorporated into phospholipid vesicles to form systems capable of energy transformation and some energy-dependent functions of the inner mitochondrial membrane. The reconstituted proteoliposomes in their structure to a certain extent resemble submitochondrial vesicles. This applies in particular to the asymmetry in the disposition of complexes I, III, and IV, as well as of F_1-ATPase and the translocases with respect to the outer and inner sides of the reconstituted membranes. The latter conclusion can only be made with reserve, since the structure of these membranes, as well as many aspects of the mitochondrial inner membrane structure, is far from being sufficiently investigated. A substantial difference between the natural and reconstituted membranes is the relative content of proteins and phospholipids. Thus the inner mitochondrial membranes contains about 30% phospholipids (see Chapter One). At the same time in reconstituted proteoliposomes there is at least a twofold excess of phospholipids (Kagawa, 1972), and in some cases it reaches a 15-fold (Leung and Hinkle, 1975) or a 20-fold value (Eytan and Racker, 1977). Furthermore, the most efficient reconstitution of a potential-generating transhydrogenase is attained, according to Rydström (1979), at a phospholipid-to-protein ratio of at least 200.

As could be seen from this chapter, no appreciable success has yet been gained in reconstituting complexes from their polypeptide components. This in great measure is due to the high hydrophobicity of many polypeptides and their propensity to spontaneous aggregation in the absence of phospholipids or detergents (see Green and Hechter, 1965).

On the whole, the experiments on the reconstitution of the inner mitochondrial membrane and its separate fragments still give no answer to the question posed in the beginning of this chapter, viz. how selective is the binding between polypeptides in the membrane. The experimental data available provide yet no grounds for claiming the exceptionally high affinity between the simplest constituent elements of the membrane to be the sole principle governing its formation. Notwithstanding, it must be acknowledged that the reconstitution of proteoliposomes carrying various components of the mitochondrial energy-transducing machinery accomplished in the laboratories of Racker (see this chapter) and Skulachev (Bakeyeva et al., 1970; Grinius et al., 1970; Jasaitis et al., 1972; Skulachev, 1972; Drachev et al., 1976a,b; Barskii et al., 1975) was a weightly contribution to the understanding of mechanisms underlying energy transformation in mitochondria.

8
Control Over Mitochondrial Assembly

As follows from Chapters One, Six, and Seven, mitochondria comprise about a hundred various enzymes (see, e.g., Munn, 1974) which are in a strictly defined way compartmentalized, integrated into supramolecular ensembles, specifically orientated in the membrane, etc. The molecular mechanisms of these processes have not yet been elucidated. Consideration of the principles governing mitochondrial formation is hindered by the fact that the organelle structural organization is far from being completely unraveled. Indeed, the exact localization has not been ascertained for all mitochondrial proteins identified to date; there is no final solution to the question of whether the enzymic complexes are the sole form of existence of a number of membrane proteins; little studied are the principles of interaction between the complexes in the membrane. Finally, we have yet no definitive knowledge of the composition of mitochondria and in particular of mitochondrial membranes. As to the assembly of the latter, here the approach predominant so far is one which is usually applied to the structural elements of bacteriophages as well as to ribosomes and soluble multienzyme complexes. Namely, it is supposed that the affinity between the protein (and lipid) membrane components is so high and their mutual binding so selective as to suffice for the formation of all the elements of mitochondrial structure. It must however be pointed out that neither the experiments on the reconstitution of the inner mitochondrial membrane (see Chapter Seven) nor the data on the formation of the membrane oligoenzyme complexes in vivo (see Chapter Six) give full grounds for such a point of view. Really, no one has yet succeeded in reconstituting the complexes from individual proteins and lipids. It is thus easily seen how far we are still from solving the problem of the complete reconstitution of the inner mitochondrial membrane.

The matter is no better with the studies into mitochondrial formation *in vivo*. The data now available testify that the assembly is a multistage spatially intricate process. No one can say how "precisely" it is carried out, i.e. whether all components reach their place of destination in different mitochondrial compartments, whether all of them find their partners in complexes, and so on. On the contrary, taking into account that errors are inevitable at every stage, that there are many stages, and that the assembly is not directed by any template (as it is the case with, e.g., ribosomes or viral particles where the template role is played by the molecules of nucleic acids), it must be *a priori* expected that, because of error amplification on the scale of the entire mitochondrion, the deviations from a structurally "ideal" organelle would be quite discernible. One of the aims of this chapter is to illustrate this issue with experimental data.

As shown in Chapter Two, mitochondria are rather labile structures rapidly degrading on being isolated from cells, first of all under the action of endogenous lytic enzymes (phospholipase A_2 and probably specific neutral proteinases) and also because of lipid oxidation induced by some membrane components (e.g., iron-sulfur proteins). With all this, mitochondria may survive even *in vitro* for a long time provided that they can perform oxidative phosphorylation in the incubation medium (see Chapter Three).

These data immediately prompt one to think that mitochondrial assembly may be controlled by a functional criterion, so that non-functioning (i.e. non-energy-transducing) organelles or their parts are subject to selective elimination ("rejection"). To demonstrate such elimination is the second aim of this chapter.

In the above context the necessity becomes evident for looking for mechanisms through which the selection can be effected. Approaches to solving this problem have been found in recent years, and the third aim of the chapter is to cover the first advances along this line.

8.1. FORMATION OF ABNORMALITIES IN THE INNER MITOCHONDRIAL MEMBRANE AND THEIR ELIMINATION DURING RESPIRATORY SYSTEM DEVELOPMENT

8.1.1. Lack of Positive Correlation between Cell Respiration and Cytochrome Content in Yeast

Galkin et al. (1975), studying the development of the respiratory system in *Saccharomyces cerevisiae* during aerobic growth in a galactose-containing medium, found that in the exponential phase there is no positive correlation between the changes in the respiration rate and in the cell cytochrome contents. According to these authors, the respiration soon reaches its maximal value which then remains constant throughout further growth (Figure 53a). At any growth moment studied the respiration is completely inhibited by cyanide or antimycin A and substantially stimulated by uncouplers (Figure 54). This means that the rate of oxygen consumption by the cells was limited by the mitochondrial respiratory chain and could serve as a measure of the development of the respiratory activity of the mitochondrial population. As to the contents of cytochromes *b*, c_1, *c*, and aa_3, they can be seen in Figure 53b to reach the peak value simultaneously with the cell respiration in the mid-exponential phase but then, most peculiarly, to gradually decrease to a level roughly corresponding to that in the initial inoculum.

These and some other data gave grounds for suggesting (Galkin et al., 1975) that at the earlier stages of respiratory development in yeast growing aerobically on galactose an excess of cytochromes is

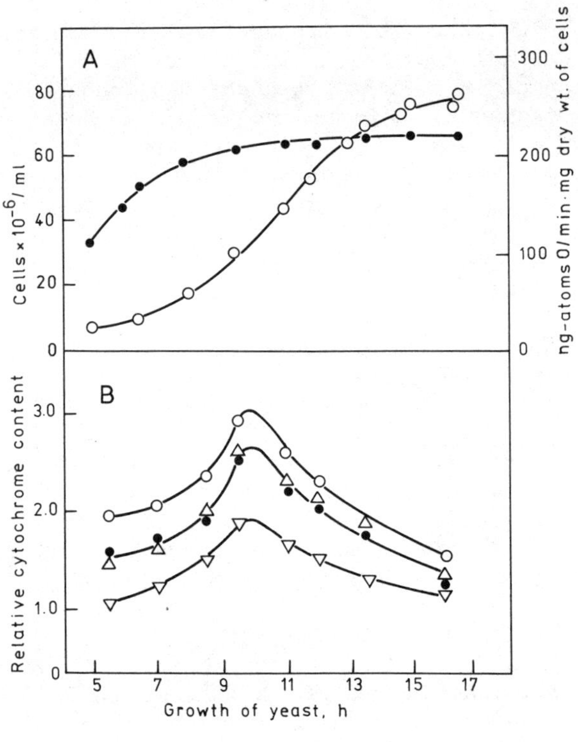

Fig. 53. Aerobic growth of Saccharomyces cerevisiae in galactose-containing medium and variations in the cell cytochrome content. (A) Growth in culture: (○) concentration of cell suspension; (●) cyanide-sensitive respiration. (B) Relative cytochrome content of the intact cells: (○) cytochrome *c*; (●) cytochrome c_1; (△) cytochrome *b*; (▽) cytochrome aa_3. Cytochrome content in the cells of the inoculum is taken equal to 1.0. From Galkin et al. (1975).

formed, which proves to be unnecessary for the maintenance of a constant respiration level and is probably eliminated in the course of further culture growth (i.e. further development of the respiratory system).

8.1.2. Involvement of Proteinases in the Control over Formation of the Mitochondrial Respiratory Chain

If elimination of a certain portion of cytochromes indeed takes place during the development of the yeast respiratory apparatus, it must involve cellular and/or mitochondrial proteinases. The validity of this suggestion can be tested in experiments with proteinase inhibitors. Use of the latter in vivo has to meet certain conditions. Firstly, inhibitors should penetrate the cell and suppress the proteinases presumably responsible for the cytochrome breakdown. Secondly, the inhibitory effect should be extended in time. Thirdly, inhibitors should be administered in such a way as not to affect appreciably the cell physiology.

Detailed studies of the effects of proteinase inhibitors on the development of the yeast respiratory system have been carried out in this laboratory (Luzikov et al., 1976; Galkin et al., 1979a,b,c). The inhibitors used were phenylmethylsulfonyl fluoride and pepstatin. The first is known to inactivate proteinase B (Saheki and Holzer, 1975; Ulane and Cabib, 1976) and carboxypeptidase Y (Hayashi et al., 1975; Wolf and Fink, 1975) localized in the yeast cell vacuoles; it also appears to inhibit the endogenous mitochondrial proteinase responsible for the breakdown of mitochondrial translation products (Kal'nov et al., 1978). Pepstatin is a potent inhibitor of the vacuolar proteinase A (Saheki and Holzer, 1975). Phenylmethylsulfonyl fluoride and pepstatin added into the growth medium of *Saccharomyces cerevisiae* were found (Bakalkin et al., 1976) to completely suppress the degradation of the total cell protein. The suppression is transient and requires therefore repeated addition of the inhibitors. Phenylmethylsulfonyl fluoride also prevents the proteolysis of mitochondrial translation products in yeast cells (Kal'nov et al., 1979a).

As shown in Figure 54b-e, addition of phenylmethylsulfonyl fluoride to the culture growing on 2% galactose (in portions of 50 μmoles per liter at intervals of 90 - 120 min), beginning from the moment corresponding to the maximal cytochrome contents, markedly retards (cytochrome *c*) or completely blocks (cytochromes *b*, c_1, and aa_3) the decrease in their levels. Similar results were obtained with pepstatin (final concentration less than 2×10^{-6} M). In both cases cell respiratory activity measured either in the presence or in the absence of uncoupler was not noticeably changed (Figure 54a).

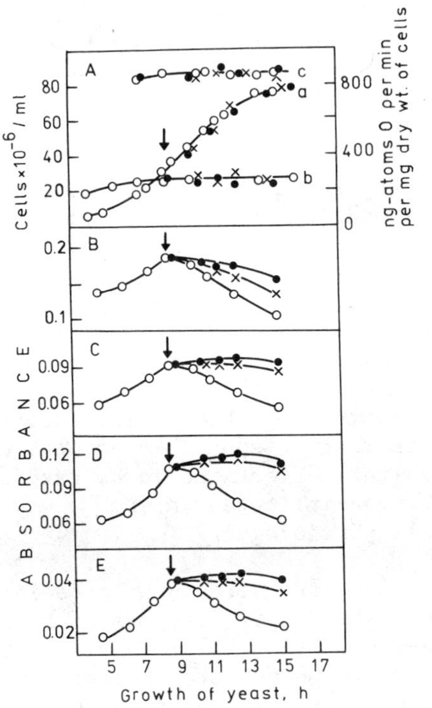

Fig. 54. Effect of proteinase inhibitors on yeast aerobic growth in galactose-containing medium. (A) Growth in culture: (a) concentration of cell suspension; (b) nonstimulated and (c) uncoupler-stimulated respiration. (B–E) Cell cytochrome content: (B) cytochrome *c*; (C) cytochrome c_1; (D) cytochrome *b*; (E) cytochrome aa_3. (o) Control; (•) in the presence of phenylmethysulfonyl fluoride; (x) in the presence of pepstatin. The arrow indicates the beginning of inhibitor administration. From Luzikov et al. (1976).

Figure 55 summarizes the data on yeast growing in a glucose medium (initial concentration 0.25%). In this case the respiratory activity increases throughout the exponential phase in parallel with the contents of cytochromes *b*, c_1, *c*, and aa_3. Cell respiration is completely inhibited by cyanide or antimycin A and significantly stimulated by uncoupler, the relation of the rates of oxygen uptake in the presence and in the absence of uncoupler being approximately the same throughout the period examined. Judging from this, the cell respiration reflects the respiratory activity of the mitochondrial population.

Introduction of phenylmethylsulfonyl fluoride or pepstatin into the yeast suspension during the exponential phase entails gradual

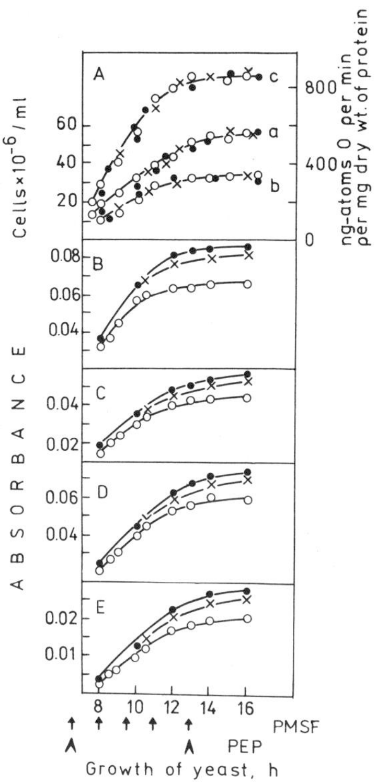

Fig. 55. Effect of proteinase inhibitors on aerobic growth of yeast in glucose-containing medium. (A) Growth in culture: (a) concentration of cell suspension; (b) nonstimulated and (c) uncoupler-stimulated respiration. (B-E) Cell cytochrome content: (B) cytochrome *c*; (C) cytochrome c_1; (D) cytochrome *b*; (E) cytochrome aa_3. (○) Control; (●) in the presence of phenylmethylsulfonyl fluoride; (x) in the presence of pepstatin. The times of inhibitor addition are indicated by arrows. From Galkin et al. (1980).

accumulation of cytochromes *b*, c_1, *c*, and aa_3 above control levels, this excess reaching 25-30% by the early stationary phase with phenylmethylsulfonyl fluoride; the effect of pepstatin is only slightly less pronounced. With all this, neither inhibitor induces any detectable changes in the cell respiratory activity (Figure 55a). Thus, under conditions when some cellular and mitochondrial proteinases are suppressed, the development of the respiratory system in yeast growing on glucose is accompanied by formation of "excess" respiratory-chain components which do not appear to contribute to the cell respiratory activity.

It must be emphasized that the experimental conditions were chosen so that neither phenylmethylsulfonyl fluoride nor pepstatin appreciably affected yeast growth. The decrease in the growth rate never exceeded 15% and was only transient, so that the cultures completely recovered within 2 h after the first addition. Some synthetic proteinase inhibitors (e.g., tosylphenylalanyl chloromethylketone and tosyllysyl chloromethylketone) are known to have a number of side effects, the main one being inhibition of protein synthesis (McIlhinney and Hogan, 1974; Pong et al., 1975; Ito, 1977). Judging from the results presented in Figures 54 and 55, phenylmethylsulfonyl fluoride had no such effects. Account here should be taken not only of the data on the biomass content but also of the data testifying to the increase in the amount of hemoproteins induced by phenylmethylsulfonyl fluoride and pepstatin. This probably has been achieved by adding the inhibitors in small portions and small total amounts. By no means unimportant was the composition of the growth media which were rich enough to support normal protein synthesis from free precursors even under conditions hindering protein catabolism.

Summing up what was said above, it can be concluded that yeast growing in the presence of small amounts of phenylmethylsulfonyl fluoride and pepstatin provide a satisfactory model for studying mitochondrial formation under conditions of selectively blocked proteolysis.

Attempts to use other inhibitors for the same purpose failed because of their side effects. In particular, the inhibitors of serine proteinases tosyllysyl chloromethylketone or tosylphenylalanyl chloromethylketone even at 0.2 mM suppressed yeast growth and only accelerated cytochrome breakdown in galactose-grown yeast. Diazoacetyl-DL-norleucine methyl ester (an inhibitor of acid pepsine-like proteinases) proved useless since it is effective only in the presence of copper ions in concentrations that completely inhibit yeast growth (Galkin, 1980).

The results available indicate that the development of the mitochondrial respiratory machinery is attended by formation in yeast cells of a certain "excess" of cytochromes that do not contribute to the mitochondrial respiratory activity and are eliminated with the participation of proteinases in the course of "improvement" of the respiratory apparatus. These "excess" entities are readily observable in galactose-grown yeast. In glucose-grown cultures they are only detectable in the presence of proteinase inhibitors. This discrepancy is probably due to that in the latter case the development of the respiratory system proceeds substantially more slowly (compare Figures 54a and 55a). Then the formation of excess cytochromes is also slower, and they do not accumulate to a detectable extent even at the usual rate of degradative processes.

It should be noted that the work the results of which were

presented in this section was initiated in 1974 (the first publication was Luzikov et al., 1976), i.e. before anything was known about the existence of precursors to some polypeptides constituting the respiratory chain and their proteolytic processing (see Chapter Six). It could have happened so that administration of proteinase inhibitors inhibited the processing and, as a consequence, suppressed the formation of the respiratory chain. By a happy chance, however, the maturases (if they are indeed proteinases! - see Section 6.3.7) proved to be insensitive to phenylmethylsulfonyl fluoride or pepstatin, and are suppressed only by chelating agents such as o-phenanthroline, EDTA, and zincon.

8.1.3. Formation of Nonfunctioning Respiratory-Chain Components and their Elimination During Differentiation of Yeast Mitochondria

A question arises about the nature of the excess cytochromes considered in the preceding section and, in particular, about their relation to the mitochondrial respiratory chain. Preliminary information along this line can be gained from analysis of low-temperature difference (reduced minus oxidized) spectra of intact cells. As follows from Figure 56, yeast grown in a glucose medium in the absence (control) or in the presence of phenylmethylsulfonyl fluoride to the early stationary phase (see Figure 55) contain an identical set of cytochromes. According to these data, phenylmethylsulfonyl fluoride does not induce the appearance of new hemoproteins or any substantial alterations in the ratios of cytochromes constituting the respiratory chain.

Figure 56 also shows that cytochromes c, c_1, and aa_3 are completely reduced by ethanol under anaerobic conditions both in control and in phenylmethylsulfonyl fluoride-grown cells. Reduction of cytochrome b in both cases reaches 75% of that obtained with dithionite. Thus the excess cytochromes accumulated in the presence of the proteinase inhibitor are reducible with physiological substrates and are therefore somehow linked with the respiratory chain.

Further studies into the nature of the excess cytochromes require detailed analysis of the activities of separate respiratory-chain segments. Such analysis can only be carried out with the entire mitochondrial population isolated from the yeast cells. Results presented below were obtained in experiments with the membrane fractions sedimented from cell homogenates at 70,000 g. These fractions quantitatively contained cytochromes b and aa_3 of the homogenates. Only minor amounts of cytochrome c and no marker enzymic activities of the inner mitochondrial membrane were detectable in 70,000 g supernatants. The amounts of cytochromes and activities of the membrane fractions were determined with respect to unit weight of cells from which the fractions were isolated. Obviously only such presentation of data allows the isolated membrane fraction to be used for studying the cell respiratory system as a whole.

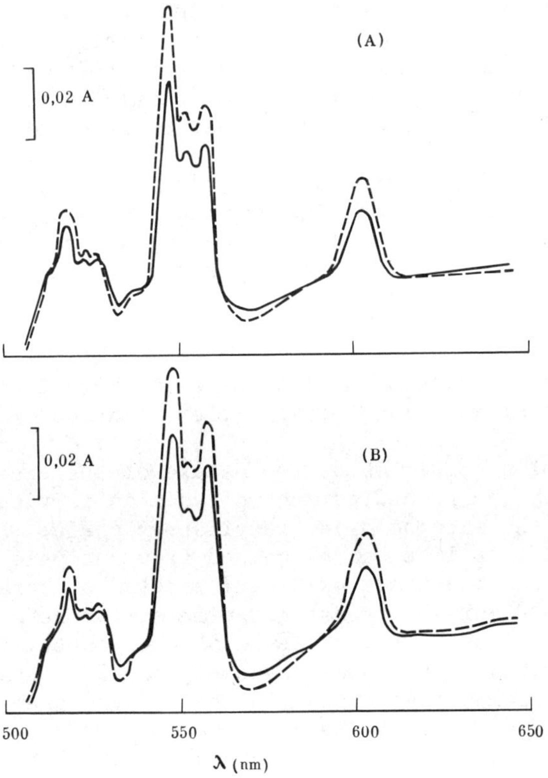

Fig. 56. Low-temperature difference spectra of intact yeast cells grown on glucose till the early stationary phase in the absence (——) and in the presence (--) of phenylmethylsulfonyl fluoride. (A) Anaerobically reduced with ethanol minus oxidized; (B) dithionite-reduced minus oxidized. From Galkin (1980).

With glucose-grown yeast, the membrane fractions were isolated in the early stationary phase from control cells and from cells treated with phenylmethylsulfonyl fluoride as described above. With galactose cultures, the membrane preparations were isolated from mid-exponential control cells possessing the highest cytochrome content as judged by whole-cell determinations (Figure 54), from stationary-phase control cells, and from cells grown from mid-exponential to stationary phase in the presence of phenylmethylsulfonyl fluoride and pepstatin.

The data on the cytochrome contents and oxidase activities in the membrane fractions, summarized in Tables 25 and 26, correspond to what has been found for intact cells (Figures 55 and 54). Indeed, the membrane fraction from glucose cells grown with phenylmethyl-

Table 25. Characteristics of Mitochondrial Populations from Glucose-Grown Yeast (Galkin et al., 1979b).

Source of preparation	Cytochrome contents (nmoles·g dry wt. cells^{-1})			Activities (μmoles substrate·min^{-1}·g dry wt. cells^{-1})			
	$c+c_1$	b	aa_3	Succinate oxidase	Ascorbate(TMPD)* oxidase	Ubiquinol oxidase	Succinate-cyt.c reductase
Control	41.7	57.8	14.4	46.0	112	44.0	81.0
PMSF**-grown	52.0	72.0	18.5	46.5	115	45.0	79.5

* TMPD, N,N,N',N'-tetramethyl p-phenylene diamine.
** PMSF, phenylmethylsulfonyl fluoride.

Table 26. Characteristics of Mitochondrial Populations from Galactose-Grown Yeast (Galkin et al., 1979a).

Source of preparation	Cytochrome contents, nmoles/g dry wt. cells			Oxidase activities, μg-atoms O/min·g dry wt. cells			
				Succinate oxidase		Ubiquinol oxidase	Ascorbate(TMPD)* oxidase
	c+c$_1$	*b*	*aa*$_3$	basal	+5μM cyt.*c*		
Mid-exponential control	48.7	59.3	13.0	26.2	38.3	95.0	278
Stationary control	26.8	39.1	8.6	27.1	38.9	94.3	261
Stationary inhibitor-grown**	43.7	57.3	12.4	26.8	37.8	95.4	263

* TMPD, N,N,N',N'-tetramethyl p-phenylene diamine.

** Proteinase inhibitors (phenylmethylsulfonyl fluoride and pepstatin) were added in portions beginning with the mid-exponential phase.

sulfonyl fluoride contains 25-30% more cytochromes $c+c_1$, bb, and aa_3 than the control one (Table 25). As to galactose-grown yeast, between the mid-exponential and early stationary phase of their growth the contents of all cytochromes fell by 35-45%, this decrease being virtually completely blocked by the proteinase inhibitors (Table 26). With all this, the proteinase inhibitors elicited no detectable changes in the succinate oxidase activity of the membrane fractions from either glucose or galactose-grown cells. These results confirm the validity of cell respiration and cell cytochrome content as criteria of the state of the mitochondrial respiratory chain in yeast. On the other hand, the similarity of the results obtained with intact cells and isolated membrane fractions testifies that the mitochondria were not differentially injured during isolation.

The following explanations can in principle be offered for the lack of surplus activity in preparations containing excess amounts of cytochromes:

(1) activity is limited by a noncytochrome component, the quantity of which does not increase in parallel with the cytochrome content;
(2) activities of all or some components are attenuated in proportion to the increase in their content in mitochondria;
(3) the "excess" respiratory-chain components are for some reason nonfunctioning.

Choice among these variants can be made on the basis of data on the partial activities of the respiratory chain.

8.1.3.1. *Activities of Respiratory-Chain Segments: Membrane Fractions from Glucose-Grown Yeast*

As follows from Table 25, the membrane fractions oxidize ubiquinol at identical rates irrespective of whether they have been isolated from control or phenylmethylsulfonyl fluoride-treated cells. In both cases the oxidation proceeds through the cytochrome chain since it is completely suppressible with cyanide or antimycin. Similarly there are no differences in ascorbate(TMPD) oxidase activities of the two preparations. The rate of this reaction can be considered as a characteristic of the respiratory-chain segment including cytochromes c and aa_3. Finally, closely similar rates of cytochrome c reduction, regardless of its concentration, are displayed by the succinate-cytochrome c reductase segments of the two preparations. Thus, proteinase inhibitors, while inducing accumulation of an "excess" of cytochromes, do not increase the activities of various respiratory-chain segments measured in intact membranes (Galkin et al., 1979b).

The next step in elucidating the causes of the inactivity of the excess respiratory-chain components in the membrane fractions was to evaluate the full potential catalytic capacity of the individual oligoenzyme complexes, which can be done after their maximal activation, for instance by freezing-thawing the membrane preparations in the presence of a nonionic detergent Lubrol WX.

Such data are available for succinate-ubiquinone reductase, ubiquinol-cytochrome *c* reductase, and cytochrome *c* oxidase segments (Galkin et al., 1979b). As shown in Figure 57, the succinate-ubiquinone reductase activity of the solubilized membrane fraction from cells grown in the presence of phenylmethylsulfonyl fluoride exceeds the corresponding value for control cells in the whole range of ubiquinone-2 concentrations. It is also evident that the proteinase inhibitor does not alter the K_m for ubiquinone. In other words, the increment in activity brought about by growth with the proteinase inhibitor is due to its increased content in mitochondria and is not associated with altered affinity of the enzyme for ubiquinone.

The maximal cytochrome *c* oxidase activity is also higher in the solubilized membrane fraction from inhibitor-grown cells (Table 27), in compliance with the higher cytochrome aa_3 content in this preparation. Kinetic analysis reveals equal affinity of this and control preparations for cytochrome *c*.

Finally, similar results were obtained for ubiquinol-cytochrome *c* reductase in activated preparations of the two membrane fractions;

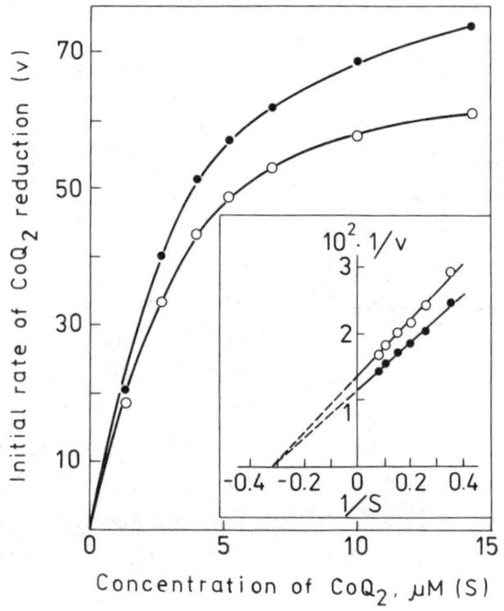

Fig. 57. Succinate-ubiquinone reductase activity of solubilized mitochondrial preparations from stationary control (○) and phenylmethylsulfonyl fluoride-grown (●) glucose yeast. Solubilization was achieved by freeze-thawing the membranes in the presence of Lubrol WX. Activity (initial rate) is expressed as μmoles ubiquinone-2 reduced per min per g dry weight cells. From Galkin et al. (1980).

Table 27. Activities of Individual Respiratory-Chain Complexes in Solubilized Membrane Preparations from Glucose Yeast (Galkin et al., 1979b).

Source of preparation	Succinate-ubiquinone reductase	Ubiquinol-cytochrome *c* reductase	Cytochrome *c* oxidase
	μmoles substrate/min·g dry wt. cells		
Control	52.2	77.0	51.5
PMSF-grown	63.4	98.3	65.1

the control one had lower activity, though this was only demonstrable at sufficiently high concentrations of both substrates (Table 27).

Thus, inhibition of proteinases leads to an increase in the activities of all respiratory-chain segments in the mitochondrial population in parallel with the increase in cytochrome content (Galkin et al., 1979b). This fact means that the first two suggestions concerning the causes of the similarity of respiratory activities in membrane fractions containing different amounts of cytochromes (see above) are hardly probable. Indeed, now there are no grounds for assuming that the mitochondrial populations of yeast grown in the absence and in the presence of phenylmethylsulfonyl fluoride have equal amounts of components limiting the rate of electron transfer in the respiratory chain. The data presented above also rule out the possibility that the activity of some components is attenuated in proportion to the increase in their content in mitochondria. The most plausible suggestion then is that the excess respiratory-chain components are normally constituted and potentially active.

8.1.3.2. *Activities of Respiratory-Chain Segments: Membrane Fractions from Galactose-Grown Yeast*

Membrane fractions isolated from galactose-grown yeast in the mid-exponential and early stationary phases displayed identical succinate oxidase, ascorbate (TMPD) oxidase, and ubiquinol oxidase activities (Table 26).

Quite another picture was observed when the activities of individual respiratory-chain complexes were measured in solubilized membrane preparations (Galkin et al., 1979a,c). The maximal succinate-ubiquinone reductase, ubiquinol-cytochrome *c* reductase, and cytochrome *c* oxidase activities of the membrane fraction isolated

Table 28. Activities of Individual Respiratory-Chain Complexes in Solubilized Membrane Preparations from Galactost Yeast (Galkin et al., 1979a).

Source of preparation	Succinate-ubiquinone reductase	Ubiquinol-cytochrome *c* reductase	Cytochrome *c* oxidase
	μmoles substrate/min·g dry wt. cells		
Mid-exponential control	44.0	77.3	56.0
Stationary control	32.5	58.1	32.5
Stationary inhibitor-grown	44.7	75.0	52.6

from mid-exponential yeast exceeded the respective values for stationary-phase yeast (Table 28). In other words, the activities of all constituent complexes of the respiratory chain decreased during culture growth. Proteinase inhibitors (phenylmethylsulfonyl fluoride and pepstatin) blocked this decrease: in their presence the activities of all segments were similar in the exponential and stationary-phase preparations. It is easy to see that these results correlate with the data on the cytochrome contents in the membrane fractions (compare Tables 26 and 28).

As follows from Figure 58, the decrease in succinate-ubiquinone reductase activity by the stationary phase is not accompanied by any changes in the K_m value; proteinase inhibitors prevent this decrease and do not affect the affinity constant. Analogous results were obtained for ubiquinol-cytochrome *c* reductase and cytochrome *c* oxidase segments of the respiratory chain (Galkin et al., 1980).

All this allows one to think that the mitochondrial population of galactose-grown yeast in the exponential growth phase includes a considerable portion of respiratory-chain components that, albeit potentially active, do not contribute to the overall respiratory activity of mitochondria (Galkin et al., 1979a,c). These components disappear in the course of further development of the yeast respiratory system but can be preserved by adding proteinase inhibitors to the growing culture.

It must be emphasized that the experiments with glucose- and galactose-grown yeast complement each other nicely. On the one hand, the data on glucose-grown cultures indicate that the accumulation of excess respiratory-chain components and their subsequent elimination are not a unique property of growth on galactose. On the other hand,

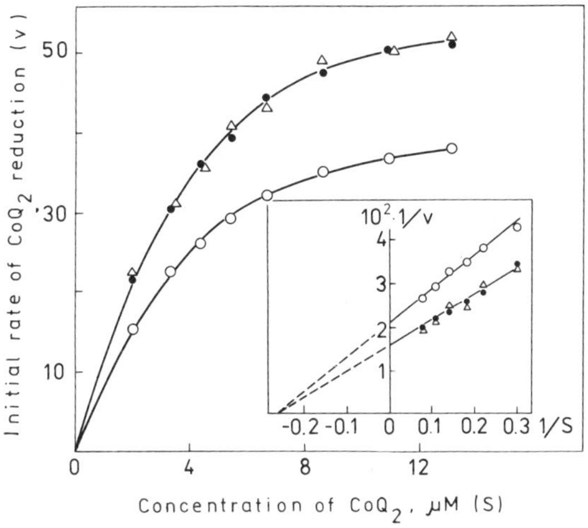

Fig. 58. Succinate-ubiquinone reductase activities of solubilized membrane fractions from yeast grown on galactose till the mid-exponential phase (△), till the stationary phase (○), and from the mid-exponential to stationary phase in the presence of phenylmethylsulfonyl fluoride (●). The mitochondrial membranes were solubilized by freeze-thawing in the presence of Lubrol WX. Initial rates are expressed as μmoles ubiquinone-2 per min per g dry weight cells. From Galkin et al. (1979a).

such results obtained in the absence of proteinase inhibitors (galactose-grown cells) give grounds for believing that in glucose-grown yeast the excess entities are also not artifacts resulting from some side effects of proteinase inhibitors.

8.1.4. Selection by a Performance Criterion as a Mechanism Controlling Mitochondrial Formation

Thus, the "excess" cytochromes accumulating in galactose yeast or in glucose yeast in the presence of proteinase inhibitors are found in normally constituted complexes of the respiratory chain which under certain conditions display catalytic activity. This appears to apply not only to cytochromes but also to other respiratory-chain elements including flavoproteins (see Galkin et al., 1980). The abnormal complexes have all the properties of usual electron carriers except their ability to participate in mitochondrial respiration. It is therefore quite likely that the excess complexes simply cannot interact with their potential partners because of improper alignment within the inner mitochondrial membrane.

Deviations from the normal arrangement of the respiratory chain were detected, for instance, during kinetic analysis of the succinate-cytochrome *c* reductase activity of the mitochondrial population from yeast in the exponential phase of growth on galactose (Galkin et al., 1980). It was found that in this preparation a large portion of the electron-donor sites is inaccessible for the exogenous acceptor, and can only be rendered reactive upon membrane labilization with a detergent. Such abnormal sites, which appear to be buried within the membrane instead of being exposed on its surface, are absent from the "mature" mitochondria of stationary-phase yeast; still, they are preserved even at this phase if the growth medium contains proteinase inhibitors.

There is yet no direct evidence for the structural defects in the mitochondria whose development is not complete. The fact is that the existing approaches to studying the mitochondrial membrane structure (in particular, the asymmetry in the disposition of individual proteins) are still of qualitative rather than quantitative character (see Chapter One). For this reason it is difficult to discern whether one third or one fourth of the total content of a certain respiratory-chain component are abnormally positioned in the membrane.

Since the abnormal respiratory-chain components are accumulated in the presence of inhibitors suppressing some cytoplasmic and mitochondrial proteinases, the proteolytic enzymes can be thought to participate in the regulation of mitochondrial formation. All the facts summarized in the preceding subsections suggest that it is the nonfunctioning entities, not contributing to the mitochondrial respiratory activity, that are selectively eliminated. Indeed, if elimination encompassed functional as well as nonfunctioning components, then the decrease in the quantity of respiratory-chain segments in galactose yeast should have been accompanied by a proportional activation of the remaining respiratory chains to keep the overall mitochondrial respiratory activity constant. In glucose-grown yeast, if addition of proteinase inhibitors entails accumulation of both functional and nonfunctioning respiratory-chain segments, it should at the same time proportionally inactivate the preexisting respiratory ensembles, again to keep the overall activity at the constant level observed experimentally.

Unfortunately, the data available cannot be used to assess the precision with which the inner mitochondrial membrane is assembled. This is in the first place caused by insufficiency in our knowledge about the membrane structural organization (it is timely to recall that the concept of the respiratory chain is structurally rather vague) and about the mechanism of electron transfer between the enzymic complexes of the respiratory chain.

One can speculate on several variants of the distribution in the membrane and mode of elimination of abnormal complexes.

(1) Let us assume that 25-30% of the total content of each complex are improperly inserted in the membrane, such "defects" being randomly scattered throughout the respiratory ensembles. If this were so, then the probability of formation of an active chain of four adjacent complexes would be low since any membrane region would be likely to carry at least one defective element of the ensemble. As follows from Chapters Three and Four, such a system is unstable.

(2) Alternatively, we may assume that during formation of the mitochondrial respiratory system 25-30% of the respiratory chains are completely eliminated, for this it being enough that each chain carried a single defective link. In this case the actual portion of the abnormally incorporated links may be significantly lower that one fourth to one third of the total content of the corresponding link in the membrane. Again, this would be true assuming that the errors are evenly distributed among all respiratory chains, and that incorrect insertion is equally probable for every complex.

(3) The next assumption is that the occurrence of defective links is an "autocatalytic" event: the probability of improper insertion of a respiratory-chain link is enhanced if the adjacent link is already incorrectly positioned in the membrane. In this case the membrane would have regions with improperly arranged individual respiratory-chain components, and regions with normal ensembles. This model allows of one fourth to one third of all components of the respiratory system being inserted into the inner mitochondrial membrane at random, while the rest form normal functional units (compare with the first suggestion).

(4) The above model does not account for the fact that, owing to a certain mobility of proteins in the membrane, "recombination" of chains is possible, so that in the simplest case two inactive chains, each containing one or more improperly inserted links, can give rise to an active chain and an inactive one in which all the initial defects are concentrated. Such recombinations may result in altered distribution of the defective links and their segregation in some regions of the membrane.

As indicated above, there is a certain correlation between the content of excess cytochromes and the surplus in the potential activity of the corresponding respiratory-chain segments. However, this correlation cannot yet be claimed to be strictly quantitative. Indeed, it is not at all obvious that all of the excess cytochromes are contained in improperly aligned complexes having potential catalytic activity. If during mitochondrial inner membrane formation errors evolve at the stage of integration of individual enzymic complexes of the respiratory chain into functional assemblies, errors are also quite possible at the stage of assembly of the complexes themselves. In such a case the membrane would contain a certain amount of "free" or "extra" polypeptides that do not carry their usual catalytic or structural functions. As will be shown in the next section, polypeptides synthesized on mitochondrial ribosomes and incorporated into the membrane under a deficiency of cytoplasmic partners are readily subject to proteolysis. With certain allowances these polypeptides may be regarded as defective elements of the membrane.

Of special interest is the question of the detailed mechanism of elimination of the excess nonfunctioning components of the mitochondrial respiratory system. Summing up the data concerning the removal from the plasma membrane of some of its components (surface antigens, receptors, etc.), Parry (1978) discerned four possible mechanisms of this process (Figure 59). The first implies the release of the component into the surrounding medium owing to direct action of a lytic enzyme (e.g., proteinase or glycosidase). The second consists in the cleavage not of the component as such but of its anchoring protein. The last two mechanisms imply shedding of some membrane components in the composition of microvesicles which detach from the membrane and are then transported either into the cell to be digested in lysosomes or into the surrounding medium. These mechanisms are to a certain extent conceivable in the case of elimination of individual components of the mitochondrial membrane.

Since the elimination of the excess components of the mitochondrial respiratory chain was suppressed by pepstatin and by phenylmethylsulfonyl fluoride, which inhibit yeast vacuolar proteinases A and B, respectively, both these enzymes could be thought to attack these components. However, taking into account the long duration of the experiments with the inhibitors (several hours), one can admit as well that proteinase A is involved indirectly, serving for activation of proteinase B (Saheki and Holzer, 1975). In either case a rather probable mechanism is the shedding of separate proteins or whole inner-membrane regions as microvesicles, with subsequent trapping and digestion in vacuoles.

It has been established that the inner mitochondrial membrane harbors proteinases capable of rapidly breaking down the mitochondrially synthesized proteins. This process is inhibited by phenylmethylsulfonyl fluoride but is insensitive to pepstatin. Though the origin of these proteinases is not yet known, they can be preliminarily assumed to be activated in the cell by a pepstatin-sensitive process. Such intramembrane "rejection" resembling the digestion of surface proteins in the plasma membranes of some cells (see references in the review by Parry, 1978) will be considered in the next section of this chapter.

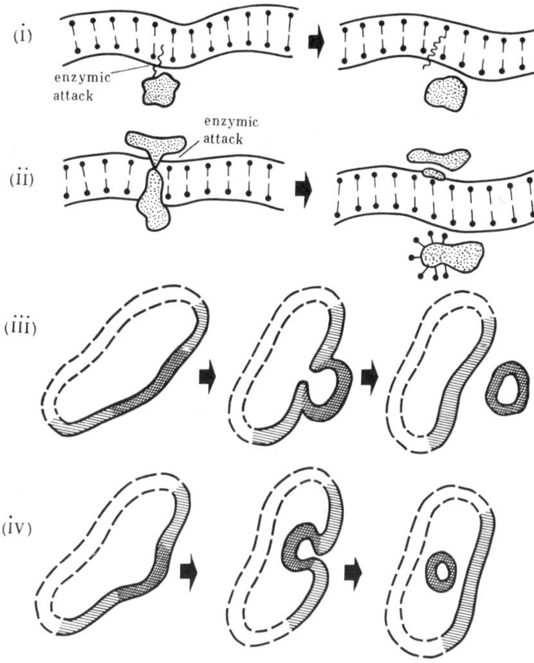

Fig. 59. Possible mechanisms of the removal of plasma membrane components. (i) Components are removed by direct lytic action. (ii) Shedding of components ensues the breakdown of anchoring proteins. (iii) Bulging of membrane sections leads to formation of vesicles with subsequent release of some components into the surrounding medium. (iv) Invagination of membrane sections may likewise lead to vesicle formation inside the cell. Adapted from Parry (1978).

8.2. MEMBRANE-BOUND PROTEINASES OF YEAST MITOCHONDRIA AND THEIR POSSIBLE INVOLVEMENT IN THE REGULATION OF MITOCHONDRIAL FORMATION

8.2.1. Proteolysis of Mitochondrial Translation Products in Yeast Cells

Some notions on how the selection of mitochondrial membrane components can actually be performed evolved on the basis of experiments in which the breakdown of intramitochondrially made polypeptides was studied. Such data were obtained with the use of a method specially designed in this laboratory (Bakalkin et al., 1978) to overcome two major difficulties in estimating the real rates of protein degradation in vivo: (1) recycling of the labeled amino acids and (2) postincorporation of the label retained in the cells after washing in the form of amino acyl-tRNA or in slowly equilibrating compartments.

The data on the proteolysis of mitochondrial translation products in Saccharomyces cerevisiae are depicted in Figure 60 (Kal'nov et al., 1979a). It can be seen that [^{3}H]leucine pulse-incorporated into mitochondrial protein in the presence of cycloheximide is lost from mitochondria already within one hour in the early exponential phase of growth in a glucose medium. Under such conditions the half-life of the mitochondrial translation products only amounts to about 20 min. This is indicative of an exceptionally high activity of the proteinase responsible for this process.

As shown in Figure 60, the proteolysis of mitochondrial translation products is suppressed by phenylmethylsulfonyl fluoride and to a lesser extent by leupeptine, a naturally occurring proteinase peptide inhibitor (Umezawa, 1973); at the same time it is insensitive to pepstatin. It is of interest that the action of phenylmethylsulfonyl fluoride is transient, which is probably caused by some conversions of the compound in the cell that do not allow its concentration to be maintained high enough for a long time, so that if the proteinase is rapidly renewed, after a while the newly synthesized molecules would be uninhibited.

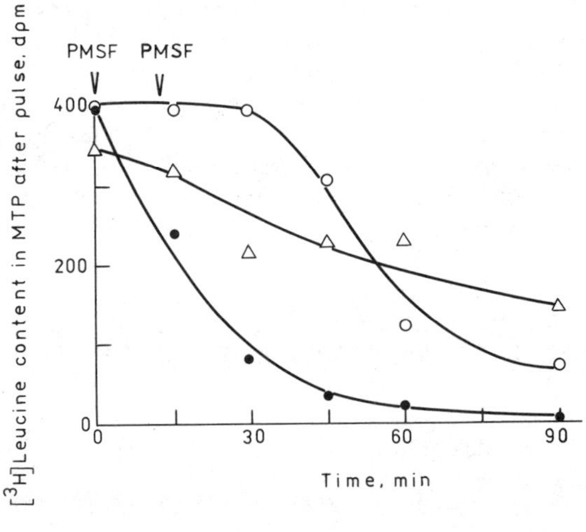

Fig. 60. Effect of proteinase inhibitors on the proteolysis of mitochondrial translation products in yeast cells. (●) Control; (○) in the presence of phenylmethylsulfonyl fluoride (PMSF) (added at 50 μmoles per liter during the pulse and at moments indicated by arrows; (△) in the presence of leupeptin (2 mg per liter). From Kal'nov et al. (1979a).

It must be noted that such rates of degradation of mitochondrial translation products are not entirely unprecedented. Constantino and Attardi (1977) made an attempt to estimate the stability of polypeptides synthesized on mitochondrial ribosomes in HeLa cells. These authors monitored the decrease in the content of labeled leucine incorporated into these polypeptides in the presence of cycloheximide, and found that the bulk of the label associated with two polypeptides out of eight is already lost in 4 h. This time, although clearly overestimated, is still in good correspondence with the data presented in Figure 61 for the early stationary phase of yeast growth.

Analysis of the efficiency of various proteinase inhibitors in suppressing the cleavage of mitochondrial translation products in yeast cells gives some ideas on the identity of the proteinases responsible. According to the data of Lenney (1975), phenylmethylsulfonyl fluoride inactivates proteinase (endopeptidase) B and carboxypeptidase Y which are vacuolar enzymes. At the same time, neither enzyme is sensitive to peptide inhibitors leupeptin and pepstatin. The proteinase cleaving the mitochondrial translation products, on the contrary, is partially suppressed by leupeptin. If one also takes into account that the mitochondrial proteolysis is unaffected by pepstatin, which inhibits proteinase A, it can be concluded that at least the three vacuolar proteinases (A, B, and Y) do not participate in the breakdown of polypeptides released from mitochondrial ribosomes.

It can be added that the proteolysis of mitochondrial translation products is retarded in the early stationary phase of yeast growth (Figure 61) as well as under glucose repression (Figure 64). At the same time, both the rate of degradation of total cell protein and the activities of vacuolar proteinases A and B increase in stationary-phase yeast (Bakalkin et al., 1976). During glucose repression these activities were also found to be elevated (Novikova et al., 1982). The lack of correlation between the proteolysis rate of intramitochondrially synthesized polypeptides and the activities of vacuolar proteinases indicates that these enzymes are not involved in the breakdown of mitochondrial translation products.

As just mentioned, the mitochondrial translation products are degraded far more rapidly in exponential than in stationary glucose-grown yeast. In other words, the proteolysis is more effective in the developing mitochondria of young cells having low respiratory activity than in mitochondria of mature cells displaying highly active oxidative phosphorylation. The mechanism of the regulation of mitochondrial intramembrane proteolysis will be considered in a special section of this chapter.

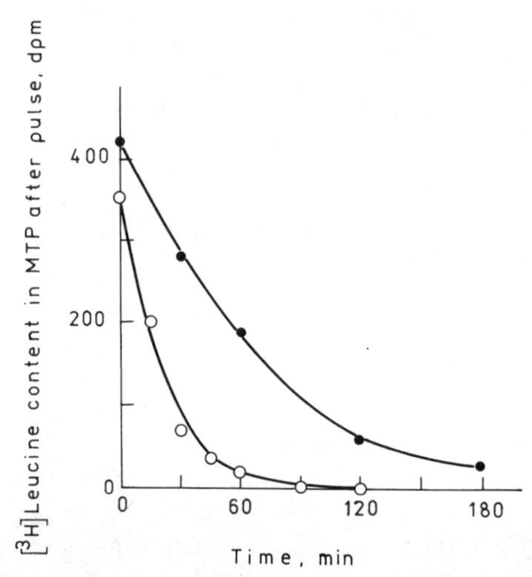

Fig. 61. Breakdown of mitochondrial translation products in exponential (○) and stationary (●) phases of yeast growth. From Kal'nov et al. (1979a).

Further, in this section we have been dealing with the proteolysis of mitochondrial translation products synthesized in the presence of an inhibitor of cytoplasmic translation, cycloheximide. Such conditions upset the balance between the mitochondrial and cytoplasmic protein syntheses. As a result, after exhaustion of the pool of free cytoplasmic polypeptides their partners released from mitochondrial ribosomes might accumulate in the membrane without entering any enzymic complexes. In this case it is a matter not of "routine" proteolysis of mitochondrially made polypeptides but rather of the breakdown of artificially induced abnormal elements of the inner membrane. Since the half-life of the mitochondrial translation products synthesized in the presence of cycloheximide is much shorter in the exponential than in the stationary phase of yeast growth, it may be suggested that the probability of elimination of such polypeptides, which may be abnormally bound with the membrane, is higher for the young cells with incompletely formed mitochondria.

The idea of the rapid and selective cleavage of mitochondrial translation products made in the presence of cycloheximide was earlier expressed by von Rücker et al. (1973). It was based on some indirect data obtained in experiments with Neurospora crassa. These authors regarded such polypeptides as extra entities not integrated into functional membrane complexes; particular proteins referred to were subunits I, II, and III of cytochrome *c* oxidase (von Rücker, 1976). The breakdown of mitochondrial translation products was ascribed to a hypothetical mitochondrial proteinase.

8.2.2. Proteolytic System of Yeast Mitochondria Responsible for the Breakdown of Intramitochondrially Made Polypeptides

Research conducted in this laboratory has indeed shown that yeast mitochondria harbor a proteolytic system which is highly effective in digesting the mitochondrial translation products.

As mentioned in Chapter Two, Wheeldon et al. (1974) observed rapid degradation of a portion of polypeptides synthesized by isolated liver mitochondria. The half-life of these polypeptides in the presence of ATP in the incubation medium was about 20 min. It was supposed that degraded proteins are nascent polypeptides released from mitochondrial ribosomes. It remained unknown, however, whether the proteolysis is performed by mitochondrial proteinases or by cytoplasmic enzymes probably contaminating the mitochondrial fraction.

Kal'nov et al. (1978) in this laboratory obtained similar data for Saccharomyces cerevisiae mitochondria. As shown in Figure 62, incubation of isolated mitochondria in the presence of ATP and succinate results in the loss of [^{3}H]leucine previously incorporated into mitochondrial protein in vivo in the presence of cycloheximide. The process is shown not to be associated with the action of cytoplasmic proteinases. Indeed, the mitochondrial fraction did not catalyze the cleavage of Azocoll or hemoglobin in standard assays for proteinases A and B which can be regarded as vacuole markers. Furthermore, the proteolysis of mitochondrial translation products is unaffected by antisera raised against these enzymes (Kal'nov et al., 1979b). Both these facts indicate a lack of appreciable contamination of the mitochondrial fraction by vacuolar enzymes, or at least their inability to digest mitochondrial translation products. The proteolysis was not stimulated and was even inhibited by detergents (Triton X-100 and sodium dodecyl sulfate), in contrast to that observed by Michel et al. (1976) who reported cleavage of mitochondrial proteins by a proteinase residing in specific granules of Neurospora crassa cells; these vesicles cosedimented with mitochondria during isolation and could only be removed by sucrose density gradient centrifugation.

A strong argument in favor of the existence of an endogenous mitochondrial proteinase responsible for the breakdown of mitochondrially made polypeptides (Kal'nov et al., 1978, 1979b) is the observation of this process not only in isolated yeast mitochondria but also in inside-out submitochondrial vesicles derived therefrom. This indicates a rather tight binding of the proteinase to the inner mitochondrial membrane.

Table 29 summarizes the data on the inhibitor sensitivity of the mitochondrial proteinase tested with different synthetic and naturally occurring compounds (Kal'nov et al., 1978, 1979b). First of all

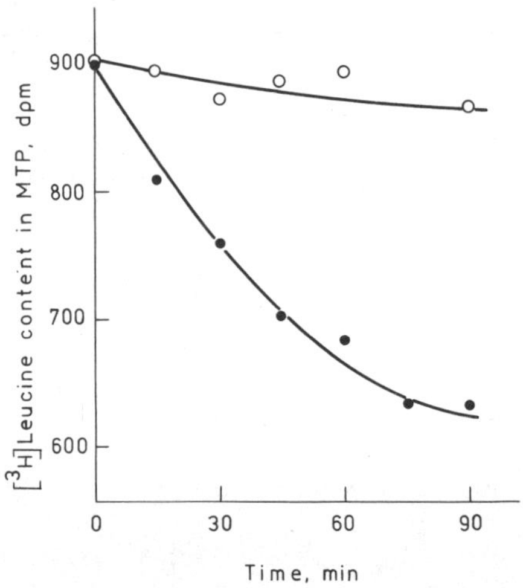

Fig. 62. Degradation of mitochondrial translation products in isolated yeast mitochondria. (●) In the presence of ATP and succinate; (o) without ATP and succinate. From Kal'nov et al. (1978).

it should be noted that, like the in vivo proteolysis (see Figure 60), in vitro breakdown of mitochondrial translation products is inhibited by phenylmethylsulfonyl fluoride and leupeptin, but insensitive to pepstatin. Hence a reasonable conclusion is that both processes are due to the same enzyme. A peculiar feature distinguishing the mitochondrial proteinase from proteinases A and B (Table 29) is its sensitivity to chloramphenicol. Proteolysis in isolated mitochondria is suppressed by inhibitors of trypsin and chymotrypsin (phenylmethylsulfonyl fluoride, chymostatin, and antipain) and by a sulfhydryl-blocking reagent, p-chloromercuriphenyl sulfonate.

A necessary step in elucidating the nature and function of the proteolytic system of the inner mitochondrial membrane is the isolation and purification of its components. Experiments with mitochondrial inner membrane preparations (ultrasonic submitochondrial vesicles) from Saccharomyces cerevisiae showed that they can hydrolyze at an appreciable rate ^{3}H-labeled cytochrome *c*, and N-acetyl-L-tyrosine ethyl ester conventionally used as a substrate for chymotrypsin-like proteinases. Hydrolysis of these substrates is suppressed by the same inhibitors as the breakdown of intramitochondrially made polypeptides (Table 29). Hence, these substrates can be used as test ones in the search for proteinases responsible for the proteolysis in mitochondria.

The first progress in isolating membrane-bound proteinases from yeast mitochondria was achieved by subjecting Triton or sodium

Table 29. Effects of Various Inhibitors on Yeast Proteinases A and B, on the Proteolysis of Mitochondrial Translation Products in Isolated Yeast Mitochondria, and on the Cytochrome *c* Hydrolase and ATEE Hydrolase Activities of Ultrasonic Submitochondrial Particles (compiled from Kal'nov et al., 1979, and Novikova et al., 1981).

		Inhibition, %				
Inhibitor	Concn.	Proteinase A	Proteinase B	Mitochondrial proteolysis	Cytochrome *c* hydrolase	ATEE** hydrolase
PMSF*	1 mM	0	100	85	44	82
pCMPS***	1 mM	0	100	75	65	95
PMSF + pCMPS		nd****	nd	nd	95	nd
Pepstatin	1 μg/ml	100	0	0	nd	nd
Antipain	2 μg/ml	0	100	95	61	40
Chymostatin	2 μg/ml	0	100	83	nd	15
Leupeptin	2 μg/ml	5	0	55	50	48
PMSF + leupeptin		nd	nd	nd	63	nd
Chloramphenicol	4 mg/ml	0	0	95	nd	nd

* PMSF, phenylmethylsulfonyl fluoride; ** ATEE, acetyltyrosine ethyl ester;
*** pCMPS, p-chloromercuriphenyl sulfonate; **** nd, not determined.

dodecyl sulfate extracts from submitochondrial vesicles to polyacrylamide gel electrophoresis (Zubatov et al., 1983). Such extracts contain up to 90% of the cytochrome *c* hydrolase activity of the initial vesicles. Proteinases were located by carrying out cytochrome *c* hydrolysis directly in the gels (see a preliminary communication by Novikova et al., 1981). As can be seen in Figure 63, three bands with cytochrome *c* hydrolase activity are reliably detected, having R_f of 0.95, 0.96, and 1.00 with respect to bromophenol blue.

Analysis of the cytochrome *c* hydrolase isoenzymes extracted from the gel showed that they all have a molecular weight about 17,000 by the data of sodium dodecyl sulfate-polyacrylamide gel electrophoresis but differ in isoelectric points (pI 4.0, 4.2, and 4.4).

In addition to these proteinases, the Triton or sodium dodecyl sulfate extracts from submitochondrial vesicles contain enzymes capable of hydrolyzing synthetic substrates of serine proteinases: Nα-benzoyl-DL-arginine 4-nitroanilide and Nα-benzoyl-DL-arginine naphthylamide. These substrates are hydrolyzed in polyacrylamide gels in zones with R_f of 0.35 and 0.43 with respect to bromophenol blue. These enzymes, like the cytochrome *c* hydrolase isoenzymes, have similar molecular weights (19,000 by the data of sodium dodecyl sulfate-polyacrylamide gel electrophoresis) but different pI values (4.3 and 4.7). An analogous situation has been observed in the studies of Bacillus subtilis metalloendopeptidase (Vaganova et al., 1976), for which four molecular forms have been found having close molecular weights.

It is yet too early to speak about the true number of proteinases localized in the inner mitochondrial membrane. However, their multiplicity is already evident, and this indicates that the proteolysis of polypeptides in the membrane may be differentiated.

In connection with the material on the proteolytic processing of cytoplasmically and mitochondrially made precursors to some mitochondrial proteins, which in some cases appears to take place in the inner mitochondrial membrane (see Chapter Six), it is of interest to ascertain whether those maturases and the hydrolases described in this section have anything in common. Unfortunately, the only maturases studied so far were localized in the mitochondrial matrix. As indicated in Section 6.3.7, they are metalloenzymes inhibited by o-phenanthroline or EDTA but insensitive to synthetic inhibitors of serine proteinases and a number of peptide inhibitors of microbial origin. The cytochrome *c* hydrolase activity of submitochondrial vesicles, on the contrary, is suppressible by phenylmethylsulfonyl fluoride and p-chloromercuriphenyl sulfonate but unaffected by o-phenanthroline or EDTA (Novikova et al., 1981; Zubatov et al., unpublished). Maturases are incapable of digesting protein substrates (Boehni et al., 1980; Miura et al., 1982). One of them, performing the processing of ornithine carbamoyltransferase precursor

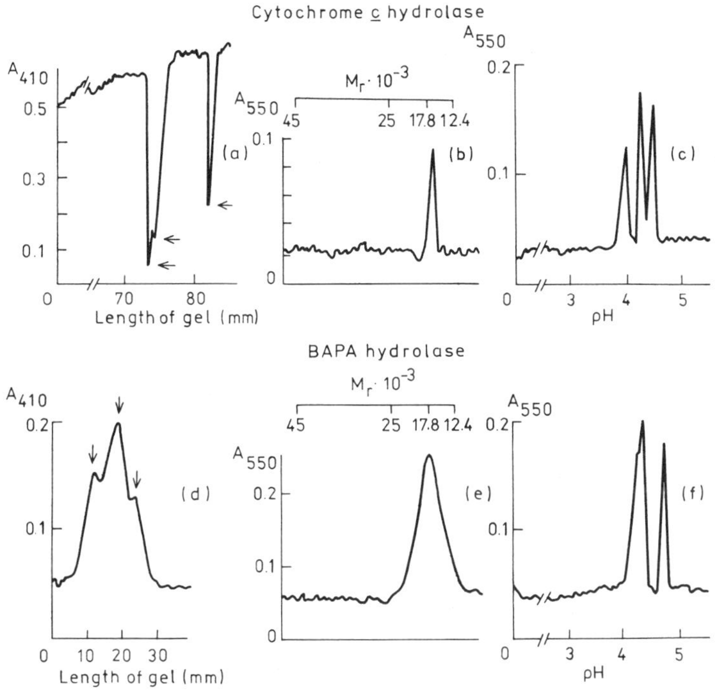

Fig. 63. Multiplicity of the proteinases of the yeast mitochondrial inner membrane. (a,b,c) Data for the isoenzymes of cytochrome *c* hydrolase extractable from submitochondrial particles with 0.05% sodium dodecyl sulfate. (d,e,f) Analogous data for BAPA (BANA) hydrolase isoenzymes. (a,d) Polyacrylamide gel electrophoresis of extracts from submitochondrial particles; arrows show the bands having the corresponding activity. (b,e) Molecular-weight determination: bands from (a,d) were excised, extracted with 0.05% sodium dodecyl sulfate, and subjected to sodium dodecyl sulfate-polyacrylamide gel electrophoresis. (c,f) Determination of isoelectric points: the pooled extracts from (a,d) were subjected to isoelectric focusing.

in rat liver mitochondria, was found to have a molecular weight of 108,000 and pI 5.5. These characteristics substantially differ from those for the above-described hydrolases of yeast mitochondria.

8.2.3. Regulation of the Breakdown of Mitochondrial Translation Products in Yeast

As already mentioned, the rate of the breakdown of intramitochondrially made polypeptides in yeast cells substantially varies,

Table 30. Correlation Between the Breakdown Rate of Mitochondrial Translation Products in Yeast and the Fluidity of the Mitochondrial Inner Membrane.

Conditions of yeast growth	Breakdown of mitochondrial translation products	Cytochrome *c* hydrolase activity* of submitochondrial particles	Temperatures of phase transitions in the membrane**	Coefficient K*** for membrane phospholipids
Exponential phase, aerobic	+++++	4.9	2°C - 5°C	8.6
Stationary phase, aerobic	+++	12.0	5°C -15°C	6.5
Stationary phase, 2-h glucose repression	+	10.0	13°C -23°C	4.0
Stationary phase, anaerobic	++	12.0	13°C -29°C	1.5

*Expressed as per cent cytochrome *c* hydrolyzed per hour per 0.25 mg protein.
**Assayed with the paramagnetic probe 2-(3-carboxypropyl)-4,4-dimethyl-2-tridecyl-3-oxazolidinyloxyl.
***Ratio of unsaturated to saturated fatty acids.

depending on the phase of culture growth (Figure 61). What may be the mechanism of this phenomenon? A working hypothesis on that score evolved from studying the proteolysis of mitochondrial translation products during yeast glucose repression. It has been recently shown in this laboratory (Novikova et al., 1982) that transfer of yeast grown aerobically on glucose to the stationary phase into a medium with 10% glucose entails pronounced activation of vacuolar proteinases A and B. Although by that time convincing proof was available that the proteolysis of intramitochondrially made polypeptides in yeast is performed by the proteolytic system of the inner mitochondrial membrane (see the preceding section), it seemed of interest to find out whether the increase in proteinase A and B activities during glucose repression would be accompanied by accelerated breakdown of mitochondrial translation products.

Figure 64 demonstrates that it is not so. On the contrary, during glucose repression the proteolysis of mitochondrially synthesized polypeptides in the yeast cells gradually slows down and in 2 h virtually completely stops. This phenomenon (Luzikov et al., 1983a) appears to be due neither to changes in the overall proteolytic activity of the inner mitochondrial membrane nor to variations in the set of the mitochondrial translation products. A more plausible cause of the retarded breakdown of these polypeptides is the altered physical state of the membrane. The results of experiments with lipid-soluble spin-labeled probes together with the analysis of the fatty acid composition of lipid molecules showed that glucose repression increases the order and decreases the mobility of hydrocarbon chains of the phospholipid molecules in the inner mitochondrial membrane, in all probability because of the enhanced content of saturated fatty acyls in the phospholipids. Such changes could well hinder the interaction between proteinases and mitochondrial translation products, for instance by attenuating the rates of their lateral and rotational diffusion in the membrane.

There are grounds for suggesting that the correlation between the rate of the breakdown of mitochondrially made polypeptides and the mobility of lipid molecules in the inner mitochondrial membrane has a universal character (Luzikov et al., 1983b). This follows from the data presented in Figure 65 and Table 30. It is easy to see that the highest rate of proteolysis observed in yeast in the early exponential phase of aerobic growth corresponds to the lowest temperatures of phase transitions in the membrane of submitochondrial vesicles and the highest value of the coefficient K (the ratio of the total amounts of unsaturated to saturated fatty acids in the phospholipid fraction of submitochondrial vesicles). In anaerobically grown yeast, as well as under glucose repression, the low breakdown rate agrees with the high phase-transition temperatures and low K values. Such relationships become explicable if one takes into account that the temperature of the phase transition in the membrane is related to the mobility of hydrocarbon chains of the phospholipid molecules, and

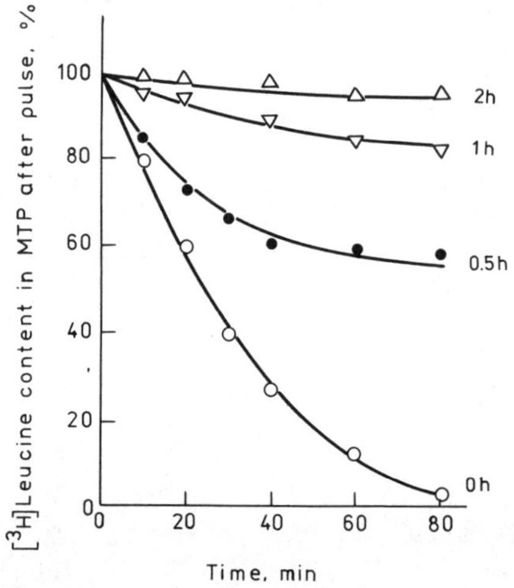

Fig. 64. Changes in the breakdown rate of mitochondrial translation products during yeast glucose repression in the early stationary phase of growth. Periods of repression preceding the pulse of [^{3}H]leucine are indicated in the Figure.

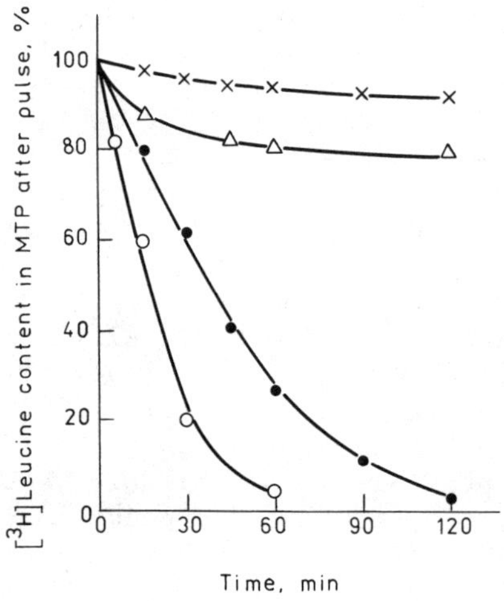

Fig. 65. Degradation of mitochondrial translation products in yeast under different cultivation conditions. (o) Exponential phase of aerobic growth on glucose; (•) early stationary phase of aerobic growth; (x) 2 h of glucose repression in the stationary phase; (△) early stationary phase of anaerobic growth.

that this mobility in no small measure determines the ability of membrane proteins to undergo rotational and lateral diffusion (McConnel, 1976). The latter circumstance, as already pointed out, may be of paramount importance for the proteolysis of mitochondrial translation products in the inner mitochondrial membrane.

Of course, the observed correlation (see Table 30) is only qualitative; furthermore, it has so far been observed only for a limited set of physiological conditions.

8.2.4. Some General Aspects of Proteolytic Control of Mitochondrial Inner Membrane Assembly

As follows from the three preceding subsections, polypeptides synthesized on mitochondrial polysomes under blocked cytoplasmic translation in yeast are subject to rapid breakdown which under certain conditions may be virtually complete within 1 h (half-life of 15 min). For comparison it can be mentioned that the degradation of the total cell protein under the same conditions proceeds at a rate not exceeding 9% per hour (Bakalkin et al., 1976).

It was shown in Chapter Six that integration of polypeptides into enzymic complexes of the inner mitochondrial membrane is an intricate multistage process including the proteolytic modification of protein precursors, attachment of prosthetic groups, transmembrane transport and lateral diffusion of polypeptides in search of their partners in complexes. In particular, only the maturation of the cytochrome c_1 precursor as such, not to mention its insertion into complex III of the respiratory chain, in yeast cells takes more than 20 min*. If the rates of various stages of maturation and insertion into the membrane of the mitochondrially made polypeptides are of the same order, the assembly of the complexes proceeds not much faster than the polypeptide cleavage.

As discussed above, the high breakdown rates appear to be characteristic of polypeptides not bound to their cytoplasmically produced partners because of the lack of the latter. Since the polypeptides under consideration are constituents of some enzymic complexes of the inner mitochondrial membrane (complexes III and IV of the respiratory chain, H^+-ATPase), the assembly of the latter may be under effective control exercised by the membrane proteolytic system.

There are grounds for believing that a second level of control also exists, which is effective on whole complexes and can be regarded as control over mitochondrial inner membrane assembly by a performance criterion. Indeed, evidence is available that membrane

* See Addendum.

formation is accompanied by appearance of nonfunctioning complexes which are selectively eliminated during mitochondrial differentiation into respiratory-competent coupled organelles. Moreover, it has been shown how such "excess" elements of the inner mitochondrial membrane can be purposefully accumulated. Unfortunately, the mechanism of elimination of nonfunctioning enzymic complexes is still obscure, except that some stage of this rejection involves proteolytic enzymes suppressible by phenylmethylsulfonyl fluoride and, directly or indirectly, by pepstatin.

All this taken together allows the conclusion that in mitochondrial assembly the principle of self-assembly, operative during formation of multisubunit proteins, ribosome particles, viruses, bacterial flagella, microtubules, and other subcellular structures (see a review by Bouck and Brown, 1976), is supplemented with the principle of selection by a performance criterion (Luzikov, 1973, 1979, 1981): any elements that for some reason fail to perform properly are eliminated, all errors occurring during the assembly being thus rejected.

9
A Broader Look at the Prospects for Further Studies

On acquaintance with the eight foregoing chapters it is easy to conclude that the pivot of the book is the idea that the mitochondrion should be regarded as a dynamic system which retains its composition and structure only as long as it performs its functions in energy transformation. It is to be hoped that the experimental data considered in Chapters Three and Four suffice for accepting this point of view. The basic issue has at least two implications which allow one to look from a new angle at some aspects of mitochondrial biogenesis.

Firstly, it becomes evident that the extent of mitochondrial expansion in the cell may be regulated not only by the nuclear and mitochondrial genomes but on the substrate level as well. This means that "overproduction" of, for instance, respiratory-chain components would result in exhaustion of the intracellular stores of oxidizable substrates etc.; this in its turn would moderate or stop the functioning of the energy-transformation system with subsequent elimination of a portion of the membrane material. No experiments have yet been conducted to test this suggestion directly*.

Secondly, a thought inevitably occurs about the existence of control over mitochondrial assembly underlied by the principle of selection by a performance criterion. The first steps made to test this idea so far pertain only to the assembly of the respiratory chain. The approach proposed is obviously applicable to elucidating other aspects of mitochondrial formation, including the assembly of the inner mitochondrial membrane as a whole and assortment of "soluble" proteins in the organellar compartments.

---

* See Addendum.

It is only natural to ask oneself: How does all this bear on other organelles? To date, a comparatively detailed study has been made of chloroplast degradation. Harnischfeger (1972) and Siegenthaler (1972) put forward generally similar schemes, according to which degradation in vitro sets in with the cleavage of thylakoid galactolipids and phospholipids under the action of an endogenous galactolipase (Anderson et al., 1974; Shaw et al., 1976). Free fatty acids released in this process induce chloroplast swelling and suppress their photochemical reactions (see, e.g., McCarty and Jagendorf, 1965; Molotkowsky and Zheskova, 1965; Constantopoulos and Kenyon, 1968; Siegenthaler, 1970, 1974). Finally, in illuminated chloroplasts free fatty acids undergo peroxidation. Unlike the above authors, Hoshina et al. (1975) believed that the primary cause of light-induced chloroplast swelling and impairment of the photochemical reactions is the conversion of phosphatidylcholine to lysophosphatidylcholine with ensuing sensitization of thylakoid membranes to the action of free fatty acids and lipid peroxides (Hoshina and Nishida, 1975). As can be seen, the mechanisms of chloroplast degradation have much in common with what has already been said about mitochondrial aging.

For chloroplasts the suggestion appears to hold true that the impairment of their photochemical functions entails organelle degradation. In particular, darkness is known to induce chloroplast swelling (Mlodzianowski and Ponitka, 1973), a decrease in the content of chlorophylls, RNA, and protein, a loss of the ability to perform Hill's reaction and other functions (Choundhury and Biswal, 1979, 1980). One of the possible explanations for these alterations is that in the dark the synthesis of chlorophylls *a* and *b* is hindered first of all; the decline of their level in chloroplasts resulting from normal catabolism lowers the resistance of the light-harvesting

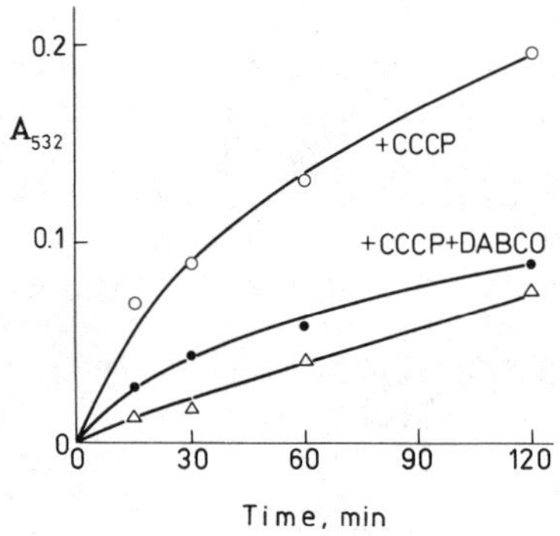

Fig. 66. Effect of carbonylcyanide m-chlorophenylhydrazone (CCCP) on lipid peroxide formation in illuminated chloroplasts. (△) Without additions; (○) with CCCP; (●) with CCCP and a quencher of singlet molecular oxygen 1,4-diazobicyclo-[2,2,2]octane (DABCO). After Takahama and Nishimura (1975).

chlorophyll *a/b* protein to proteolysis, as supposed by Bennett (1981). The same may apply to other proteins of thylakoid membranes. However, it cannot be ruled out that the primary effect of darkness is de-energization of thylakoid membranes which initiates the whole chain of degenerative alterations considered above.

The connection between chloroplast stability and functional state is more graphically revealed in experiments on the stimulation of lipid photooxidation with inhibitors of electron transfer. As shown by Takahama and Nishimura (1975), carbonylcyanide m-chlorophenylhydrazone, by interrupting electron transfer on the oxidizing side of photosystem II (or I), accelerates malondialdehyde accumulation in illuminated chloroplasts (Figure 66). Similar effects are produced by another inhibitor - 3-(4,4-dichlorophenyl)-1,1-dimethylurea - and by washing chloroplasts with Tris solution.

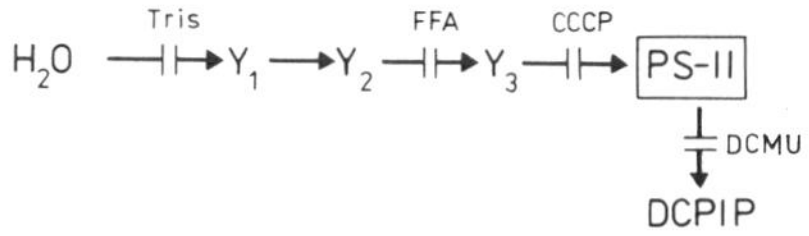

These data imply that any natural disturbances in the electron-transfer system, if such occurred during formation of thylakoid membranes, would inevitably result in their labilization by accumulating lipid peroxides, so that they would be subsequently degraded in the course of chloroplast development.

The idea that cell organelles may exist as prostructures in some conditions, convert into mature structures in other conditions, and go back into the rudimentary state upon return to the initial conditions has in recent years begun to acquire a universal character. In particular, experimental data along this line have been obtained for glyoxysomes. Kudielka et al. (1982) found that cultured anise cells growing in a sucrose medium contain precursor vesicles which possess catalase but are devoid of key glyoxysomal enzymes, isocitrate lyase and malate synthetase. When sugar in the growth medium is replaced by acetate, these structures develop into mature glyoxysomes by being replenished with derepressed glyoxysomal enzymes. Transfer of cells into fresh sucrose medium again induces formation of "nonspecialized" microbodies. The question of how the enzymes characteristic of mature glyoxysomes are therewith selectively degraded is still enigmatic. Obviously the study of such degradation mechanisms is an indispensable step in understanding the biogenesis of glyoxysomes, as well as of any other organelle*.

Dynamic transformations of intracellular structures require the existence of powerful, selectively acting lytic systems. Of extreme importance in this context are the data recently obtained in Lloyd's laboratory with synchronous cultures of microorganisms. It has been shown, in particular (Edwards et al., 1982), that during the cell

* See Addendum.

cycle of Acanthamoeba castellanii the content of mitochondrial ATPase undergoes oscillation with a period of about an hour, a two-fold decrease from the maximal level being attained in less than half an hour, i.e. at a rate comparable with that observed for the enzyme synthesis (see Figure 67)*.

These data once again return us to the idea of the paramount role of proteinases in adjusting the content of individual enzymes or multienzyme systems (including the membrane ones) in the cell. It may be timely to mention the early work of Schimke et al. (1965) who showed that tryptophan and N-methyltryptophan enhance the resistance of tryptophan pyrrolase to proteolysis. A correlation was established between the ability of various tryptophan analogs to stabilize the enzyme *in vitro* and to promote its accumulation in rat liver. This was supposed to be due to attenuation of the enzyme breakdown in the cells, with the rate of its synthesis being constant.

These concepts in certain measure extend to the multienzyme system of oxidative demethylation in the endoplasmic reticulum of liver cells. It has long been known (see, e.g., Ernster and Orrenius, 1965) that phenobarbital induces accumulation in microsomes of NADPH-cytochrome *c* reductase and cytochrome P-450 which are constituents of this system. As this takes place, the level of cytochrome b_5 remains approximately constant whereas a number of enzymic activities including glucose-6-phosphatase, NADH-cytochrome *c* reductase, and ATPase markedly decrease. While the increment in cytochrome

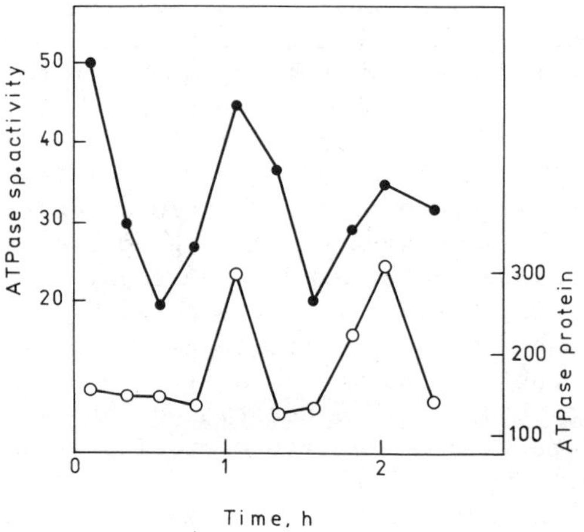

Fig. 67. Changes in the content of the ATPase protein and ATPase activity during synchronized growth of Acanthamoeba castellanii. (●) ATPase specific activity (milliunits per mg protein); (○) content of the ATPase protein (immunological units per mg protein). After Edwards et al. (1982).

* See Addendum.

P-450 in the presence of phenobarbital may be due to excessive synthesis of the corresponding mRNA (Adesnik et al., 1981), the behavior of NADPH-cytochrome *c* reductase and cytochrome b_5 results at least partly from an almost complete cessation of their intracellular breakdown (Kuriyama et al., 1969). It must be noted that phenobarbital causes only a slight decrease in the rate of degradation of the total microsomal protein, which is indicative of selective retardation of the digestion of particular proteins, probably owing to their stabilization by metabolizing the drug (see Figure 68). The mechanism of protein breakdown in the endoplasmic reticulum is not yet clear. In this connection it seems reasonable to pay attention to the neutral proteinases tightly associated with the microsomal fraction, which have been found in various tissues (Sogawa and Takahashi, 1978, 1979; Sogawa et al., 1981).

Selective stabilization of certain enzymes of the endoplasmic reticulum upon addition of phenobarbital somehow extends to the phospholipid components of the membrane. Indeed, a number of works have demonstrated that the rate of microsomal phospholipid breakdown decreases upon drug administration to animals (Holtzman and Gillette, 1968; Stein and Stein, 1969; Infante et al., 1971).

Thus, the situation with the oxidative demethylation system can be regarded as a basically new case of the regulation of enzyme contents at the substrate level. Firstly, the stabilization in all probability encompasses not one but a whole group of enzymes; secondly, a stabilization of proteins is attended by stabilization of the lipid components of the membrane. On acquaintance with this book, such events should not seem very surprising. Indeed, it could be seen in Chapter Three that electron transfer through an uncoupled respiratory chain enhances the resistance of the entire multienzyme system not only to trypsin or chymotrypsin, but to exogenous phospho-

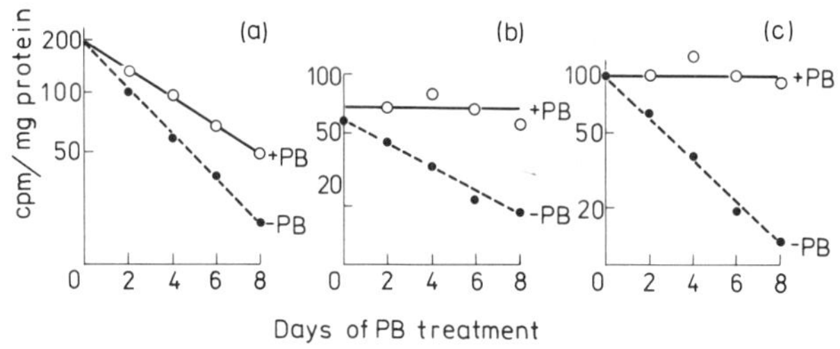

Fig. 68. Effect of phenobarbital on the degradation rates of total microsomal protein (a), NADPH-cytochrome *c* reductase (b), and cytochrome b_5 (c). Decrease of specific radioactivity in these fractions was followed after pulse of [^{14}C-guanidino]L-arginine; (-PB) without phenobarbital, (+PB) in the presence of phenobarbital. After Kuriyama et al. (1969).

lipase A_2 as well. Similar observations have been made for the coupled oxidative phosphorylation system.

In the case of mitochondria, we have to deal with the next level of organization of intracellular structures, which is marked not only by the presence of highly specialized membrane but also by a highly differentiated distribution of soluble proteins among the organellar compartments. Taking into account that the mitochondrial compartments are intimately interlinked functionally, and hence depend on each other for survival (see Chapter Three), it is not unreasonable to conceive of the mitochondrion as a giant multienzyme whose content is under substrate control, as in the trivial case of tryptophan pyrrolase.

Another field of application of the rule establishing the connection between organelle stability and the level of its metabolic activity (let us not forget that this rule can be considered proven only for mitochondria) is intracellular topography, i.e. regularities in the mutual arrangement of various intracellular structures. Summarizing the information available on the localization of mitochondria in the cell, Lehninger (1964) made some generalizations on that score. One is that mitochondria are frequently located in the vicinity of substrate sources. Another is that mitochondria are often observed near those regions of the cytoplasm where a need evolves for mitochondrially produced ATP. How can these phenomena be explained? No concrete mechanisms have so far been put forward. In the light of the material covered in Chapters Three and Four it would be tempting to suggest that the mitochondrial structures survive only in those parts of the cell where they can actively oxidize substrates and transduce energy, while in other regions they are forced out of existence, being destabilized by the conditions unfavorable for oxidative phosphorylation. An alternative explanation to the concentration of mitochondria around the source of substrates would be some sort of chemotaxis; such a mechanism, however, has no real evidence for it.

Not only Lehninger's generalizations about mitochondrial disposition in the cytoplasm have irrefutable grounds; close analogies have been observed in recent years. In particular, Lord (1980) indicates in his review that in leaf cells peroxisomes are often found appressed to the chloroplast outer membrane. To understand this phenomenon, it might help to take into account that peroxisomes convert glycolate produced by chloroplasts into glycerate, this being the main metabolic function of leaf peroxisomes. Likewise, fat-utilizing glyoxysomes in seedlings are during gluconeogenesis intimately associated with spherosomes which store triglycerides. It is of interest that upon exhaustion of lipid reserves glyoxysomes are degraded (Lord, 1980).

It would be of course naive to claim that peroxisomes and glyoxysomes are found preferentially in proximity to substrate sources only because there they are stabilized while performing their metabolic functions. Nevertheless, with all its naiveté this idea carries a constructive principle and has its merit already in being amenable to experimental testing. Actually, this book was in no small measure prompted by a wish to outline one of the possible ways of solving problems of this kind.

The reader has no doubt noticed that, proceeding through the book, he was more and more often pushed off the solid ground of facts towards the quicksand of speculation; lest he should be utterly carried away into the world of fantasy, I end my narration.

Addendum

Page 15

Later Ogata et al. (Ogata, K., Shimomura, Y., and Ozawa, T., 1982, Biochem.Intern. 4:621-627) obtained complex I comprising only nine types of subunits (molecular weights 60,000, 50,000, 48,000, 43,000, 28,000, 25,500, 20,000, 12,500, and 7800, in 2:2:1:1:2:1:2:2:1 stoichiometry). This preparation of molecular weight about 480,000 contained one FMN molecule, 26 atoms of nonheme iron, and 2-3 molecules of tightly bound phospholipids.

Page 186

Recently larger precursors were reported for the β subunit of propionyl-coenzyme A carboxylase of rat liver mitochondria (Kraus, J.P., Kalousek, F., and Rosenberg, L.E., 1983, J.Biol.Chem. 258:7245-7248), for the flavoprotein subunit of sweet-potato mitochondrial succinate dehydrogenase (Hattori, T., Iwasaki, Y., Sakajo, S., and Asahi, S., 1983), Biochem.Biophys.Res.Commun. 113:235-240), for the creatine phosphokinase subunit of canine myocardial mitochondria (Perryman, M.B., Strauss, A.W., Olson, J., and Roberts, R., 1983, Biochem.Biophys.Res.Commun. 110:967-972), and for the protein inhibitor of yeast mitochondrial F_1-ATPase (Yoshida, Y., Hashimoto, T., and Tagawa, K., 1983, J.Biochem. (Tokyo) 94:283-290).

Page 190

Data have since become available on the amino acid sequences of some precursors to mitochondrial proteins. The complete primary structure of the imported form of the proteolipid (ninth) subunit of mitochondrial H^+-ATPase from Neurospora crassa was established by molecular cloning and sequencing of its mRNA (Viebrock, A., Perz, A., and Sebald, W., 1982, EMBO J. 1:565-571). Nucleotide sequencing showed that the precursor, as compared to the mature proteolipid (an extremely hydrophobic protein of 81 amino acid residues), has an NH_2-terminal presequence of 66 amino acid residues containing 12 basic and no acidic side chains. This must ensure the high polarity of the precursor, which precludes its aggregation and nonspecific "sticking" to various intracellular structures and favors its translocation into the mitochondrial interior charged negatively with respect to the surrounding medium.

Yoshida et al. (Yoshida, Y., Hashimoto, T., Hase, T., Matsubara, H., and Tagawa, K., 1983, J.Biochem. (Tokyo) 94:291-297) isolated mRNA coding for the ATPase protein inhibitor, and used it to synthesize the precursor of the inhibitor with several labeled amino acids ([^{35}S]methionine, [^{3}H]serine, [^{3}H]leucine, and [^{3}H]glycine) in separate reactions *in vitro*. Each precursor preparation was subjected to 20-28 consecutive cycles of Edman degradation. This made it possible to determine the position of these amino acids at the NH_2-terminal region of the precursor. Comparison of this partial amino acid sequence with the one known for the mature inhibitor revealed the existence of an extra peptide comprising 21 amino acid residues.

Sannia et al. (Sannia, G., Colombo, M., Palomba, R., and Marino, G., 1983, Biochem.Intern. 6:731-736) isolated the precursor of mitochondrial aspartate aminotransferase synthesized in a cell-free system containing [^{35}S]methionine, and determined the positions of the first two methionine residues at the NH_2 end of the polypeptide chain. They turned out to occupy positions 1 and 33 whereas in the mature enzyme the first methionine is in position 9. Hence it was concluded that the precursor exceeds the mature protein by 24 amino acid residues.

Page 191

To date at least eight more mitochondrial polypeptides are known to be synthesized in yeast cells both on free and on mitochondria-bound polysomes (Suissa, M., and Schatz, G., 1982, J.Biol.Chem. 257:13048µ13055).

Page 194

Reid, G.A., and Schatz, G. (1982, J.Biol.Chem. 257:13056-13061), using a yeast cytoplasmic petite (rho$^-$) and adding carbonylcyanide m-chlorophenylhydrazone to the growth medium, have been able to show that under such conditions the cytoplasmically made mitochondrial polypeptides may accumulate in considerable amounts. These polypeptides are localized in the cytosol as precursors, and the amount of each reaches 0.015% of the total cell protein. Such data were obtained for the β subunit of F_1-ATPase, cytochrome b_2, and cytochrome c_1. As follows from pulse-chase experiments, the precursors are degraded in the cytosol at different rates. The half-life of the most labile (the cytochrome c_1 precursor) is only 10 min.

Page 196

Gasser et al. (Gasser, S.M., Daum, G., and Schatz, G., 1982, J.Biol.Chem. 257:13034-13041) presented data on the import of five mitochondrial proteins of different localization: cytochrome b_2, F_1-ATPase β subunit, isopropyl malate synthetase, cytochrome c_1, and subunit V of cytochrome *c* oxidase. All these polypeptides synthesized in a cell-free system were imported into isolated mitochondria and found in mature form in proper compartments: cytochrome b_2 in the intermembrane space, the ATPase subunit and isopropyl malate synthetase in the matrix, cytochrome c_1 and subunit V of cytochrome *c* oxidase in the membrane fraction. However, addressing of the polypeptides proved to be preferential rather than exclusive. More distinct results were obtained for cytochrome b_2 and F_1-ATPase β subunit with mitoplasts.

Page 199

Zwizinski et al. (Zwizinski, C., Schleyer, M., and Neupert, W., 1983, J.Biol.Chem. 258:4071-4074) recently showed that the precursor form of Neurospora crassa mitochondrial ADP/ATP carrier binds with the mitochondrial surface even when the membrane potential is abolished. Consequently, this first stage of interaction of the precursor with the receptor site does not depend on the potential. Similar data were obtained earlier in a study of the synthesis of the phage M13 coat protein and its insertion into the host-cell plasma membrane (Date et al., 1980).

Page 201

A neutral metalloendopeptidase capable of processing the β subunit of F_1-ATPase was isolated from hypotonic extracts of yeast mitochondria as a soluble complex of molecular weight 115,000 (McAda, P.C., and Douglas, M.G., 1982, J.Biol.Chem. 257:3177-3182). Sodium dodecyl sulfate-polyacrylamide gel electrophoresis revealed one major (molecular weight of 59,000) and several minor polypeptides in the endopeptidase preparation. The enzyme was insensitive to inhibitors of serine proteinases but was ibhibited by EDTA and o-phenanthroline.

Page 204

*Later it was shown that the 37,000-molecular-weight product appears owing to nonspecific events and does not represent a bona fida intermediate (Argan, C., Lusty, C.J., and Shore, G., 1983, J.Biol.Chem. 258:6667-6670).

Page 204

**More detailed studies of the processing and import into mitochondria of precursors to cytochrome c_1 (Ohashi, A., Gibson, J., Gregor, I., and Schatz, G., 1982, J.Biol.Chem. 257:13042-13047) and cytochrome b_2 (Reid, G.A., Yonetani, T., and Schatz, G., 1982, J.Biol.Chem. 257:13068-13074; Daum, G., Gasser, S.M., and Schatz, G., 1982, J.Biol.Chem. 257:13075-13080) showed that maturation of cytochrome b_2 in yeast cells is slower than of cytochrome c_1 and takes about an hour to be practically complete (90% conversion). The precursor of cytochrome *c* peroxidase is processed even more slowly.

Page 207

Later Suissa, M., and Schatz, G. (1982, J.Biol.Chem. 257:13048-13055) studied the distribution in yeast cells of 12 mRNAs for imported mitochondrial proteins localized in the matrix (citrate synthetase, isopropyl malate synthetase), in the inner membrane (α, β, and γ subunits of F_1-ATPase, subunits V and VI of cytochrome *c* oxidase, cytochrome c_1, and subunit I of the cytochrome bc_1 complex), in the intermembrane space (cytochrome b_2 and cytochrome *c* peroxidase), and in the outer membrane (the 29,000-dalton polypeptide). They found that under certain experimental conditions mRNAs for only three of the above proteins (cytochrome *c* oxidase subunits V and VI and the 29,000-dalton polypeptide) are present exclusively in free polysomes, whereas other mRNAs are distributed between free polysomes and those associated with mitochondria. These data pointed to the dual (co-and-posttranslational) mechanism of import of at least some polypeptides into mitochondria. It is yet impossible to quantitatively assess the relative contribution of these two pathways; in all probability it may be controlled by the rate of translation.

Page 287

Precursors of cytochrome b_2 and cytochrome *c* peroxidase in yeast cells maturate still more slowly (Reid, G.A., Yonetani, T., and Schatz, G., 1982, J.Biol.Chem. 257:13068-13074).

Page 289

Recently this thesis was confirmed (Abramova, N.B., Korzh, V.P., and Neifakh, A.A., 1983, Biokhimiya 49:1279-1286): it was shown that microinjection of isolated mitochondria into developing loach embryos does not increase oxygen consumption. Furthermore, a portion of the mitochondrial material containing both endogenous and injected mitochondria was degraded.

Page 291

Something like this happens to peroxisomes of methanol-grown yeast *Hansenula polymorpha* when the organism is transferred from methanol-containing to glucose-containing medium (see references in Van Dijken, J.P., Veenhuis, M., and Harder, W., 1982, Ann.N.Y.Acad. Sci. 356:200-216). Under such conditions peroxisomes, which harbor the enzymes responsible for utilization of methanol, are selectively degraded with the participation of the cell vacuolar apparatus. Here we can again ask ourselves how the cell knows that it is the peroxisomes that must be eliminated and not, for instance, mitochondria.

Page 292

Another example of the significant role of proteolysis in controlling the content of membrane proteins is the data for acetyl choline receptors of muscle cells. As found by Hyman, C., and Floehner, S.C. (1983, J.Cell Biol. 96:1316-1324), introduction of leupeptine, an inhibitor of lysosomal proteinases, to a mouse muscle-cell line results in substantial suppression of the receptor breakdown and in a consequent 2-3-fold increase in their content in the cell. The excess receptors originate in all probability from the cell surface and accumulate in a cellular compartment with lysosomal characteristics. These data resemble the above-described results of our earlier studies (see Chapter Eight) which demonstrated that addition of proteinase inhibitors to growing yeast leads to preservation or accumulation of nonfunctioning components of the mitochondrial respiratory chain (Galkin et al., 1980).

References

Aas, M., and Bremer, J., 1968, Biochim.Biophys.Acta, 164:157-166.

Abracham, A. K., Vedeler, A., Reistad, R., and Phil, A., 1981, Exptl. Cell Res., 132:265-272.

Ackrell, B. A. C., Ball, M. B., and Kearney, E. B., 1980, J.Biol. Chem., 255:2761-2769.

Ades, I. Z., and Butow, R. A., 1980a, J.Biol.Chem., 255:9918-9924.

Ades, I. Z., and Butow, R. A., 1980b, J.Biol.Chem., 255:9925-9935.

Adesnik, M., Bar-Nun, S., Maschio, F., Zunich, M., Lippman, A., and Bard, E., 1981, J.Biol.Chem., 256:10340-10345.

Alberti, K. G. M. M., and Bartley, W., 1963, Biochem.J., 87:104-114.

Alberti, K. G. M. M., and Bartley, W., 1965, Biochem.J., 95:641-656.

Alberti, K. G. M. M., and Bartley, W., 1969, Biochem.J., 111:763-776.

Aleksandrowicz, Z., Swierczyński, J., and Wrzokowa, T., 1973, Biochim.Biophys.Acta, 305:59-66.

Alexandre, A., Rossi, C. R., Carignani, G., and Rossi, C. S., 1975, FEBS Lett., 52:107-110.

Alfonso, M., Kandrach, M. A., and Racker, E., 1981, J.Bioenerg. Biomembr., 13:375-391.

Anderson, W. M., and Fisher, R. R., 1978, Arch.Biochem.Biophys., 187:180-190.

Anderson, W. M., and Fisher, R. R., 1981, Biochim.Biophys.Acta, 635:194-199.

Anderson, M. M., McCarty, R. E., and Zimmer, E. A., 1974, Plant Physiol., 53:699-704.

Anderson, S., Bankier, A. T., Barrell, B. G., de Bruijn, M. H. L., Coulson, A. R., Drouin, J., Eperon, I. C., Nierlish, D. P., Roe, B. A., Sanger, F., Schreier, P. H., Smith, A. J. H., Staden, R., and Young, I. G., 1981, Nature, 290:457-465.

Anderson, S., de Bruijn, M. H. L., Coulson, A. R., Eperon, I. C., Sanger, F., and Young, I. G., 1982, J.Mol.Biol., 156:683-717.

Aquila, H., Eierman, W., Babel, W., and Klingenberg, M., 1978, Eur.J. Biochem., 85:549-560.

Aquila, H., Misra, D., Eulitz, M., and Klingenberg, M., 1982, Hoppe-Seyler's Z.Physiol.Chem., 363:345-349.

Arias, I. M., Doyle, D., and Schimke, R. T., 1969, J.Biol.Chem., 244: 3303-3315.

Arnold, C. G., Schimmer, O., Schötz, F., and Bathelt, H., 1972, Arch. Microbiol., 81:50-67.

Aroskar, V. A., and Avadhani, N. G., 1979, Biochem.Biophys.Res. Commun., 91:17-23.
Artsatbanov, V. Yu., Konstantinov, A. A., and Skulachev, V. P., 1977, Dokl.Akad.Nauk SSSR 237:461-464.
Arutjunjan, A. M., Kamensky, Y. A., Milgrö̈m, E., Surkov, S., Konstantinov, A. A., and Sharomov, Y. A., 1978, FEBS Lett., 95:40-44.
Aschenbrenner, V., Druyan, R., Albin, R., and Rabinowitz, M., 1970 Biochem.J., 119:157-160
Ashford, T. P., and Porter, K. R., 1962, J.Cell Biol., 12:198-202.
Ashhurst, D. E., 1965, J.Cell Biol., 24:497-499.
Ashwell, M. A., and Work, T. S., 1968, Biochem.Biophys.Res.Commun., 32:1006-1012.
Ashwell, M. A., and Work, T. S., 1970, Annu.Rev.Biochem., 39:251-290.
Astin, A. M., Haslam, J. M., and Woods, R. A., 1977, Biochem.J., 166: 275-285.
Attardi, G., Constantino, P., England, J., Lynch, D., Murphy, W., Ojala, D., Posakony, J., and Storrie, B., 1975, in: "Genetics and Biogenesis of Mitochondria and Chloroplasts", P. S. Perlman, C. W. Birky, and T. J. Byers, eds., pp.3-65, Ohio State University Press, Columbus.
Autor, A. P., 1982, J.Biol.Chem., 257:2713-2718.
Awasthi, Y. C., Chuang, T. F., Keenan, T. W., and Crane, F. L., 1970a, Biochem.Biophys.Res.Commun., 39:822-832.
Awasthi, Y. C., Ruzicka, F. J., and Crane, F. L., 1970b, Biochim. Biophys.Acta, 203:233-248.
Aziz, L. E., Chien, S. -M., Patel, H. V., and Freeman, K. B., 1981, FEBS Lett., 133:127-129.
Azzi, A., 1980, Biochim.Biophys.Acta, 594:231-252.
Baginsky, M. L., and Hatefi, Y., 1969, J.Biol.Chem., 244:5313-5319.
Bailey, E., Taylor, C. B., and Bartley, W., 1967, Biochem.J., 104: 1026-1032.
Bakalkin, G. Ya., Kalnov, S. L., Zubatov, A. S., and Luzikov, V. N., 1976, FEBS Lett., 63:218-221.
Bakalkin, G. Ya., Kalnov, S. L., Galkin, A. V., Zubatov, A. S., and Luzikov, V. N., 1978, Biochem.J., 170:569-576.
Bakeyeva, L. E., Grinius, L. L., Jasaitis, A. A., Kuliene, V. V., Levitsky, D. O., Liberman, E. A., Severina, I. I., and Skulachev, V. P., 1970, Biochim.Biophys.Acta, 216:13-21
Baldwin, A. N., and Berg, P., 1966, J.Biol.Chem., 241:831-838.
Ball, A. J. S., and Tustanoff, E. R., 1970, Biochim.Biophys.Acta, 199:476-489.
Ball, M. B., Bell, R. L., and Capaldi, R. A., 1977, FEBS Lett., 83: 99-102.
Bandlow, W., 1972, Biochim.Biophys.Acta, 282:105-122.
Banerjee, R. K., Shertzer, H. G., Kanner, B. I., and Racker, E., 1977, Biochem.Biophys.Res.Commun., 75:772-778.
Banno, Y., Morris, H., and Katunuma, N., 1978, J.Biochem., (Tokyo), 83:1545-1554.
Barskii, E. L., Drachev, L. A., Kaulen, A. D., Kondrashin, A. A., Liberman, E. A., Ostroumov, S. A., Samuilov, V. P., Semenov,

A. Yu., Skulachev, V. P., and Jasaitis, A. A., 1975, Bioorg. Khim., 1:113-126.
Bartelink, A. K. M., and De Kort, C. A. D., 1973, Biochem.J., 136: 795-802.
Bartley, W., and Birt, L. M., 1970, in: "Essays in Cell Metabolism", W. Bartley, H. L. Kornberg, and J. R. Quayle, eds., pp.1-44, Wiley Interscience, London - New York - Sydney - Toronto.
Bartley, W., and Tustanoff, E. R., 1966, Biochem.J., 99:599-603.
Baum, H., Silman, H. I., Rieske, J. S., and Lipton, S. H., 1967, J.Biol.Chem., 242:4876-4887.
Beattie, D. S., 1968, Biochem.Biophys.Res.Commun., 30:57-62.
Beattie, D. S., 1969, Biochem.Biophys.Res.Commun., 35:721-727.
Beattie, D. S., and Stuchell, R. N., 1970, Arch.Biochem.Biophys., 139:291-297.
Beattie, D. S., Basford, R. E., and Koritz, S. R., 1967a, J.Biol. Chem., 242:4584-4586.
Beattie, D. S., Basford, R. E., and Koritz, S. R., 1967b, Biochemistry, 6:3099-3106.
Beattie, D. S., Patton, G. M., and Stuchell, R. N., 1970, J.Biol. Chem., 245:2177-2184.
Beattie, D. S., Clejan, L., Chen, Y. -S., Lin, C. -I. P., and Sidhu, A., 1981, J.Bioenerg.Biomembr., 13:357-373.
Beechey, R. B., 1974, in: "Membrane ATPases and Transport Processes", R. J. Bronk, ed., Biochem.Soc.Spec.Publ., 4:41-62, Biochemical Society, London.
Bell, R. L., and Capaldi, R. A., 1976, Biochemistry, 15:996-1101.
Bell, R. L., Sweetland, J., Ludwig, B., and Capaldi, R. A., 1979, Proc.Natl.Acad.Sci.USA, 76:741-745.
Bennett, J., 1981, Eur.J.Biochem., 118:61-70.
Berden, J. A., and Slater, E. C., 1970, Biochim.Biophys.Acta, 216: 237-249.
Berden, J. A., and Slater, E. C., 1972, Biochim.Biophys.Acta, 256: 199-215.
Berden, J. A., and Voorn-Brouwer, M. M., 1978, Biochim.Biophys.Acta, 501:424-439.
Berezney, R., Awasthi, Y. C., and Crane, F. L., 1970a, J.Bioenergetics, 1:457-465.
Berezney, R., Awasthi, Y. C., Funk, L. K., and Crane, F. L., 1970b, J.Bioenergetics, 1:445-456.
Bertina, R. M., and Out, T. A., 1973, 9th Intern.Congr.Biochem., Stockholm, Abstr., p.236.
Bertina, R. M., Schrier, P. I., and Slater, E. C., 1973, Biochim. Biophys.Acta, 305:503-518.
Bertrand, H., and Werner, S., 1977, Eur.J.Biochem., 79:599-606.
Betz, H., and Weiser, U., 1976, Eur.J.Biochem., 70:385-395.
Bhat, N. K., Niranjan, B. G., and Avadhani, N. G., 1982, Biochemistry 21:2452-2460.
Bibb, M. J., van Etten, R. A., Wright, C. T., Walberg, M. W., and Clayton, D. A., 1981, Cell, 26:167-180.
Biewald, R., and Buse, G., 1982, Hoppe-Seyler's Z.Physiol.Chem., 363: 1141-1153.

Bill, K., Broger, C., and Azzi, A., 1982, Biochim.Biophys.Acta, 679: 28-34.
Bingham, R. W., and Campbell, P. N., 1972, Biochem.J., 126:211-215.
Birchmeier, W., Kohler, C. E., and Schatz, G., 1976, Proc.Natl.Acad. Sci.USA, 73:4334-4338.
Birkmayer, G. D., and Bücher, Th., 1969, FEBS Lett., 5:28-30.
Bisson, R., Azzi, A., Gutweniger, H., Colonna, R., Montecucco, C., and Zanotti, A., 1978a, J.Biol.Chem., 253:1874-1880.
Bisson, R., Gutweniger, H., and Azzi, A., 1978b, FEBS Lett., 92:219-222.
Bisson, R., Montecucco, C., Gutweniger, H., and Azzi, A., 1979, J.Biol.Chem., 254:9962-9965.
Bisson, R., Jacobs, B., and Capaldi, R. A., 1980, Biochemistry, 19: 4173-4178.
Blobel, G., 1980, Proc.Natl.Acad.Sci.U.S.A., 77:1496-1500.
Blobel, G., and Dobberstein, B., 1975, J.Cell Biol., 67:835-851.
Boehni, P., Gasser, S., Leaver, C., and Schatz, G., 1980, in: "The Organization and Expression of the Mitochondrial Genome", A. M. Kroon and S. Saccone, eds., pp.423-433, Elsevier/North-Holland Biomedical Press, Amsterdam - New York - Oxford.
Boime, I., Smith, E. E., and Hunter, F. E., 1970, Arch.Biochem. Biophys., 139:425-443.
Bojanovski, D., Schlimme, E., Wang, C. S., and Alaupovic, P., 1975, 10th FEBS Meeting Abstr. No 1156, Paris.
Bojanovski, D., Schlimme, E., Wang, C. S., and Alaupovic, P., 1976, Eur.J.Biochem., 71:539-548.
Bolotina, I. A., Markovich, D. S., Volkenstein, M. V., and Zavodsky, P., 1967, Biochim.Biophys.Acta, 132:260-270.
Bomstein, R., Goldberger, R., and Tisdale, H., 1961, Biochim.Biophys. Acta, 50:527-543.
Bonitz, S. G., Coruzzi, G., Thalenfeld, B. E., Tzagoloff, A., and Macino, G., 1980, J.Biol.Chem., 255:11927-11941.
Boonman, J. C. P., van Beek, G. G. M., Muijsers, A. O., and van Gelder, B. F., 1979, Molec.Cell Biol., 26:183-192.
Boquet, P., and Pappenheimer, A. M., 1976, J.Biol.Chem., 251:5770-5778.
Borgese, N., and Meldolesi, J., 1980, J.Cell Biol., 85:501-515.
Borst, P., 1972, Annu.Rev.Biochem., 41:333-376.
Borst, P., and Grivell, L. A., 1971, FEBS Lett., 13:73-88.
Borst, P., and Kroon, A. M., 1969, Intern.Rev.Cytol., 26:108-190.
Bouck, G. B., and Brown, D. L., 1976, Annu.Rev.Plant Physiol., 27: 71-94.
Boutry, M., and Coffeau, A., 1982, Eur.J.Biochem., 125:471-477.
Boulay, F., Brandolin, G., Lauquin, G. J. M., Jollès, J., Jollès, P., and Vignais, P. V., 1979, FEBS Lett., 98:161-164.
Boxer, D. H., Feckl, J., and Klingenberg, M., 1977, FEBS Lett., 73: 43-46.
Boyer, P. D., Luchsinger, W. W., and Falcone, A. B., 1956, J.Biol. Chem., 223:405-421.
Bracker, C. E., and Grove, S. N., 1971, Protoplasma, 73:15-34.
Bragin, E. O., Sorokovoi, V. I., Kogan, E. M., and Vladimirov, Yu. A., 1975, Vopr.Med.Khim., 21:150-152.

Bragin, E. O. Dergunov, A. D., Neugodova, G. L., Sorokovoi, V. I., and Vladimirov, Yu. A., 1977a, Vopr.Med.Khim., 23:673-677.
Bragin, E. O., Sorokovoi, V. I., Chernikov, V. P., Kogan, E. M., and Vladimirov, Yu. A., 1977b, Vopr.Med.Khim., 23:297-302.
Brambl, R., 1980, J.Biol.Chem., 255:7673-7680.
Brandolin, G., Meyer, C., Defaye, G., Vignais, P. M., and Vignais, P. V., 1974, FEBS Lett., 46:149-153.
Brdiczka, D., Pette, D., Brunner, G., and Miller, F., 1968, Eur.J. Biochem., 5:294-304.
Brega, A., and Baglioni, C., 1971, Eur.J.Biochem., 22:415-422.
Brenner-Holzach, O., and Raaflaub, J., 1954, Helv.Physiol.Acta, 12: 242-252.
Brierley, G. P., 1976, Mol.Cell.Biochem., 10:41-62.
Brierley, G. P., and Jurkowitz, M., 1976, Biochem.Biophys.Res. Commun., 68:82-88.
Brierley, G. P., Merola, A. J., and Fleischer, S., 1962, Biochim. Biophys.Acta, 64:218-228.
Brierley, G. P., Murer, E., and Bachmann, E., 1964, J.Biol.Chem., 239-2706-2712.
Brierley, G. P., Jurkowitz, M., and Chávez, E., 1977a, Biochem. Biophys.Res.Commun., 74:235-241.
Brierley, G. P., Jurkowitz, M., Chávez, E., and Jung, D. W., 1977b, J.Biol.Chem., 252:7932-7939.
Briggs, M. M., and Capaldi, R. A., 1978, Biochem.Biophys.Res.Commun., 80:553-559.
Briggs, M. M., Kamp, P. -F., Robinson, N. C., and Capaldi, R. A., 1975, Biochemistry, 14:5123-5128.
Bronnikov, G. I., and Evtodienko, Yu. V., 1977, Biokhimiya, 42:952-954.
Brooks, J. C., and Senior, A. E., 1971, Arch.Biochem.Biophys., 147: 467-470.
Brooks, J. C., and Senior, A. E., 1972, Biochemistry, 11:4675-4678.
Brown, G. G., and Beattie, D. S., 1977, Biochemistry, 16:4449-4454.
Brown, G. G., and Beattie, D. S., 1978, Biochim.Biophys.Acta, 538: 173-187.
Browning, K. S., and RajBhandary, U. L., 1982, J.Biol.Chem., 257: 5253-5256.
Bruni, A., and Racker, E., 1968, J.Biol.Chem., 243:962-971.
Brunner, G., and Bücher, T., 1970, FEBS Lett., 6:105-108.
Brunner, G., and Neupert, W., 1968, FEBS Lett., 1:153-155.
Bucher, J. R., and Penniall, R., 1975, FEBS Lett., 60:180-184.
Buetow, D. E., and Wood, W. M., 1978, Subcell.Biochem., 5:1-85.
Bulos, B., and Racker, E., 1968, J.Biol.Chem., 243:3891-3900.
Bunn, C. L., Mitchell, C. H., Lukins, H. B., and Linnane, A. W., 1970, Proc.Natl.Acad.Sci. U.S.A., 67:1233-1240.
Burke, J. B., and Beattie, D. S., 1974, Arch.Biochem.Biophys., 164: 1-11.
Burstein, C., Kandrach, A., and Racker, E., 1971a, J.Biol.Chem., 246: 4083-4089.
Burstein, C., Loyter, A., and Racker, E., 1971b, J.Biol.Chem., 246: 4075-4082.

Buse, G., and Steffens, G., 1978, Hoppe-Seyler's Z. Physiol.Chem., 359:1005-1009.
Buse, G., Steffens, G. J., Steffens, G. C. M., Sacher, R., and Erdweg, M., 1980, in: "First European Bioenergetics Conference", pp.41-42, Eogesta Center, Urbino, Italy.
Butler, W. H., and Judah, J. D., 1970, Biochem.J., 118:883-886.
Butow, R. A., Lopez, I. C., Chang, H. -P., and Farrelly, F., 1980, in: "The Organization and Expression of the Mitochondrial Genome", A. M. Kroon and C. Saccone, eds., pp.195-205, Elsevier/North-Holland Biomedical Press, Amsterdam - New York - Oxford.
Cabral, F., and Schatz, G., 1978, J.Biol.Chem., 253:4396-4401.
Campbell, P. N., and Blobel, G., 1976, FEBS Lett., 72:215-226.
Campbell, M. T., Sutton, R., and Pollak, J. K., 1982, Eur.J.Biochem., 125:401-406.
Calvayrac, R., van Leute, F., and Butow, R. A., 1971, Science, 173: 252-254.
Capaldi, R. A., 1973, Biochem.Biophys.Res.Commun., 53:1331-1337.
Capaldi, R. A., 1974, Arch.Biochem.Biophys., 163:99-105.
Capaldi, R. A., Komai, H., and Hunter, D. R., 1973, Biochem.Biophys. Res.Commun., 55:655-659.
Capaldi, R. A., Sweetland, J., and Merli, A., 1977, Biochemistry, 16:5707-5710.
Capeillere-Blandin, C., and Ohnishi, T., 1982, Eur.J.Biochem., 122: 403-413.
Caplan, A. I., and Greenawalt, J. W., 1966, J.Cell Biol., 36:15-31.
Carafoli, E., and Gazzotti, P., 1970, Biochem.Biophys.Res.Commun., 39:842-846.
Cartledge, T. G., and Lloyd, D., 1973, Biochem.J., 132:609-621.
Cartledge, T. G., Lloyd, D., Erecińska, M., and Chance, B., 1972, Biochem.J., 130:739-747.
Case, G. D., and Leigh, J. S., 1976, Biochem.J., 160:769-784.
Case, G. D., Ohnishi, T., and Leigh, J. S., 1976, Biochem.J., 160: 785-795.
Casey, R. P., Thelen, M., and Azzi, A., 1980, J.Biol.Chem., 255: 3994-4000.
Cattell, K. J., Lindop, C. R., Knight, I. G., and Beechey, R. B., 1971, Biochem.J., 125:169-177.
Catterall, W. A., and Pedersen, P. L., 1971, J.Biol.Chem., 246: 4987-4994.
Catterall, W. A., and Pedersen, P. L., 1974, in: "Membrane ATPases and Transport Processes", R. J. Bronk, ed., pp.63-88, Biochem. Soc.Spec.Publ., 4, Biochemical Society, London.
Catterall, W. A., Coty, W. A., and Pedersen, P. L., 1973, J.Biol. Chem., 248:7427-7431.
Cerletti, N., and Schatz, G., 1979, J.Biol.Chem., 254:7746-7751.
Cerletti, P., Giordano, M. G., Giovenco, M. A., Barra, D., and Strom, R., 1966, Biochim.Biophys.Acta, 122:352-355.
Chaix, P., 1961, in: "Haematin Enzymes", J. E. Falk, R. Lemberg, and R. K. Morton, eds., pp.225-232, Pergamon Press, Oxford - London - New York - Paris.

Chan, S. H. P., and Barbour, B. L., 1976, Biochem.Biophys.Res.Commun. 72:499-506.

Chan, S. H. P., and Tracy, R. P., 1978, Eur.J.Biochem., 89:595-605.

Chan, S. H. P., Love, B., and Stotz, E., 1970, J.Biol.Chem., 245: 6669-6674.

Chance, B., 1972, FEBS Lett., 23:3-20.

Chance, B., Parsons, D. F., and Williams, G. R., 1963, Science, 143: 136-139.

Chance, B., Wilson, D. F., Dutton, P. L., and Erecińska, M., 1970, Proc.Natl.Acad.Sci. U.S.A., 66:1175-1182.

Chang, T. -M., and Penefsky, H. S., 1973, J.Biol.Chem., 248:2746-2754.

Chapman, C., and Bartley, W., 1968, Biochem.J., 107:455-465.

Chappell, J. B., 1968, Br.Med.Bull., 24:150-157.

Chappell, J. B., and Crofts, A. R., 1965, Biochem.J., 95:707-716.

Chappell, J. B., and Greville, G. D., 1958, Nature, 182:813-814.

Chappell, J. B., and Greville, G. D., 1959, Nature, 183:1737-1738.

Chappell, J. B., and Greville, G. D., 1960, Biochim.Biophys.Acta, 38:483-494.

Chappell, J. B., and Greville, G. D., 1961, Nature, 190:502-504.

Chappell, J. B., and Greville, G. D., 1963, Biochem.Soc.Symp., 23:39-65.

Chappell, J. B., and Perry, S. V., 1954, Nature, 173:1094-1095.

Chapeville, F., and Haenni, A. -L., 1974, Biosynthèse des Protéines, Hermann Collection, Paris Methodes.

Chávez, E., Jung, D. W., and Brierley, G. P., 1977, Biochem.Biophys. Res.Commun., 75:69-75.

Chefurka, W., 1966, Biochemistry, 5:3887-3903.

Chen, Y. -S., and Beattie, D. S., 1981, Biochemistry, 20:7557-7565.

Chen, W. L., and Charalampous, F. C., 1969, J.Biol.Chem., 244:2767-2776.

Chen, W. L., and Charalampous, F. C., 1973, Biochim.Biophys.Acta, 294:329-342.

Chen, S., and Guillory, R. J., 1981, J.Biol.Chem., 256:8318-8323.

Chin, C. H., 1950, Nature, 165:926-927.

Ching, E., and Attardi, G., 1982, Biochemistry, 21:3188-3195.

Choundhury, N. K., and Biswal, U. C., 1979, Experientia, 35: 1036-1037.

Choundhury, N. K., and Biswal, U. C., 1980, Physiol.Plant., 49:43-48.

Christiansen, R. O., Loyter, A., Steensland, H., Saltzgaber, J., and Racker, E., 1969, J.Biol.Chem., 244:4428-4436.

Christophersen, B. O., 1968, Biochim.Biophys.Acta, 164:35-46.

Chua, N. -H., and Schmidt, G. W., 1978, Proc.Natl.Acad.Sci. U.S.A., 75:6110-6114.

Chua, N. -H., and Schmidt, G. W., 1979, J.Cell Biol., 81:461-483.

Chuang, T. F., Awasthi, Y. C., and Crane, F. L., 1973, J.Bioenergetics., 5:27-72.

Cintrón, N. M., and Pedersen, P. L., 1979, J.Biol.Chem., 254: 3439-3443.

Clarke, S., 1976, J.Biol.Chem., 251:1354-1363.

Clark-Walker, G. D., and Linnane, A. W., 1966, Biochem.Biophys.Res. Commun., 25:8-13.
Clark-Walker, G. D., and Linnane, A. W., 1967, J.Cell Biol., 34:1-14.
Clejan, L., Beattie, D. S., Gollub, E. G., Liu, K. -P., and Sprinson, D. B., 1980, J.Biol.Chem., 255:1312-1316.
Colli, W., Hinkle, P. C., and Pullman, M. E., 1969, J.Biol.Chem., 244:6432-6443.
Conboy, J. G., Fenton, W. A., and Rosenberg, L. E., 1982, Biochem. Biophys.Res.Commun., 105:1-7.
Connely, J. L., and Hallstrom, C. H., 1967, Biochemistry, 6:1567-1572.
Constantino, P., and Attardi, G., 1975, J.Mol.Biol., 96:291-306.
Constantino, P., and Attardi, G., 1977, J.Biol.Chem., 252:1702-1711.
Constantopoulos, G., and Kenyon, C. N., 1968, Plant Physiol., 43: 531-536.
Cooper, C., and Lehninger, A. L., 1957, J.Biol.Chem., 224:561-578.
Coote, J. L., and Work, T. S., 1971, Eur.J.Biochem., 23:564-574.
Coruzzi, G., and Tzagoloff, A., 1979, J.Biol.Chem., 254:9324-9330.
Coruzzi, G., Trembath, M. K., and Tzagoloff, A., 1979, Eur.J.Biochem. 92:279-287.
Corwin, L. M., and Lipsett, M. N., 1959, J.Biol.Chem., 234:2453-2458.
Côté, C., Solioz, M., and Schatz, G., 1979, J.Biol.Chem., 254:1437-1439.
Coty, W. A., and Pedersen, P. L., 1975, J.Biol.Chem., 250:3515-3521.
Cox, G. B., Downie, A., Langman, L., Senior, A. E., Ash, G., Fayle, D. R. H., and Gibson, F., 1981, J.Bacteriol., 148:30-42.
Crane, R. C., and Marinetti, G. V., 1979, Biochemistry, 18:2407-2414.
Crane, F. L., Glenn, J. L., and Green, D. E., 1956, Biochim.Biophys. Acta, 22:475-487.
Crane, F. L., Arntzen, C. J., Hall, J. D., Ruzicka, F. J., and Dilley, R. A., 1971, in: "Autonomy and Biogenesis of Mitochondria and Chloroplasts", N. K. Boardman, A. W. Linnane, and R. M. Smillie, eds., pp.53-69, North-Holland Publishing Comp., Amsterdam - London.
Cremona, T., and Kearney, E. B., 1963, Nature, 200:542-544.
Cremona, T., and Kearney, E. B., 1964, J.Biol.Chem., 239:2328-2334.
Criddle, R. S., and Schatz, G., 1969, Biochemistry, 8:322-334.
Criddle, R. S., Packer, L., and Shieh, P., 1977, Proc.Natl.Acad.Sci. U.S.A., 74:4306-4310.
Criddle, R. S., Johnston, R., Packer, L., Shieh, P. K., and Konishi, T., 1979, in: "Cation Flux across Biomembranes", Y. Mukohata and L. Packer, eds., pp.399-407, Acad. Press, New York - San Francisco - London.
Cross, R. J., Taggart, J. V., Covo, G. A., and Green, D. E., 1949, J.Biol.Chem., 177:655-678.
Cundliffe, E., and McQuillen, K., 1967, J.Mol.Biol., 30:137-146.
Cuzner, M. L., Davison, A. N., and Gregson, N. A., 1966, Biochem.J., 101:618-626.
Damsky, C. H., Nelson, W. M., and Claude, A., 1969, J.Cell Biol., 43:174-179.

Daniels, C. J., Bole, D. G., Quay, S. C., and Oxender, D. L., 1981, Proc.Natl.Acad.Sci. U.S.A., 78:5396-5400.
Danielson, L., and Ernster, L., 1963, Biochem.Z., 338:188-205.
Darley-Usmar, V. M., and Fuller, S. D., 1981, FEBS Lett., 135:164-166.
Das Gupta, U., and Rieske, J. S., 1973, Biochem.Biophys.Res.Commun., 54:1247-1254.
Date, T., and Wickner, W., 1981, Proc.Natl.Acad.Sci. U.S.A., 78:6106-6110.
Date, T., Zwizinski, C., Ludmerer, S., and Wickner, W., 1980, Proc.Natl.Acad.Sci. U.S.A., 77:827-831.
Datta, A., Penefsky, H. S., 1970, J.Biol.Chem., 245:1537-1544.
Davey, P. J., Yu, R., and Linnane, A. W., 1969, Biochem.Biophys.Res.Commun., 36:30-34.
Davidian, N., Pennial, R., and Elliott, E. B., 1969, Arch.Biochem.Biophys., 133:345-358.
Davidson, R. G., Cortner, J. A., Rattazzi, M. C., Ruddle, F. H., and Lubs, H. A., 1970, Science, 169:391-392.
Davis, K. A., and Hatefi, Y., 1969, Biochemistry, 8:3355-3361.
Davis, K. A., and Hatefi, Y., 1971, Biochemistry, 10:2509-2516.
Davis, K. A., Hatefi, Y., Poff, K. L., and Butler, W. L., 1972, Biochem.Biophys.Res.Commun., 46:1984-1990.
Davis, K. A., Hatefi, Y., Poff, K. L., and Butler, W. L., 1973, Biochim.Biophys.Acta, 325:341-356.
Davison, M. T., and Garland, P., 1975, J.Gen.Microbiol., 91:127-138.
Davison, M. T., and Garland, P., 1977, J.Gen.Microbiol., 98:147-153.
De Bernard, B., 1957, Biochim.Biophys.Acta, 23:510-515.
De Bernard, B., Getz, G. S., and Rabinowitz, M., 1969, Biochim.Biophys.Acta, 193:58-63.
De Córdova, C. S., Cohén, R., and González-Cadavid, N. F., 1977, Biochem.J., 166:305-313.
De Duve, C., and Wattiaux, R., 1966, Annu.Rev.Physiol., 28:435-492.
De Jong, L., Holtrop, H., and Kroon, A. M., 1980, Biochim.Biophys.Acta, 606:331-337.
Denslow, N. D., and O'Brien, T. W., 1974, Biochem.Biophys.Res.Commun., 57:9-16.
De Pierre, J. W., and Ernster, L., 1977, Annu.Rev.Biochem., 46:201-262.
De Ronde, A., van Loon, A. P. G. M., Grivell, L. A., and Kohli, J., 1980, Nature, (London), 287:361-363.
Desautels, M., and Goldberg, A. L., 1982, J.Biol.Chem., 257:11673-11679.
De Vries, H., and Kroon, A. M., 1970, Biochim.Biophys.Acta, 204:531-541.
De Winter, J. M., Vianen, G. M., and van den Bosch, H., 1982, Biochim.Biophys.Acta, 712:332-341.
Dharmalingam, K., and Jayaraman, J., 1971, Biochem.Biophys.Res.Commun., 45:1115-1118.
Dharmalingam, K., and Jayaraman, J., 1972, Biochem.J., 128:45P.
Dianoux, A. -C., Bof, M., Césarini, R., Reboul, A., and Vignais, P. V., 1976, Eur.J.Biochem., 67:61-66.

Dianoux, A. -C., Bof, M., and Vignais, P. V., 1978, Eur.J.Biochem., 88:69-77.

Di Jeso, F., 1978, 12th FEBS Meeting, Dresden, Abstr. No.1707.

Di Jeso, F., Christiansen, R. O., Steensland, H., and Loyter, A., 1969, Fed.Proc., 28:663.

Di Sabato, G., and Fonnesu, A., 1959, Biochim.Biophys.Acta, 35:358-367.

Di Sabato, G., and Kaplan, N. O., 1965, J.Biol.Chem., 240:1072-1076.

Dixon, H., Kellerman, G. M., and Linnane, A. W., 1972, Arch.Biochem. Biophys., 152:869-875.

Dobberstein, B., Blobel, G., and Chua, N. -H., 1977, Proc.Natl.Acad. Sci. U.S.A., 74:1082-1085.

Dodd, G. H., 1972, Biochem.J., 127:38P-39P.

Dooijewaard, G., Slater, E. C., Van Dijk, P. J., and De Bruin, G. J. M., 1978, Biochim.Biophys.Acta, 503:405-424.

Douglas, M. G., and Butow, R. A., 1976, Proc.Natl.Acad.Sci. U.S.A., 73:1083-1086.

Downer, N. W., Robinson, N. C., and Capaldi, R., 1976, Biochemistry, 15:2930-2936.

Drachev, L. A., Jasaitis, A. A., Kaulen, A. D., Kondrashin, A. A., Chu, L. V., Semenov, A. Yu., Severina, I. I., and Skulachev, V. P., 1976a, J.Biol.Chem., 251:7072-7076.

Drachev, L. A., Jasaitis, A. A., Mikelsaar, H., Nemecek, I. B., Semenov, A. Yu., Semenova, E. G., Severina, I. I., and Skulachev, V. P., 1976b, J.Biol.Chem., 251:7077-7082.

Druyan, R., De Bernard, B., and Rabinowitz, M., 1969, J.Biol.Chem., 244:5874-5878.

Dubiel, W., Müller, M., and Rapoport, S., 1981a, Biochem.Intern., 3:165-171.

Dubiel, W., Müller, M., Rathmann, J., Hiebsch, C., and Rapoport, S., 1981b, Acta Biol.Med.Germ., 40:625-628.

Duque-Magalhaes, M. C., and Regnier, P., 1982, Biochimie, 64:907-913.

Earle, S. R., and Fisher, R. R., 1980, J.Biol.Chem., 255:827-830.

Earle, S. R., Anderson, W. M., and Fisher, R. R., 1978, FEBS Lett., 91:21-24.

Earnshaw, M. J., and Truelove, B., 1970, Plant Physiol., 45:322-326.

Ebner, E., and Maier, K. L., 1977, J.Biol.Chem., 252:671-676.

Ebner, E., and Schatz, G., 1973, J.Biol.Chem., 248:5379-5384.

Ebner, E., Mason, T. L., and Schatz, G., 1973a, J.Biol.Chem., 248: 5369-5378.

Ebner, E., Mennucci, L., and Schatz, G., 1973b, J.Biol.Chem., 248: 5360-5368.

Edelhoch, H., Hayashi, O., and Teply, L. J., 1952, J.Biol.Chem., 197:97-104.

Edwards, S. W., and Ball, E. G., 1954, J.Biol.Chem., 209:619-633.

Edwards, S. W., and Lloyd, D., 1977, Biochem.J., 162:39-46.

Edwards, S. W., Evans, J. B., Williams, J. L., and Lloyd, D., 1982, Biochem.J., 202:453-458.

Ekholm, R., Kerstell, J., Olsson, R., Rudenstam, C. -M., and Svanborg, A., 1968, Amer.J.Cardiol., 22:312-318.

Elődi, P., and Szabolczi, G., 1959, Nature, (London), 184:56.
Emmelot, P., 1960, Nature, (London), 188:1197-1199.
Engel, W. D., Schägger, H., and von Jagow, G., 1980, Biochim.Biophys. Acta, 592:211-222.
Ephrussi, B., and Slonimski, P., 1950a, Biochim.Biophys.Acta, 6:256-267.
Ephrussi, B., and Slonimski, P., 1950b, Compt.Rend., 230:685-686.
Ephrussi, B., Slonimski, P., Yotsuyanagi, Y., and Tavlizki, J., 1956, Compt.Rend.Trav.Lab.Carlsberg,ser.physiol., 26:87-102.
Erecińska, M., Wagner, M., and Chance, B., 1973, in: "Current Topics in Bioenergetics", Vol.5, D. R. Sanadi and L. Packer, ed., pp.267-303, Acad. Press, New York - London.
Ernster, L., 1975a, FEBS Symp., 35:257-285.
Ernster, L., and Kuylenstierna, B., 1969, in: "Mitochondria: Structure and Function", L. Ernster and Z. Drahota, eds., pp.5-31, Acad. Press, London, - New York.
Ernster, L., and Löw, H., 1955, Exp.Cell Res., Suppl.3:133-153.
Ernster, L., and Nazavio, F., 1956, Acta Chem.Scand., 10:1038-1039.
Ernster, L., and Orrenius, S., 1965, Fed.Proc., 24:1190-1199.
Ernster, L., Lee, C. -P., and Janda, S., 1967, in: "Biochemistry of Mitochondria", E. C. Slater, Z. Kaniuga, and L. Wojtczak, eds., pp.29-51, Acad. Press, London - New York.
Ernster, L., Asami, K., Juntti, K., Coleman, J., and Nordenbrand, K., 1977, in: "Structure of Biological Menbranes", S. Abrahamsson and I. Pascher, eds., pp.135-156, Plenum Press, New York - London.
Esparza, M., Velours, J., and Guerin, B., 1981, FEBS Lett., 134:63-66.
Estabrook, R. W., 1961, J.Biol.Chem., 236:3051-3057.
Estabrook, R. W., 1962, Biochim.Biophys.Acta, 60:236-248.
Estrada-O, S., 1964, Arch.Biochem.Biophys., 106:498-504.
Estrada-O, S., Cárabez, A. T., and Cabeza, A. G., 1966, Biochemistry, 5:3432-3440.
Estrada-O, S., Montal, M., Célis, H., and Cárabez, A., 1970, Eur.J. Biochem., 12:227-235.
Evans, D. A., and Lloyd, D., 1967, Biochem.J., 103:22P.
Eytan, G. D., and Broza, R., 1978, J.Biol.Chem., 253:3196-3202.
Eytan, G. D., and Racker, E., 1977, J.Biol.Chem., 252:3208-3213.
Eytan, G. D., and Schatz, G., 1975, J.Biol.Chem., 250:767-774.
Eytan, G. D., Carroll, R. C., Schatz, G., and Racker, E., 1975, J.Biol.Chem., 250:8598-8603.
Eytan, G. D., Matheson, M. J., and Racker, E., 1976, J.Biol.Chem., 251:6831-6837.
Farrell, E. C., Baba, N., Brierley, G. P., and Grümer, H. -D., 1972, Lab.Invest., 27:209-213.
Fee, J. A., Choc, M. G., Findling, K. L., Lorence, R., and Yoshida, T., 1980, Proc.Natl.Acad.Sci. U.S.A., 77:147-151.
Feinstein, M. B., Fernandez, S. M., and Sha'afi, R. I., 1975, Biochim.Biophys.Acta, 413:354-370.
Ferdinand, W., Bartley, W., and Broomhead, U., 1973, Biochem.J., 134:431-436.

Fernández-Morán, H., Oda, T., Blair, P. V., and Green, D. E., 1964, J.Cell Biol., 22:63-100.
Fessenden-Raden, J. M., 1972, J.Biol.Chem., 247:2351-2357.
Fettes, I. M., Haldar, D., and Freeman, K. B., 1972, Can.J.Biochem., 50:200-209.
Fiek, C., Benz, R., Roos, N., and Brdiczka, D., 1982, Biochim. Biophys.Acta, 688:429-440.
Fikhte, B. A., and Zaichkin, E. I., 1972, Dokl.Akad.Nauk SSSR, 205: 961-962.
Fillingame, R. H., 1980, Annu.Rev.Biochem., 49:1079-1113.
Firkin, F. C., and Linnane, A. W., 1968, Biochem.Biophys.Res.Commun., 32:398-402.
Firkin, F. C., and Linnane, A. W., 1969, Exp.Cell Res., 55:68-76.
Fleisher, S., Brierley, G. P., Klouwen, H., and Slautterback, D. S., 1962, J.Biol.Chem., 237:3264-3272.
Fleisher, S., Casu, A., and Fleisher, B., 1964, Fed.Proc., 23:486.
Fleisher, S., Fleisher, B., and Stoeckenius, W., 1967, J.Cell Biol., 32:193-208.
Fletcher, M. J., and Sanadi, D. R., 1961, Biochim.Biophys.Acta., 51: 356-360.
* Flohé, L., and Schlegel, W., 1971, Hoppe-Seyler's Z.Physiol.Chem., 352:1401-1410.
Flohé, L., and Zimmerman, R., 1970, Biochim.Biophys.Acta, 223:210-213.
Frangione, B., Rosenwasser, E., Penefsky, H. S., and Pullman, M. E., 1981, Proc.Natl.Acad.Sci. U.S.A., 78:7403-7407.
Franke, W. W., and Kartenbeck, J., 1971, Protoplasma, 73:35-41.
Freitag, H., Neupert, W., and Benz, R., 1982a, Eur.J.Biochem., 123: 629-636.
Freitag, H., Janes, M., and Neupert, W., 1982b, Eur.J.Biochem., 126: 197-202.
Frey, T. G., Chan, S. H. P., and Schatz, G., 1978, J.Biol.Chem., 253: 4389-4395.
Frisell, W. R., Patwardhan, M. V., and MacKenzie, C. G., 1965, J.Biol.Chem., 240:1829-1835.
Fry, M., 1979, Biochem.Biophys.Res.Commun., 90:1119-1124.
Fukuhara, H., 1967, Proc.Natl.Acad.Sci. U.S.A., 58:1065-1072.
Fuller, S. D., Capaldi, R. A., and Henderson, R., 1979, J.Mol.Biol., 134:305-327.
Galante, Y. M., and Hatefi, Y., 1979, Arch.Biochem.Biophys., 192: 559-568.
Galante, Y. M., Stigall, D. L., Frigeri, F., and Hatefi, Y., 1977, Intern.Symp. on Membrane Bioenergetics, July 10-15, Island of Spetsai, Greece, Abstr. No.122.
Galante, Y. M., Wong, S. -Y., and Hatefi, Y., 1979, J.Biol.Chem., 254:12372-12378.
Galante, Y. M., Wong, S. -Y., and Hatefi, Y., 1981, Arch.Biochem. Biophys., 211:643-651.
Gale, E. F., and Folkes, J. P., 1953, Biochem.J., 53:493-498.
Galkin, A. V., 1980, Formation and Selective Elimination of Nonfunc-

tioning Respiratory-Chain Components during Mitochondrial Differentiation in Yeast, Ph. D. Thesis, Lomonosov State University, Moscow.
Galkin, A. V., Makhlis, T. A., Zubatov, A. S., and Luzikov, V. N., 1975, FEBS Lett., 55:42-45.
Galkin, A. V., Tsoi, T. V., and Luzikov, V. N., 1979a, FEBS Lett., 103:111-113.
Galkin, A. V., Tsoi, T. V., and Luzikov, V. N., 1979b, FEBS Lett., 105:373-375.
Galkin, A. V., Tsoi, T. V., and Luzikov, V. N., 1979c, Dokl.Akad. Nauk SSSR, 247:735-738.
Galkin, A. V., Tsoi, T. V., and Luzikov, V. N., 1980, Biochem.J., 190:145-156.
Galper, J. B., and Darnell, J. E., 1971, J.Mol.Biol., 57:363-367.
Garber, M. P., and Stepokus, P. L., 1974, J.Cell Biol., 63:24-34.
Garland, P. B., and Yates, D. W., 1967, in: "The Round Table Discussion on Mitochondrial Structure and Compartmentation", E. Quagliariello, S. Papa, E. C. Slater, and J. M. Tager, eds., pp.385-399, Adriatica Editrice, Bari.
Garland, P. B., Chance, B., Ernster, L., Lee, C. -P., and Wong, D., 1967, Proc.Natl.Acad.Sci. U.S.A., 58:1696-1702.
Garland, P. B., Haddock, B. A., and Yates, D. W., 1969, in: "Mitochondria: Structure and Function", FEBS Symposium, Vol.17, L. Ernster and Z. Drahota, eds., pp.111-126, Acad. Press, London - New York.
Garnier, J., Gaye, P., Mercier, J. -C., and Robson, B., 1980, Biochimie, 62:231-239.
Garwin, J. L., and Beckwith, J., 1982, J.Bacteriol., 149:789-792.
Gasser, S. M., Ohashi, A., Daum, G., Böhni, P. C., Gibson, J., Reid, G. A., Yonetani, T., and Schatz, G., 1982, Proc.Natl.Acad.Sci. U.S.A., 79:267-271.
Gear, A. R. L., Albert, A. D., and Bednarek, J. M., 1974, J.Biol. Chem., 249:6495-6504.
Gellerfors, P., and Lindén, M., 1981, FEBS Lett., 127:91-93.
Gellerfors, P., and Nelson, B. D., 1975, Eur.J.Biochem., 52:433-443.
Gellerfors, P., Lindén, M., and Nelson, B. D., 1976, Eur.J.Biochem., 67:463-468.
Gellerfors, P., Wielburski, A., and Nelson, B. D., 1979, FEBS Lett., 108:167-170.
George-Nascimento, C., and Poyton, R. O., 1981, J.Biol.Chem., 256: 9363-9370.
Gilmore, R., Blobel, G., and Walter, P., 1982a, J.Cell Biol., 95: 463-469.
Gilmore, R., Walter, P., and Blobel, G., 1982b, J.Cell Biol., 95: 470-477.
Girdlestone, J., Bisson, R., and Capaldi, R. A., 1981, Biochemistry, 20:152-156.
Glaser, E., Norling, B., and Ernster, L., 1977a, Intern. Symp. on Membrane Bioenergetics, July 10-15, Island of Spetsai, Greece, Abstr. No.126.

Glaser, E., Norling, B., and Ernster, L., 1977b, in: "Bioenergetics of Membranes", L. Packer, G. S. Papageorgiou and A. Trebst, eds., pp.513-526, Elsevier/North-Holland Biomedical Press, Amsterdam - New York - Oxford.

Glaser, E., Norling, B., and Ernster, L., 1980, Eur.J.Biochem., 110: 225-235.

Glaser, E., Norling, B., and Ernster, L., 1981, Eur.J.Biochem., 115: 189-196.

Goda, K., Chu, J. -W., Kimura, T., and Schaap, A. P., 1973, Biochem. Biophys.Res.Commun., 52:1300-1306.

Godinot, C., and Lardy, H. A., 1973, Biochemistry, 12:2051-2060.

Goldberger, R., Smith, A. L., Tisdale, H., and Bomstein, R., 1961, J.Biol.Chem., 236:2788-2793.

Goldberger, R., Pumphrey, A., and Smith A., 1962, Biochim.Biophys. Acta, 58:307-313.

Goldblatt, M. J., and Romani, R. J., 1977, Arch.Biochem.Biophys., 183:149-155.

Goldring, E. S., Grossman, L. I., Krupnic, D., Cryer, D. R., and Marmur, J., 1970, J.Mol.Biol., 52:323-335.

Gomez-Puyou, A., de Gomez-Puyou, M. T., Becker, G., and Lehninger, A. L., 1972, Biochem.Biophys.Res.Commun., 47:814-819.

González-Cadavid, N. F., and Campbell, P. N., 1967, Biochem.J., 102: 427-442.

González-Cadavid, N. F., Bello, E. M. A., and Raminez, J. L., 1970, Biochem.J., 118:577-586.

Graf, T., and Sebald, W., 1978, FEBS Lett., 94:218-222.

Green, A., and Watson, K., 1972, Biochem.J., 129:36P.

Green, D. E., and Hechter, O., 1965, Proc.Natl.Acad.Sci. U.S.A., 53: 318-325.

Green, D. E., and Tzagoloff, A., 1966, Arch.Biochem.Biophys., 116: 293-304.

Green, D. E., and Wharton, D. C., 1963, Biochem.Z., 338:335-348.

Green, D. E., Wharton, D. C., Tzagoloff, A., Rieske, J. S., and Brierley, G. P., 1965, in: "Oxidases and Related Redox Systems", Vol.2, T. E. King, H. S. Mason, and M. Morrison, eds., pp.1032-1060, John Wiley and Sons, Inc., New York - London - Sydney.

Green, D. E., Allmann, D. W., Bachmann, E., Baum, H., Kopaczyk, K., Korman, E. F., Lipton, S., MacLennan, D. H., McConnell, D. G., Perdue, J. F., Rieske, J. S., and Tzagoloff, A., 1967, Arch. Biochem.Biophys., 119:312-335.

Green, R. C., and O'Brien, P. J., 1970, Biochim.Biophys.Acta, 197: 31-39.

Griffith, D. E., and Wharton, D. C., 1961, J.Biol.Chem., 235:1850-1856.

Grigolava, I. V., and Konstantinov, A., 1977, FEBS Lett., 78:36-40.

Grimes, G. W., Mahler, H. R., and Perlman, P. S., 1974, J.Cell Biol. 61:565-574.

Grinius, L. L., Jasaitis, A. A., Kadziauskas, Yu. P., Liberman, E. A., Skulachev, V. P., Topali, V. P., Tsofina, L. M., and Vladimirova, M. A., 1970, Biochim.Biophys.Acta, 216:1-12.

Grinius, L. L., Guds, T. I., and Skulachev, V. P., 1971, J.Bioenergetics, 2:101-113.
Grisolia, S., Knecht, E., Hernández-Yago, J., and Wallace, R., 1980, in: "Protein Degradation in Health and Disease", pp.167-184, Ciba Foundation 75, Excerpta Medica/Elsevier/North-Holland.
Groot, G. S. P., 1974, in: "Th Biogenesis of Mitochondria", pp.443-452, Acad. Press.
Groot, G. S. P., Kovác, L., and Schatz, G., 1971, Proc.Natl.Acad.Sci. U.S.A., 68:308-311.
Gross, S. R., McCoy, M. T., and Gilmore, E. B., 1968, Proc.Natl.Acad. Sci. U.S.A., 61:253-260.
Gross, N. J., Getz, G. S., and Rabinowitz, M., 1969, J.Biol.Chem., 244:1552-1562.
Grossman, S., Cobley, J., Hogue, P. K., Kearney, E. B., and Singer, T. P., 1973, Arch.Biochem.Biophys., 158:744-753.
Grossman, A. R., Bartlett, S. G., Schmidt, G. W., Mullet, J. E., and Chua, N. -H., 1982, J.Biol.Chem., 257:1558-1563.
Guerin, M., and Napias, C., 1978, Biochemistry, 17:2510-2516.
Haas, R., and Heinrich, P. C., 1979, Eur.J.Biochem., 96:9-15.
Haas, R., Nagasawa, T., and Heinrich, P. C., 1977, Biochem.Biophys. Res.Commun., 74:1060-1065.
Haas, R., Heinrich, P. C., Tesch, R., and Witt, I., 1978, Biochem. Biophys.Res.Commun., 85:1039-1046.
Haas, R., Heinrich, P. C., and Sasse, D., 1979, FEBS Lett., 103: 168-171.
Hackenberg, H., Riccio, P., and Klingenberg, M., 1978, Eur.J. Biochem., 88:373-378.
Hackenbrock, C. R., 1966, J.Cell Biol., 30:269-297.
Hackenbrock, C. R., 1975, Arch.Biochem.Biophys., 170:139-148.
Hackenbrock, C. R., 1976, in: "Structure of Biological Membranes", 34th Nobel Foundation Symposium, S. Abrahamson and I. Pasher, eds., pp.199-234, Plenum Publ. Comp., New York.
Hackenbrock, C. R., and Hammon, K., 1975, J.Biol.Chem., 250: 9185-9197.
Hackenbrock, C. R., and Miller, K. J., 1975, J.Cell Biol., 65: 615-630.
Hackenbrock, C. R., Höchli, M., and Chau, R. M., 1976, Biochim. Biophys.Acta, 455:466-484.
Haddock, B. A., Yates, D. W., and Garland, P. B., 1970, Biochem.J., 119:565-573.
Hadváry, P., and Kadenbach, B., 1973, Eur.J.Biochem., 39:11-20.
Haid, A., Schweyen, R. J., Bechmann, H., Kaudewitz, F., Solioz, M., and Schatz, G., 1979, Eur.J.Biochem., 94:451-464.
Hallermayer, G., Zimmerman, R., and Neupert, W., 1977, Eur.J. Biochem., 81:523-532.
Hanes, D. M., and Tappel, A. L., 1971, Biochim.Biophys.Acta, 245: 42-53.
Hanson, D. K., Miller, D. H., Mahler, H. R., Alexander, N. J., and Perlman, P. S., 1979, J.Biol.Chem., 254:2480-2490.
Hanstein, W. G., and Hatefi, Y., 1970, Arch.Biochem.Biophys., 138: 87-95.

Hanstein, W. G., Davis, K. A., Ghalambor, M. A., and Hatefi, Y., 1971, Biochemistry, 10:2517-2524.
Hare, J. F., and Crane, F. L., 1974, Subcell.Biochem., 3:1-25.
Hare, J. F., and Hodges, R., 1982a, J.Biol.Chem., 257:3575-3580.
Hare, J. F., and Hodges, R., 1982b, Biochem.Biophys.Res.Commun., 105: 1250-1256.
Hare, J. F., Ching, E., and Attardi, G., 1980, Biochemistry, 19: 2023-2030.
Harman, J. W., and Feigelson, M., 1952a, Exp.Cell Res., 3:47-58.
Harman, J. W., and Feigelson, M., 1952b, Exp.Cell Res., 3:509-525.
Harmey, M. A., and Neupert, W., 1979, FEBS Lett., 108:385-389.
Harmey, M. A., Hallermayer, G., Korb, H., and Neupert, W., 1977, Eur. J.Biochem., 81-533-544.
Harmon, H. J., and Crane, F. L., 1976, Biochim.Biophys.Acta, 440: 45-58.
Harmon, H. J., and Sharrock, M., 1978, Biochim.Biophys.Acta, 503: 56-66.
Harmon, H. J., Hall, J. D., and Crane, F. L., 1974, Biochim.Biophys. Acta, 344:119-155.
Harnischfeger, G., 1972, Planta, 104:316-328.
Hartzell, C. R., Beinert, H., van Gelder, B. F., and King, T. E., 1978, Methods Enzymol., 53:54-66.
Haslam, J. M., Perkins, M., and Linnane, A. W., 1973, Biochem.J., 134:935-947.
Hatase, O., and Oda, T., 1972, J.Biochem. (Tokyo), 71:759-765.
Hatefi, Y., 1968, Proc.Natl.Acad.Sci. U.S.A., 60:733-740.
Hatefi, Y., 1978, Method Enzymol., 53:11-14.
Hatefi, Y., and Galante, Y. M., 1978, in: "Energy Conservation in Biological Membranes", G. Schäfer and M. Klingenberg, eds., pp.19-30, Springer-Verlag, Berlin - Heidelberg - New York.
Hatefi, Y., and Galante, Y. M., 1980, J.Biol.Chem., 255:5530-5537.
Hatefi, Y., and Hanstein, W. G., 1970, Arch.Biochem.Biophys., 138: 73-86.
Hatefi, Y, and Stempel, K. E., 1967, Biochim.Biophys.Acta, 26: 301-308.
Hatefi, Y., and Stempel, K. E., 1969, J.Biol.Chem., 244:2350-2357.
Hatefi, Y., and Stiggall, D. L., 1978, Methods Enzymol., 53:21-27.
Hatefi, Y., Haavik, A. G., and Jurtshuk, P., 1961, Biochim.Biophys. Acta, 52:106-118.
Hatefi, Y., Haavik, A. G., and Griffith, D. E., 1962a, J.Biol.Chem., 237:1676-1680.
Hatefi, Y., Haavik, A. G., and Griffith, D. E., 1962b, J.Biol.Chem., 237:1681-1685.
Hatefi, Y., Haavik, A. G., Fowler, L. D., and Griffith, D. E., 1962c, J.Biol.Chem., 237:2661-2669.
Hatefi, Y., Stiggall, D. L., Galante, Y., and Hanstein, W. G., 1974, Biochem.Biophys.Res.Commun., 61:313-321.
Hatefi, Y., Yagi, T., Phelps, D. C., Wong, S. -Y., Vik, S. B., and Galante, Y. M., 1982, Proc.Natl.Acad.Sci. U.S.A., 79: 1756-1760.

Hawley, E. S., and Greenawalt, J. W., 1970, J.Biol.Chem., 245: 3574-3583.
Hawley, E. S., and Greenawalt, J. W., 1975, Eur.J.Biochem., 54: 585-601.
Hayashi, H., and Capaldi, R. A., 1972, Biochim.Biophys.Acta, 282: 166-173.
Hayashi, R., Bai, Y., and Hata, T., 1975, J.Biochem. (Tokyo), 77: 1313-1318.
Heilporn, V., and Limbosch, S., 1971, Biochim.Biophys.Acta, 240: 94-108.
Heinrich, P. C., Raydt, G., Puschendorf, B., and Jusić, M., 1976, Eur.J.Biochem., 62:37-43.
Heinrich, P. C., Northemann, W., and Schmelzer, E., 1981, Acta Biol. Med.Germ., 40:1451-1463.
Helinski, D. R., and Cooper, C., 1960, J.Biol.Chem., 235:3573-3579.
Henderson, R., Capaldi, R. A., and Leigh, J. S., 1977, J.Mol.Biol., 112:631-648.
Hennig, B., and Neupert, W., 1981, Eur.J.Biochem., 121:203-212.
Hensgens, L. A. M., Grivell, L. A., Borst, P., and Bos, J. L., 1979, Proc.Natl.Acad.Sci. U.S.A., 76:1663-1667.
Henslee, J. G., and Srere, P. A., 1979, J.Biol.Chem., 254:5488-5497.
Henson, C. P., Perlman, P., Weber, C. N., and Mahler, H. R., 1968, Biochemistry, 7:4445-4454.
Heron, C., Ragan, C. I., and Trumpower, B. L., 1978, Biochem.J., 174:791-800.
Hershko, A., and Ciechanover, A., 1982, Annu.Rev.Biochem., 51: 335-364.
Hinkle, P., and Pullman, M. E., 1968, Fed.Proc., 27:648.
Hinkle, P., Kim, J. J., and Racker, E., 1972, J.Biol.Chem., 247: 1338-1339.
Höchli, L., and Hackenbrock, C. R., 1978, Biochemistry, 17:3712-3719.
Hoffmann, H. -P., and Avers, C. J., 1973, Science, 181:749-751.
Hofmann, H. -D., and Kadenbach, B., 1976, FEBS Lett., 66:27-30.
Höjeberg, B., and Rydström, J., 1977, Biochim.Biophys.Res.Commun., 78:1183-1190.
Holtzman, J. L., and Gillette, J. R., 1968, J.Biol.Chem., 243: 3020-3028.
Honjo, I., Ozawa, K., Kitamura, O., Sakai, A., and Ohsawa, T., 1968, J.Biochem. (Tokyo), 63:311-320.
Hoppel, C., and Cooper, C., 1968, Biochem.J., 107:367-375.
Horie, S., and Morrison, M., 1964, J.Biol.Chem., 239:1438-1441.
Horstman, L. L., and Racker, E., 1970, J.Biol.Chem., 245:1336-1344.
Hoshina, S., and Nishida, K., 1975, Plant Cell Physiol., 16:475-484.
Hoshina, S., Kaji, T., and Nishida, K., 1975, Plant Cell Physiol., 16:465-474.
Houstek, J., Kopecký, J., Svoboda, P., and Drahota, Z., 1982, J.Bioenerg.Biomembr., 14:1-13.
Howard, C. F., 1970, J.Biol.Chem., 245:462-468.
Howell, N., Zuiches, C. A., and Munkres, K. D., 1971, J.Cell Biol., 50:721-736.

Huber, C. T., and Morrison, M., 1973, Biochemistry, 12:4274-4282.
Hubert, P., Crémel, G., Rendon, A., Sacko, B., and Waksman, A., 1979, Biochemistry, 18:3119-3126.
Hüllsmann, W. C., Elliott, W. B., and Slater, E. C., 1960, Biochim. Biophys.Acta, 39:267-276.
Hundt, E., Trapp, M., and Kadenbach, B., 1980, FEBS Lett., 115:95-99.
Hunter, F. E., and Ford, L., 1955, J.Biol.Chem., 216:395-403.
Hunter, F. E., Davis, J., and Carlat, L., 1956, Biochim.Biophys.Acta, 20:237-242.
Hunter, F. E., Levy, J. F., Fink, J., Schutz, B., Guerra, F., and Hurwitz, A., 1959a, J.Biol.Chem., 234:2176-2186.
Hunter, F. E., Malison, R., Bridgers, W. F., Schutz, B., and Atchison, A., 1959b, J.Biol.Chem., 234:693-699.
Hunter, F. E., Geblicki, J. M., Hoffstein, P. E., Weinstein, J., and Scott, A., 1963, J.Biol.Chem., 238:828-835.
Hunter, F. E., Scott, A., Hoffstein, P. E., Geblicki, J. M., Weinstein, J., and Schneider, A., 1964a, J.Biol.Chem., 239: 614-620.
Hunter, F. E., Scott, A., Hoffstein, P. E., Guerra, F., Weinstein, J., Schneider, A., Schutz, B., Fink, J., Ford, L., and Smith, E., 1964b, J.Biol.Chem., 239:604-613.
Ibrahim, N. G., Burke, J. P., and Beattie, D. S., 1973a, FEBS Lett., 29:73-76.
Ibrahim, N. G., Stuchell, R. N., and Beattie, D. S., 1973b, Eur.J. Biochem., 36:519-527.
Infante, R., Petit, D., Polonovski, J., and Caroli, J., 1971, Experientia, 27:640-642.
Inouye, M., and Halegoua, S., 1980, CRC Crit.Rev.Biochem., 7:339-371.
Ip, M. M., Chee, P. Y., and Swick, R. W., 1974, Biochim.Biophys.Acta, 354:29-38.
Ito, K., 1977, J.Bacteriol, 132:1021-1023.
Iyengar, M. R., and Iyengar, C. L., 1980, Biochemistry, 19:2176-2182.
Jackl. G., and Sebald, W., 1975, Eur.J.Biochem., 54:97-106.
Jacobs, E. E., and Sanadi, D. R., 1960, J.Biol.Chem., 235:531-534.
Jacobus, W. E., and Lehninger, A. L., 1973, J.Biol.Chem., 248: 4803-4810.
Jacq, C., Lazowska, J., and Slonimski, P. P., 1980, in: "The Organization and Expression of the Mitochondrial Genome", A. M. Kroon and C. Saccone, eds., pp.139-152, Elsevier/North-Holland Biomedical Press, Amsterdam - New York - Oxford.
Jasaitis, A. A., Energy Transformation in Mitochondria (in Russian), Izd., AN SSSR-VINITI, Moscow.
Jasaitis, A. A., Nemecek, I. B., Severina, I. I., Skulachev, V. P., and Smirnova, S. M., 1972, Biochim.Biophys.Acta, 275:485-490.
Jaussi, R., Sonderegger, P., Flückiger, J., and Christen, Ph., 1982, J.Biol.Chem., 257:13334-13340.
Jayaraman, J., Cotman, C., Mahler, H. R., and Sharp, C. W., 1966, Arch.Biochem.Biophys., 116:224-251.
Jeffreys, A. J., and Craig, I. W., 1976, Eur.J.Biochem., 68:301-311.
Jeng, A. Y., and Shamoo, A. E., 1980, J.Biol.Chem., 255:6897-6903.

Johnson, C. L., and Schwartz, A., 1969, J.Pharmacol.Exp.Ther., 167: 365-373.
Johnson, C. L., Goldstein, M. A., and Schwartz, A., 1973, Mol. Pharmacol., 9:360-371.
Jollow, D., Kellerman, G. M., and Linnane, A. W., 1968, J.Cell Biol., 37:221-230.
Jones, O. T. G., 1979, in: "The Porphyrins", D. Dolphin, ed., Vol. VI, pp.179-232, Acad.Press, New York - San Francisco - London.
Jones, M. S., and Jones, O. T. G., 1969, Biochem.J., 113:507-514.
Josefsson, L. -G., and Randall, L. L., 1981, J.Biol.Chem., 256:2504-2507.
Julliard, J. H., and Gautheron, D. C., 1978, Biochim.Biophys.Acta, 503:223-237.
Jusić, M., Seifert, S., Weiss, E., Haas, R., and Heinrich, P. C., 1976, Arch.Biochem.Biophys., 177:355-363.
Kadenbach, B., 1969, Eur.J.Biochem., 10:312-317.
Kadenbach, B., 1970, Eur.J.Biochem., 12:3920-398.
Kadenbach, B., 1971, Biochem.Biophys.Res.Commun., 44:724-730.
Kadenbach, B., and Hadváry, P., 1970, Eur.J.Biochem., 32:343-349.
Kadenbach, B., and Merle, P., 1981, FEBS Lett., 135:1-11.
Kagawa, Y., 1972, Biochim.Biophys.Acta, 265:297-338.
Kagawa, Y., and Racker, E., 1966a, J.Biol.Chem., 241:2467-2474.
Kagawa, Y., and Racker, E., 1966b, J.Biol.Chem., 241:2475-2482.
Kagawa, Y., and Racker, E., 1971, J.Biol.Chem., 246:5477-5487.
Kagawa, Y., Kandrach, A., and Racker, E., 1973, J.Biol.Chem., 248:676-684.
Kalf, G. F., and Faust, A. S., 1969, Arch.Biochem.Biophys., 134: 103-112.
Kal'nov, S. L., Serebryakova, N. V., Zubatov, A. S., and Luzikov, V. N., 1978, Biokhimiya, 43:662-667.
Kal'nov, S. L., Novikova, L. A., Zubatov, A. S., and Luzikov, V. N., 1979a, FEBS Lett., 101:355-358.
Kal'nov, S. L., Novikova, L. A., Zubatov, A. S., and Luzikov, V. N., 1979b, Biochem.J., 182:195-202.
Kanner, B. I., Serrano, R., Kandrach, M. A., and Racker, E., 1976, Biochem.Biophys.Res.Commun., 69:1050-1056.
Kaschnitz, R. M., and Hatefi, Y., 1975, Arch.Biochem.Biophys., 171: 292-304.
Katan, M. B., Pool, L., and Groot, G. S. P., 1976, Eur.J.Biochem., 65:95-105.
Katki, A. G., and Myers, C. E., 1980, Biochem.Biophys.Res.Commun., 96:85-91.
Katunuma, N., Kominami, E., Kobayashi, K., Banno, Y., Suzuki, K., Chichibu, K., Hamaguchi, Y., and Katsunuma, T., 1975, Eur.J. Biochem., 52:37-50.
Kaufman, B. T., and Kaplan, N. O., 1960, Biochim.Biophys.Acta, 39: 332-342.
Kawajiri, K., Harano, T., and Omura, T., 1977a, J.Biochem. (Tokyo), 82:1403-1416.
Kawajiri, K., Harano, T., and Omura, T., 1977b, J.Biochem. (Tokyo), 82:1417-1423.

Kazakova, T. B., Gachava, M. M., and Markosyan, K. A., 1969, Biokhimiya, 34:1000-1007.
Kazakova, T. B., Golubkov, V. I., and Ermakova, S. A., 1972, Mol. Biol., 6:597-605.
Kazakova, T. B., Mel'nikova, M. P., and Sverdlov, E. D., 1977, Mol. Biol., 11:1029-1038.
Kearney, E. B., Watari, H., Singer, T. P., and Basinski, D. H., 1962, Fed.Proc., 21:251.
Keddie, F. M., and Barajas, L., 1969, J.Ultrastruct.Res., 29:260-275.
Keilin, D., and Hartree, E. F., 1939, Proc.Royal Soc., Ser.B 127: 167-191.
Kellems, R. E., and Butow, R. A., 1972, J.Biol.Chem., 247:8043-8050.
Kellems, R. E., Allison, V. F., and Butow, R. A., 1974, J.Biol.Chem., 249:3297-3303.
Kellems, R. E., Allison, V. F., and Butow, R. A., 1975, J.Cell Biol., 65:1-14.
Kellerman, G. M., Biggs, D. G., and Linnane, A. W., 1969, J.Cell Biol., 42:378-391.
Kellerman, G. M., Griffith, D. E., Hansby, J. E., Lamb, A. J., and Linnane, A. W., 1971, in: "Autonomy and Biogenesis of Mitochondria and Chloroplasts", N. K. Boardman, A. W. Linnane, and R. M. Smillie, eds., pp.346-359, North-Holland, Amsterdam - London.
Keyhani, E., 1973, J.Cell Biol., 58:480-484.
Khesin, R. B., Astaurova, O. B., Shemyakin, M. F., Kamzolova, S. G., and Manyakov, V. F., 1967, Mol.Biol., 1:736-752.
Kielley, R. K., 1952, Cancer Res., 12:124-128.
Kim, I. -Ch., and Beattie, D. S., 1973, Eur.J.Biochem., 36:509-518.
King, T. E., 1963, J.Biol.Chem., 238:4037-4051.
King, T. E., 1966, Advances in Enzymol., 28:155-236.
King, T. E., 1973, in: "Oxidases and Related Redox Systems", Vol. 2, T. E. King, H. S. Mason, and M. Morrison, eds., pp.816, 819, University Park Press, Baltimore - London - Tokyo.
King, T. E., Nickel, K. S., and Jensen, D. R., 1964, J.Biol.Chem., 239:1989-1994.
King, T. E., Godman, G. C., and King, D. W., 1972, J.Cell Biol., 53: 127-142.
Kiselev, A. V., and Konstantinov, A. A., 1976, Bioorg.Khim., 2: 253-258.
Kislev, N., and Eisenstadt, J. M., 1972, Eur.J.Biochem., 31:226-229.
Kleiner, D., and von Jagow, G., 1972, FEBS Lett., 20:229-232.
Kleitke, B., and Wollenberger, A., 1968, FEBS Lett., 1:187-190.
Klingenberg, M., 1968, in: "Biological Oxidations", T. P. Singer, ed., pp.3-54, Interscience Publications, New York - London - Sydney.
Klingenberg, M., 1970, FEBS Lett., 6:145-154.
Klingenberg, M., and Buchholz, M., 1970, Eur.J.Biochem., 13:247-252.
Klingenberg, M., and Pfaff, E., 1966, in: "Regulation of Metabolic Processes in Mitochondria", B.B.A. Library, Vol. 7, J. M. Tager et al., eds., pp.180-201, Amsterdam - London - New York.

Klingenberg, M., Riccio, P., and Aquila, H., 1978, Biochim.Biophys. Acta, 503:193-210.
Knowles, A. F., and Penefsky, H. S., 1972a, J.Biol.Chem., 247:6617-6623.
Knowles, A. F., and Penefsky, H. S., 1972b, J.Biol.Chem., 247:6624-6630.
Knowles, J. K. C., and Tait, A., 1972, Mol.Gen.Genet., 117:53-59.
Koch, G., 1976, J.Biol.Chem., 251:6097-6107.
Kolansky, D. M., Conboy, J. G., Fenton, W. A., and Rosenberg, L. E., 1982, J.Biol.Chem., 257:8467-8471.
Kolarov, J., and Klingenberg, M., 1975, FEBS Lett., 45:320-323.
Kolarov, J., Subík, J., and Kovác, L., 1972, Biochim.Biophys.Acta, 267:465-478.
Komai, H., and Capaldi, R. A., 1973, FEBS Lett., 30:273-276.
Komai, H., Hunter, D. R., and Takahashi, Y., 1973, Biochem.Biophys. Res.Commun., 53:82-89.
Konstantinov, A. A., Maslov, S. P., Severina, I. I., and Skulachev, V. P., 1975, Biokhimiya, 40;401-407.
Kopaczyk, K., Asai, J., Allmann, D. W., Oda, T., and Green, D. E., 1968a, Arch.Biochem.Biophys., 123:602-621.
Kopaczyk, K., Asai, J., and Green, D. E., 1968b, Arch.Biochem. Biophys., 126:358-379.
Korb, H., and Neupert, W., 1978, Eur.J.Biochem., 91:609-620.
Korner, A., and Tarver, H., 1957, J.Gen.Physiol., 41:219-231.
Kotel'nikova, A. V., and Zvyagil'skaya, R. A., 1973, Biochemistry of Yeast Mitochondria (in Russian), Izd.Nauka, Moscow.
Kovác, L., 1972, FEBS Lett., 22:270-272.
Kovác, L., and Weissova, K., 1968, Biochim.Biophys.Acta, 153:55-59.
Kovác, L., Hrusovská, E., and Smigán, P., 1970, Biochim.Biophys.Acta, 205:520-523.
Kozlov, I. A., and Chernyak, B. V., 1976, Dokl.Akad.Nauk SSSR, 231: 222-225.
Kozlov, I. A., and Mikelsaar, H. N., 1974, FEBS Lett., 43:212-214.
Kozlov, I. A., and Skulachev, V. P., 1977, Biochim.Biophys.Acta, 463:29-89.
Kozlov, I. A., and Skulachev, V. P., 1982, Current Topics in Membranes and Transport, 16:285-301.
Kozlov, Yu. P., Danilov, V. S., Kagan, V. E., and Sitkovskii, M. V., 1972, Free-Radical Lipid Oxidation in Biological Membranes, (in Russian), Moscow University Press, Moscow.
Krämer, R., and Klingenberg, M., 1977, FEBS Lett., 82:363-367.
Kramer, R. A., Du Bois, R. N., Simpson, E. R., Anderson, C. M., Kashiwagi, K., Lambeth, J. D., Jefcoate, C. R., and Waterman, M. R., 1982, Arch.Biochem.Biophys., 215:478-485.
Krause, W., and Halangk, W., 1977, Acta Biol.Med.Germ., 36:381-387.
Kröger, A., and Klingenberg, M., 1970, Vitamins and Hormones, 28: 533-574.
Kroon, A. M., and De Vries, H., 1971, in: "Autonomy and Biogenesis of Mitochondria and Chloroplasts", N. K. Boardman, A. W. Linnane, and R. M. Smillie, eds., pp.318-327, North-Holland, Amsterdam - London.

Kroon, A. M., and Jansen, R. J., 1968, Biochim.Biophys.Acta, 155: 629-632.
Kropp, E. S., and Wilson, J. E., 1970, Biochem.Biophys.Res.Commun., 38:74-79.
Kudielka, R. A., Kock, H., and Theimer, R. R., 1981, FEBS Lett., 136: 8-12.
Kumar, C. C., and Padmanaban, G., 1980, J.Biol.Chem., 255:11130-11134
Kumar, S. A., Rao, N. A., Felton, S. P., and Huennekens, F. M., 1968, Arch.Biochem.Biophys., 125:436-448.
Küntzel, H., 1969, Nature (London), 222:142-146.
Kuriyama, Y., and Luck, D. J. L., 1973a, J.Mol.Biol., 73:425-537.
Kuriyama, Y., and Luck, D. J. L., 1973b, J.Cell Biol., 59:776-784.
Kuriyama, Y., Omura, T., Siekevitz, P., and Palade, G. E., 1969, J.Biol.Chem., 244:2017-2026.
Kuzela, S., and Grecná, E., 1969, Experientia, 25:776-777.
Kuzela, S., Krempaský, V., Kolarov, J., and Ujházy, V., 1975, Eur.J. Biochem., 58:483-491.
Kuzela, S., Liciaková, K., and Lakota, J., 1980, FEBS Lett., 114:197-201.
Lamb, A. J., Clark-Walker, G. D., and Linnane, A. W., 1968, Biochim. Biophys.Acta, 161:415-427.
Lambeth, D. O., and Lardy, H. A., 1971, Eur.J.Biochem., 22:355-363.
Lambeth, D. O., Lardy, H. A., Senior, A. E., and Brooks, J. C., 1971, FEBS Lett., 17:330-332.
Lambowitz, A. M., Chua, N. -H., and Luck, D. J. L., 1976, J.Mol. Biol., 107:223-253.
Lambowitz, A. M., La Polla, R. J., and Collins, R. A., 1979, J.Cell Biol., 82:17-31.
Landriscina, C., Papa, S., Coratelli, P., Mazzarella, L., and Quagliariello, E., 1970, Biochim.Biophys.Acta, 205:136-147.
Lansman, R. A., Rowe, M. J., and Woodward, D. O., 1974, Eur.J. Biochem., 41:15-23.
La Polla, R. J., and Lambowitz, A. M., 1977, J.Mol.Biol., 116:189-205.
Lauquin, G., and Vignais, P. V., 1973, Biochim.Biophys.Acta, 305: 534-556.
Lauquin, G., Lunardi, J., and Vignais, P. V., 1976, Biochimie, 58: 1213-1220.
Lederman, M., and Attardi, G., 1973, J.Mol.Biol., 78:275-283.
Lee, Ch. -pu, 1971, in: "Probes of Structure and Function of Macromolecules and Membranes", Vol. I: Probes and Membrane Function, B. Chance, Ch. -pu Lee, and J. K. Blasie, eds., pp.417-426, Acad. Press, New York - London.
Lee, Ch. -pu, Sottocasa, G. L., and Ernster, L., 1967, in: "Methods in Enzymology", Vol. X, R. W. Estabrook and M. Pullman, eds., pp.33-37, Academic Press, New York - London.
Lee, Ch. -pu, Johanson, B., and King, T. E., 1969, Biochem.Biophys. Res.Commun., 35:243-248.
Lee, Ch. -pu, Johanson, B., and King, T. E., 1971, in: "Probes of Structure and Function of Macromolecules and Membranes", Vol.

I: Probes and Membrane Function, B. Chance, Ch. -pu Lee, and J. K. Blasie, eds., pp.401-406, Acad. Press, New York - London
Lehninger, A. L., 1955, The Harvey Lectures, Ser. XLIX, pp.176-215.
Lehninger, A. L., 1959a, J.Biol.Chem., 234:2187-2195.
Lehninger, A. L., 1959b, J.Biol.Chem., 234:2465-2471.
Lehninger, A. L., 1962, Physiol.Rev., 42:467-517.
Lehninger, A. L., 1964, The Mitochondrion, W. A. Benjamin, Inc., New York - Amsterdam.
Lehninger, A. L., 1971, Biochem.Biophys.Res.Commun., 42:312-318.
Lehninger, A. L., and Gotterer, G. S., 1960, J.Biol.Chem., 235:PC8.
Lehninger, A. L., and Ray, B. L., 1957, Biochim.Biophys.Acta, 26:643-644.
Lehninger, A. L., and Remmert, L. F., 1959, J.Biol.Chem., 234:2459-2464.
Lehninger, A. L., and Schneider, M., 1959, J.Biophys.Biochem.Cytol., 5:109-116.
Lehninger, A. L., Ray, B. L., and Scheider, M., 1959, J.Biophys. Biochem.Cytol., 5:97-108.
Lehninger, A. L., Carafoli, E., and Rossi, C., 1967, Advances in Enzymol., 29:259-320.
Lemberg, M. R., 1969, Physiol.Rev., 49:48-121.
Lenaz, G., and MacLennan, D. H., 1966, J.Biol.Chem., 241:5260-5265.
Lenney, J. F., 1975, J.Bacteriol., 122:1265-1273.
Leonard, K., Wingfield, P., Arad, T., and Weiss, H., 1981, J.Mol. Biol., 149:259-274.
Lester, R. L., and Hatefi, Y., 1958, Biochim.Biophys.Acta, 29:103-112.
Leung, K. H., and Hinkle, P. C., 1975, J.Biol.Chem., 250:8467-8471.
Lévy, M., Toury, R., and André, J., 1967, Biochim.Biophys.Acta, 135:599-613.
Lévy, M., Toury, R., Sauner, M. -T., and André, J., 1969, in: "Mitochondria: Structure and Function", FEBS Symposium, Vol. 17, pp.33-42, Acad. Press, London - New York.
Lewin, A. S., Gregor, I., Mason, T. L., Nelson, N., and Schatz, G., 1980, Proc.Natl.Acad.Sci. U.S.A., 77:3998-4002.
Lewis, S. E., and Fowler, K. S., 1960, Biochim.Biophys.Acta, 38:564-565.
Li, Y., Leonard, K., and Weiss, H., 1981a, Eur.J.Biochem., 116:119-205.
Li, Y., de Vries, S., Leonard, K., and Weiss, H., 1981b, FEBS Lett., 135:277-280.
Liberman, E. A., Khachatryan, G. I., Tsofina, L. M., Elizarova, G. V., and Saks, V. A., 1980, Biokhimiya, 45:418-423.
Lima, M. S., Nachbaur, G., and Vignais, P. M., 1968, C.R.Acad.Sci. Paris, Ser. D, 266:739-742.
Lin, L. -F. H., and Beattie, D. S., 1978, J.Biol.Chem., 253:2412-2418
Lin, L. -F. H., Kim, I. -C., and Beattie, D. S., 1974, Arch.Biochem. Biophys., 160:458-464.
Lin, L. -F. H., Clejan, L., and Beattie, D. S., 1978, Eur.J.Biochem., 87:171-179.

Lindén, M., Gellerfors, P., and Nelson, B. D., 1982, FEBS Lett., 141: 189-192.
Lindenmayer, A., and Estabrook, R. W., 1958, Arch.Biochem.Biophys., 78:66-82.
Linnane, A. W., 1965, in: "Oxidases and Related Redox Systems", T. E., King, H. S. Mason, and M. Morrison, eds., pp.1102-1121, John Wiley and Sons, New York - London - Sydney.
Linnane, A. W., Biggsm, D. R., Huang, M., and Clark-Walker, G. D., 1968, in: "Aspects of Yeast Metabolism", A. K. Mills, ed., pp.217-242, Blackwell Scientific Publications, Oxford - Edinburgh.
Linnane, A. W., Haslam, J. M., Lukins, H. B., and Nagley, P., 1972, Annu.Rev.Microbiol., 26:163-198.
Lipsett, M. N., and Corwin, L. M., 1959, J.Biol.Chem., 234:2448-2452.
Lipsky, N. G., and Pedersen, P. L., 1981, J.Biol.Chem., 256:8652-8657.
Lipsky, N. G., and Pedersen, P. L., 1982, J.Biol.Chem., 257:1473-1481.
Little, C., and O'Brien, P. J., 1968, Biochem.Biophys.Res.Commun., 31:145-150.
Lizardi, P. M., and Luck, D. J. L., 1972, J.Cell Biol., 54:56-74.
Lloyd, D., 1974, The Mitochondria of Microorganisms, Acad. Press, London - New York - San Francisco.
Lloyd, D., and Edwards, S. W., 1976, Biochem.J., 160:335-342.
Lloyd, D., and Edwards, S. W., 1977, Biochem.J., 162:581-590.
Longo, G. P., and Scandalios, J. G., 1969, Proc.Natl.Acad.Sci. U.S.A. 62:104-111.
Lord, J. M., 1980, Subcell.Biochem., 7:171-211.
Lötscher, H. R., Winterhalter, K. H., Carafoli, E., and Richter, C., 1979, Proc.Natl.Acad.Sci. U.S.A., 76:4340-4344.
Love, B., and Auer, H. E., 1970, Biochem.Biophys.Res.Commun., 41: 1437-1442.
Love, B., Chan, S. H. P., and Stotz, E., 1970, J.Biol.Chem., 245: 6664-6668.
Luck, D. J. L., 1963, J.Cell Biol., 16:483-499.
Luck, D. J. L., 1965, J.Cell Biol., 24:461-470.
Ludwig, B., and Schatz, G., 1980, Proc.Natl.Acad.Sci. U.S.A., 77: 196-200.
Ludwig, B., Downer, N. W., and Capaldi, R. A., 1979, Biochemistry, 18:1401-1407.
Ludwig, B., Prochaska, L., and Capaldi, R. A., 1980, Biochemistry, 19:1516-1523.
Lukyanenko, A. I., Berestovskii, G. N., and Evtodienko, Yu. V., 1980, Biofizika, 25:82-86.
Lustig, A., Levens, D., and Rabinowitz, M., 1982, J.Biol.Chem., 257: 5800-5808.
Lusty, C. J., Machinist, J. M., and Singer, T. P., 1965, J.Biol.Chem. 240:1804-1810.
Luzikov, V. N., 1973a, Stabilization of the Enzymic Systems of the

Inner Mitochondrial Membrane and Related Problems, D.Sc.Thesis, Moscow State University, Moscow.
Luzikov, V. N., 1973b, Subcell.Biochem., 2:1-31.
Luzikov, V. N., 1979, FEBS Symp., 53:133-142.
Luzikov, V. N., 1981, FEBS Lett., 125:131-133.
Luzikov, V. N., and Romashina, L. V., 1972, Biochim.Biophys.Acta, 267:37-47.
Luzikov, V. N., Rakhimov, M. M., Saks, V. A., and Berezin, I. V., 1967, Biokhimiya, 32:1234-1241.
Luzikov, V. N., Rakhimov, M. M., and Berezin, I. V., 1968, Biokhimiya 33:1115-1123.
Luzikov, V. N., Saks, V. A., and Berezin, I. V., 1969, Biokhimiya 34:874-877.
Luzikov, V. N., Zubatov, A. S., and Rainina, E. I., 1970, FEBS Lett., 11:233-236.
Luzikov, V. N., Romashina, L. V., and Voznaya, N. M., 1971a, FEBS Lett., 13:261-264.
Luzikov, V. N., Saks, V. A., and Kupriyanov, V. V., 1971b, Biochim. Biophys.Acta, 253:46-57.
Luzikov, V. N., Zubatov, A. S., Rainina, E. I., and Bakeyeva, L. E., 1971c, Biochim.Biophys.Acta, 245:321-334.
Luzikov, V. N., Kupriyanov, V. V., and Makhlis, T. A., 1973a, J.Bioenergetics, 4:521-532.
Luzikov, V. N., Zubatov, A. S., and Rainina, E. I., 1973b, J.Bioenergetics, 5:129-149.
Luzikov, V. N., Makhlis, T. A., and Galkin, A. V., 1976, FEBS Lett., 69:108-110.
Luzikov, V. N., Novikova, L. A., Tikhonov, A. K., and Zubatov, A. S., 1983a, submitted for publication.
Luzikov, V. N., Novikova, L. A., Zubatov, A. S., and Tikhonov, A. K., 1983b, submitted for publication.
Maccecchini, M. -L., Rudin, Y., Blobel, G., and Schatz, G., 1979a, Proc.Natl.Acad.Sci. U.S.A., 76:343-349.
Maccecchini, M. -L., Rudin, Y., and Schatz, G., 1979b, J.Biol.Chem., 254:7468-7471.
MacFarlane, M. G., and Spencer, A. G., 1953, Biochem.J., 54:569-575.
MacGregor, C. H., and McElhaney, G. E., 1981, J.Bacteriol., 148:551-558.
Machinist, J. M., and Singer, T. P., 1965, J.Biol.Chem., 240:3182-3190.
Machleidt, W., and Werner, S., 1979, FEBS Lett., 107:327-330.
Macino, G., and Tzagoloff, A., 1979, J.Biol.Chem., 254:4617-4623.
Macino, G., and Tzagoloff, A., 1980, Cell, 20:507-517.
MacLennan, D. H., and Asai, J., 1968, Biochem.Biophys.Res.Commun., 33:441-447.
MacLennan, D. H., and Tzagoloff, A., 1966, J.Biol.Chem., 241:1933-1937.
MacLennan, D. H., and Tzagoloff, A., 1968, Biochemistry, 7:1603-1610.
MacLennan, D. H., Lenaz, G., and Szarkowska, L., 1966, J.Biol.Chem., 241:5251-5259.

MacLennan, D. H., Smoly, J. M., and Tzagoloff, A., 1968, J.Biol.Chem. 243:1589-1597.
Maddaiah, V. T., Sharma, R. K., Balachandar, V., Rezvani, I., Collip, P. J., and Chen, Sh. -Y., 1973, J.Biol.Chem., 248:4263-4268.
Mager, J., 1960, Biochim.Biophys.Acta, 38:150-152.
Mahler, H. R., and Perlman, P. S., 1971, Biochemistry, 10:2979-2990.
Mahler, H. R., Sarkar, N. K., Vernon, L. R., and Alberty, R. A., 1952, J.Biol.Chem., 199:585-597.
Mahler, H. R., Perlman, P. S., and Mehrotra, B. D., 1971, in: "Autonomy and Biogenesis of Mitochondria and Chloroplasts", N. K. Boardman, A. W. Linnane, and R. M. Smillie, eds., pp.492-511, North-Holland, Amsterdam - London.
Mahler, H. R., Bastos, R. N., Flury, U., Lin, C. C., and Phan, S. H., 1975, in: "Genetics and Biogenesis of Mitochondria and Chloroplasts", P. S. Perlman, C. W. Birky, and T. J. Byers, eds., pp.66-116, Ohio State University Press, Columbus.
Mairouch, H., and Godinot, C., 1977, Proc.Natl.Acad.Sci.U.S.A., 74: 4185-4189.
Malmström, B. G., 1974, Quart.Rev.Biophys., 6:389-431.
Malviya, A. N., Parsa, B., Yodaiken, R. E., and Elliott, W. B., 1968, Biochim.Biophys.Acta, 162:195-209.
Manella, C. A., and Bonner, W. D., 1975, Biochim.Biophys.Acta, 413: 213-225.
Mansurova, S. E., Shakhov, Yu. A., and Kulaev, I. S., 1977, FEBS Lett., 74:31-34.
Marahiel, M. A., Imam, G., Nelson, P., Pieniazek, N. J., Stepien, P. P., and Küntzel, H., 1977, Eur.J.Biochem., 76:345-354.
Marchant, R., and Smith, D. G., 1968, J.Gen.Microbiol., 50:391-397.
Marra, E., Doonan, S., Saccone, C., and Quagliariello, E., 1977, Biochem.J., 164:685-691.
Marra, E., Doonan, S., Saccone, C., and Quagliariello, E., 1978, Eur.J.Biochem., 83:427-435.
Marra, E., Passarella, S., Doonan, S., and Quagliariello, P., 1980, FEBS Lett., 122:33-36.
Marres, C. A. M., and Slater, E. C., 1977, Biochim.Biophys.Acta, 462:531-548.
Marver, H. S., Collins, A., Tschudy, D. P., and Rechcigh, M., 1966, J.Biol.Chem., 241:4323-4329.
Marzoev, A. I., and Vladimirov, Yu. A., 1977a, Byull.Eksp.Biol.Med., 84:426-428.
Marzoev, A. I., and Vladimirov, Yu. A., 1977b, Byull.Eksp.Biol.Med., 84:565-567.
Marzuki, S., Cobon, G. S., Haslam, J. M., and Linnane, A. W., 1975a, Arch.Biochem.Biophys., 169:577-590.
Marzuki, S., Cobon, G. S., Crowfoot, P. D., and Linnane, A. W., 1975b, Arch.Biochem.Biophys., 169:591-600.
Mason, T. L., and Schatz, G., 1973, J.Biol.Chem., 248:1355-1360.
Matlib, M. A., and O'Brien, P. J., 1975, Arch.Biochem.Biophys., 167: 193-202.
Matsuura, S., Arpin, M., Hannum, C., Margoliash, E., Sabatini, D. D.,

and Morimoto, T., 1981, Proc.Natl.Acad.Sci.U.S.A., 78:4368-4372.
Matthaei, J. H., and Nierenberg, M. W., 1961, Proc.Natl.Acad.Sci. U.S.A., 47:1580-1588.
Mayer, R. J., and Hübscher, G., 1971, Biochem.J., 124:491-500.
Mayer, G., Ullrich, U., and Staudinger, H., 1968, Hoppe-Seyler's Z.Physiol.Chem., 349:459-464.
McCarty, R. E., and Jagendorf, A. T., 1965, Plant Physiol., 40:725-735.
McCay, P. B., Gibson, D. D., Fong, K. -L., and Hornbrook, K. R., 1976, Biochim.Biophys.Acta, 431:459-468.
McClary, D. O., and Bowers, W. D., 1968, J.Ultrastruct.Res., 25:37-45.
McConnel, H. M., 1976, in: "Spin labeling theory and Applications", L. G. Besliner, ed., pp.515-561, Academic Press, New York.
McIlhinney, A., and Hogan, B. L. M., 1974, Biochim.Biophys.Acta, 372:358-365.
McIntyre, J. O., Bock, H. -G. O., and Fleisher, S., 1978, Biochim. Biophys.Acta, 513:255-267.
McKay, R., Druyan, R., Getz, G. S., and Rabinowitz, M., 1969, Biochem.J., 114:455-461.
McKnight, R. C., and Hunter, F. E., 1966, J.Biol.Chem., 241:2757-2765
Meisel', M. N., Biryuzova, V. I., Volkova, T. M., Malatyan, M. N., and Medvedeva, G. A., 1964, in: "Electron and Fluorescent Microscopy of the Cell", (in Russian), pp.3-15, Izd. Nauka, Moscow - Leningrad.
Meisner, H. M., 1970, Biochim.Biophys.Acta, 205:27-34.
Meldolesi, J., Corte, G., Pietrini, G., and Borgese, N., 1980, J.Cell Biol., 85:516-526.
Mellors, A., Tappell, A. L., Sawant, P. L., and Desai, I. D., 1967, Biochim.Biophys.Acta, 143:299-309.
Melnick, R. L., and Packer, L., 1971, Biochim.Biophys.Acta, 253:503-508.
Mendel-Hartvig, I., and Nelson, B. D., 1978, FEBS Lett., 92:36-40.
Merle, P., and Kadenbach, B., 1980, Eur.J.Biochem., 105:499-507.
Merli, A., Capaldi, R. A., Ackrell, B. A. C., and Kearney, E. B., 1979, Biochemistry, 18:1393-1400.
Meyer, U. A., and Schmid, R., 1973, Fed.Proc., 32:1649-1655.
Mian, F. A., Küenzi, M. T., and Halvorson, H. O., 1973, J.Bacteriol., 115:876-881.
Michel, R., and Neupert, W., 1973, Eur.J.Biochem., 36:53-67.
Michel, R., Liebl. A., Hartmann, A., and Neupert, W., 1976, Hoppe-Seyler's Z.Physiol.Chem., 357:415-426.
Michel, R., Wachter, E., and Sebald, W., 1979, FEBS Lett., 101:373-376.
Mihara, K., Omura, T., Harano, T., Brenner, S., Fleisher, S., Rajagopalan, K. V., and Blobel, G., 1982, J.Biol.Chem., 257:3355-3358.
Millis, A. J. T., and Suyama, Y., 1972, J.Biol.Chem., 247:4063-4073.
Minakami, S., Cremona, T., Ringler, R. L., and Singer, T. P., 1963, J.Biol.Chem., 238:1529-1537.

Minakami, S., Schindler, F. J., and Estabrook, R. W., 1964, J.Biol. Chem., 239:2042-2048.
Miralles, V., Felipo V., Hernández-Yago, J., and Grisolia, S., 1982, Biochem.Biophys.Res.Commun., 107:1028-1036.
Mitchell, P., 1966, Chemiosmotic Coupling in Oxidative and Photosynthetic Phosphorylation, Glynn Research Laboratories, Bodmin.
Mitchell, P., 1972, in: "Mitochondria/Biomembranes", FEBS, Eighth meeting, Vol. 28, Van Den Bergh, P. Borst, L. L. M. Van Deenen, J. C. Riemersma, E. C. Slater, and J. M. Tager, eds., pp.353-370, North-Holland/American Elsevier.
Mitchell, P., 1976, J.Theor.Biol., 62:327-367.
Mitchell, P., and Moyle, J., 1969, Eur.J.Biochem., 9:149-165.
Mittnacht, S., Sherman, S. C., and Farber, J. L., 1979, J.Biol.Chem., 254:9871-9878.
Miura, S., Mori, M., Amaya, Y., and Tatibana, M., 1982, Eur.J. Biochem., 122:641-647.
Mlodzianowski, F., and Ponitka, A., 1973, Z.Pflanzenphysiol., 69: 13-25.
Mockel, J., 1972, Biochim.Biophys.Acta, 277:628-638.
Molotkovsky, Y. G., and Zheskova, I. M., 1965, Biochem.Biophys.Res. Commun., 20:411-415.
Moreadith, R. W., and Jacobus, W. E., 1982, J.Biol.Chem., 257:899-905.
Mori, M., Miura, S., Tatibana, M., and Cohen, P. P., 1980, J.Biochem. (Tokyo), 88:1829-1836.
Mori, M., Morita, T., Ikeda, F., Amaya, Y., Tatibana, M., and Cohen, P. P., 1981a, Proc.Natl.Acad.Sci. U.S.A., 78:6056-6060.
Mori, M., Morita, T., Miura, S., and Tatibana, M., 1981b, J.Biol. Chem., 256:8263-8266.
Morita, T., Mori, M., Tatibana, M., and Cohen, P. P., 1981, Biochem. Biophys.Res.Commun., 99:623-629.
Morita, T., Miura, S., Mori, M., and Tatibana, M., 1982, Eur.J. Biochem., 122:501-509.
Morré, D. J., and Ovtracht, L., 1977, Intern.Rev.Cytol., Suppl. 5, pp.61-188.
Morré, D. J., Merritt, W. D., and Lembi, C. A., 1971, Protoplasma, 73:43-49.
Morré, D. J., Kartenbeck, J., and Franke, W. W., 1979, Biochim. Biophys.Acta, 559:71-152.
Morton, D. J., Hoppel, C., and Cooper, C., 1968, Biochem.J., 107: 377-380.
Mounolou, J. C., Jakob, H., and Slonimski, P. P., 1966, Biochem. Biophys.Res.Commun., 24:218-224.
Mueckler, M. M., Himeno, M., and Pitot, H. C., 1982, J.Biol.Chem., 257:7178-7180.
Müller, M., Dubiel, W., Rathman, J., and Rapoport, S., 1980, Eur.J. Biochem., 109:405-410.
Munkres, K. D., 1968, Ann.N.Y.Acad.Sci., 151:294-306.
Munkres, K. D., Giles, N. H., and Case, M. E., 1965, Arch.Biochem. Biophys., 109:397-403.

Munn, E. A., 1974, The Structure of Mitochondria, Acad. Press, London - New York.
Nabi, N., and Omura, T., 1980, Biochem.Biophys.Res.Commun., 97:680-686.
Nachbaur, J., and Vignais, P. M., 1968, Biochem.Biophys.Res.Commun., 33:315-320.
Nachbaur, J., Colbeau, A., and Vignais, P. M., 1969, FEBS Lett., 3: 121-124.
Nachbaur, J., Colbeau, A., and Vignais, P. M., 1972, Biochim.Biophys. Acta, 274:426-446.
Nagata, I., Furuya, E., Yoshida, Y., Kanaseki, T., and Tagawa, T., 1975, J.Biochem.(Tokyo), 78:1353-1364.
Nagley, P., and Linnane, A. W., 1970, Biochem.Biophys.Res.Commun., 39:989-996.
Nan Sen Tseng, L., and Cunningham, L., 1966, Biochemistry, 5:144-149.
Nass, M. M. K., 1969, J.Mol.Biol., 42:521-529.
Nass, S., 1969, Intern.Rev.Cytol., 25:55-129.
Neal, W. K., Hoffmann, H. -P., Avers, C. J., and Price, C. A., 1970, Biochem.Biophys.Res.Commun., 38:414-422.
Nelson, B. D., and Gellerfors, P., 1974, Biochim.Biophys.Acta, 357: 358-364.
Nelson, N., and Schatz, G., 1979, Proc.Natl.Acad.Sci.U.S.A., 76: 4365-4369.
Nelson, N., Nelson, H., and Racker, E., 1972, J.Biol.Chem., 247: 7657-7662.
Neubert, D., Wojtczak, A. B., and Lehninger, A. L., 1962, Proc.Natl. Acad.Sci.U.S.A., 48:1651-1658.
Neupert, W., and Ludwig, G. D., 1971, Eur.J.Biochem., 19:523-532.
Neupert, W., and Schatz, G., 1981, TIBS, 6:1-4.
Neupert, W., Brdiczka, D., and Bücher, Th., 1967, Biochem.Biophys. Res.Commun., 27:488-493.
Neupert, W., Sebald, W., Schwab, A. J., Massinger, P., and Bücher, Th., 1969a, Eur.J.Biochem., 10:589-591.
Neupert, W., Sebald, W., Schwab, A. J., Pfaller, A., and Bücher, Th., 1969b, Eur.J.Biochem., 10:585-588.
Neupert, W., Massinger, P., and Pfaller, A., 1971, in: "Autonomy and Biogenesis of Mitochondria and Chloroplasts", N. K. Boardman, A. W. Linnane, and R. Smillie, eds., pp.328-338, North-Holland Amsterdam - London.
Newman, H. A. I., Gordesky, S. E., Hoppel, C., and Cooper, C., 1968, Biochem.J., 107:381-385.
Nishibayashi-Yamashita, H., Cunningham, C., and Racker, E., 1972, J.Biol.Chem., 247:698-704.
Nordenbrand, K., Hundal, T., Carlsson, C., Sandri, G., and Ernster, L., 1977, Intern.Symp. on Membrane Bioenergetics, July 10-15, Island of Spetsai, Greece, Abstr. No 108.
Norum, K. R., Farstad, M., and Bremer, J., 1966, Biochem.Biophys. Res.Commun., 24:797-804.
Novikoff, A., 1963, in: "Ciba Foundation Symposium on Lysosomes", A. V. S. Reuck and M. P. Cameron, eds., Discussion, pp.301-302, J. & A. Churchill, Ltd., London.

Novikoff, A. B., and Essner, E., 1962, J.Cell Biol., 15:140-146.
Novikoff, L. A., Zubatov, A. S., and Luzikov, V. N., 1981, FEBS Lett., 135:245-248.
Novikova, L. A., Zubatov, A. S., and Luzikov, V. N., 1982, Biokhimiya 47:284-289.
Nygaard, A. P., 1953, J.Biol.Chem., 204:655-663.
Nygaard, A. P., and Sumner, J. B., 1953, J.Biol.Chem., 200:723-729.
Ohnishi, T., 1979, in: "Membrane Proteins in Energy Transduction", R. A. Capaldi, ed., pp.1-87, Marcel Dekker, New York and Basel.
Ohnishi, T., Leigh, J. S., Winter, D. B., Lim, J., and King, T. E., 1974a, Biochem.Biophys.Res.Commun., 61:1026-1035.
Ohnishi, T., Winter, D. B., Lim, J., and King, T. E., 1974b, Biochem. Biophys.Res.Commun., 61:1017-1025.
Ohnishi, T., Cartledge, T. G., and Lloyd, D., 1975, FEBS Lett., 52: 90-94.
Ohnishi, T., Lim, J., Winter, D. B., and King, T. E., 1976a, J.Biol. Chem., 251:2105-2109.
Ohnishi, T., Salerno, J. C., Winter, D. B., Lim, J., Yu, C. A., Yu. L., and King, T. E., 1976b, J.Biol.Chem., 251:2094-2104.
Ojala, D., Merkel, C., Gelfand, R., and Attardi, G., 1980, Cell, 22: 393-403.
Ojala, D., Montoya, J., and Attardi, G., 1981, Nature (London), 290: 470-474.
Okamoto, H., Yamamoto, S., and Nozaki, M., 1967, Biochem.Biophys.Res. Commun., 26:309-314.
Okunuki, K., 1961, Advances in Enzymol., 23:29-82.
Ono, H., and Ito, A., 1982, Biochem.Biophys.Res.Commun., 107:258-264.
Orme-Johnson, N. R., Hansen, R. E., and Beinert, H., 1971, Biochem. Biophys.Res.Commun., 45:871-878.
Orme-Johnson, N. R., Hansen, R. E., and Beinert, H., 1974, J.Biol. Chem., 249:1928-1939.
Osafune, T., 1973, J.Electron Microscopy, 22:51-61.
Ozawa, K., Kitamura, O., Ohsawa, T., Murata, T., and Honjo, I., 1969, J.Biochem.(Tokyo), 66:361-367.
Packer, L., 1972, J.Bioenergetics, 3:115-127.
Packer, L., and Utsumi, K., 1969, Arch.Biochem.Biophys., 131:386-403.
Packer, L., Williams, M. A., and Criddle, R. S., 1973, Biochim. Biophys.Acta, 292:92-104.
Palade, G., 1975, Science, 189:347-358.
Palmer, J. W., Schmid, P. C., Pfeiffer, D. R., and Schmid, H. H. O., 1981, Arch.Biochem.Biophys., 211:674-682.
Palmieri, F., Genchi, G., Stipani, I., Francia, F., and Quagliariello E., 1974, in: "Membrane Proteins in Transport and Phosphorylation", (G. F. Azzone, M. E. Klingenberg, E. Quagliariello, and N. Siliprandi, eds., pp.245-256, North-Holland/American Elsevier, Amsterdam - London - New York.
Paltauf, F., and Schatz, G., 1969, Biochemistry, 8:335-339.
Panchenko, L. F., Brusov, O. S., Gerasimov, A. M., and Lokteva, T. D., 1975, FEBS Lett., 55:84-87.

Panfili, E., Sandri, G., Sottocasa, G. L., 1974, Acta Vitamin.Enzymol 28:323-330.
Panfili, E., Sandri, G., Sottocasa, G. L., Lunazzi, G., Liut, G., and Graziosi, G., 1976, Nature (London), 264:185-186.
Pappenheimer, A. M., 1977, Annu.Rev.Biochem., 46:69-94.
Pappu, A. S., Fatterpaker, P., and Sreenivasan, A., 1978, Biochem.J., 172-349-352.
Parce, J. W., Cunningham, C. C., and Waite, M., 1978, Biochemistry, 17:1634-1639.
Parce, J. W., Spach, P. I., and Cunningham, C. C., 1980, Biochem.J., 188:817-822.
Parisi, B., and Cella, R., 1971, FEBS Lett., 14:209-213.
Parkes, J. G., Thompson, W., 1970, Biochim.Biophys.Acta, 196:162-169.
Parry, G., 1978, Subcell.Biochem., 5:261-326.
Parsons, D. F., 1963, Science, 140:985-987.
Parsons, D. F., and Williams, G. R., 1967, in: "Methods in Enzymology", Vol.X, R. W. Estabrook and M. E. Pullman, eds., pp.443-447, Acad. Press, New York - London.
Parsons, D. F., and Yano, Y., 1967, Biochim.Biophys.Acta, 135: 362-364.
Parsons, D. F., Williams, G. R., Thompson, W., Wilson, D., and Chance, B., 1967, in: "The Round Table Discussion on Mitochondrial Structure and Compartmentation", E. Quagliariello, S. Papa, E. C. Slater, and J. M. Tager, eds., pp.29-70, Adriatica Editrice, Bari.
Passarella, S., and Quagliariello, E., 1976, Biochimie., 58:989-1001.
Passarella, S., Marra, E., Doonan, S., and Quagliariello, E., 1980, Biochem.J., 192:649-658.
Passarella, S., Marra, E., Doonan, S., Languino, L. R., Saccone, C., and Quagliariello, E., 1982, Biochem.J., 202:353-362.
Pedersen, P. L., 1975, J.Bioenergetics, 6:243-275.
Peeters-Joris, C., Vandervoorde, A. -M., and Baudhuin, P., 1975, Biochem.J., 150:31-39.
Penefsky, H. S., 1979, Advances in Enzymol., 49:223-280.
Penn, N. W., 1961, Biochim.Biophys.Acta, 53:490-494.
Pennial.R., and Davidian, N., 1968, FEBS Lett., 1:38-41.
Perkins, M., Haslam, J. M., and Linnane, A. W., 1973, Biochem.J., 134:923-934.
Perlman, P. S., and Mahler, H. R., 1970, J.Bioenergetics, 1:113-138.
Perlman, P. S., and Mahler, H. R., 1974, Arch.Biochem.Biophys., 162: 248-271.
Perlman, S., and Penman, S., 1970, Biochem.Biophys.Res.Commun., 40: 941-948.
Petrushka, E., Quastel, J. H., and Scholefield, P. G., 1959, Canad., J.Biochem.Physiol., 37:989-998.
Pfaff, E., and Schwalbach, K., 1967, in: "The Round Table Discussion on Mitochondrial Structure and Compartmentation", E. Quagliariello, S. Papa, E. C. Slater, and J. M. Tager, eds., pp.346-350, Adriatica Editrice, Bari.
Pfeifer, P. M., and McCay, P. B., 1972, J.Biol.Chem., 247:6763-6769.

Phan, S. H., and Mahler, H. R., 1976a, J.Biol.Chem., 251:257-269.
Phan, S. H., and Mahler, H. R., 1976b, J.Biol.Chem., 251:270-276.
Pharo, R. L., Sordahl, L. A., Vyas, S. R., and Sanadi, D. R., 1966, J.Biol.Chem., 241:4771-4780.
Pharo, R. L., Sordahl, L. A., Edelhoch, H., and Sanadi, D. R., 1968, Arch.Biochem.Biophys., 125:416-428.
Pickett, C. B., Montisano, D., Eisner, D., and Cascarano, J., 1980, Exp.Cell Res., 128:343-352.
Plattner, H., and Schatz, G., 1969, Biochemistry, 8:339-343.
Plattner, H., Salpeter, M. M., Saltzgaber, J., and Schatz, G., 1970, Proc.Natl.Acad.Sci.U.S.A., 66:1252-1259.
Polakis, E. S., Bartley, W., and Meek, G. A., 1965, Biochem.J., 97: 298-302.
Pong, S. -S., Nuss, D. N., and Koch, G., 1975, J.Biol.Chem., 250: 240-245.
Poole, R. K., Lloyd, D., and Chance, B., 1974, Biochem.J., 138:201-210.
Poyton, R. O., and Groot, G. S. P., 1975, Proc.Natl.Acad.Sci.U.S.A., 72:172-176.
Poyton, R. O., and McKemmie, E., 1979a, J.Biol.Chem., 254:6763-6771.
Poyton, R. O., and McKemmie, E., 1979b, J.Biol.Chem., 254:6772-6780.
Poyton, R. O., and Schatz, G., 1975a, J.Biol.Chem., 250:752-761.
Poyton, R. O., and Schatz, G., 1975b, J.Biol.Chem., 250:762-766.
Pressman, B. C., 1970, in: "Membranes of Mitochondria and Chloroplasts", E. Racker, ed., pp.213-250, Van Nostrand Reinhold, Co., New York - Cincinnati - Toronto - London - Melbourne.
Prestipino, G., Ceccarelli, D., Conti, F., and Carafoli, E., 1974, FEBS Lett., 45:99-103.
Price, C. A., Fonnesu, A., and Davies, R. E., 1956, Biochem.J., 64: 754-768.
Priegnitz, A., and Wojtczak, L., 1961, Biochim.Biophys.Acta, 48: 585-587.
Prochaska, L., Bisson, R., and Capaldi, R. A., 1980, Biochemistry, 19:3174-3179.
Pullman, M. E., and Monroy, G. C., 1963, J.Biol.Chem., 238:3762-3769.
Pullman, M. E., Penefsky, H. S., Datta, A., and Racker, E., 1960, J.Biol.Chem., 235:3322-3329.
Quagliariello, E., and Palmieri, F., 1968, Eur.J.Biochem., 4:20-27.
Quagliariello, E., Landriscina, C., Coratelli, P., 1968, Biochim. Biophys.Acta, 164:12-24.
Quagliariello, E., Papa, S., Meijer, A. J., and Tager, J. M., 1969, in: "Mitochondria: Structure and Function", FEBS Symposium, Vol.17, pp.335-346, Acad. Press, London - New York.
Raaflaub, J., 1953, Helv.Physiol.Acta, 11:157-165.
Rabinowitz, M., 1968, Bull.Soc.Chim.Biol., 50:311-363.
Rabinowitz, M., and Swift, H., 1970, Physiol.Rev., 50:376-427.
Rabinowitz, M., Getz, G. S., Casey, J., and Swift, H., 1969, J.Mol. Biol., 41:381-400.
Racker, E., 1962, Proc.Natl.Acad.Sci. U.S.A., 48:1659-1663.
Racker, E., 1965, Mechanisms in Bioenergetics, Acad. Press, New York - London.

Racker, E., 1970, in: "Essays in Biochemistry", P. N. Campbell and F. Dickens, eds., pp.1-22, Acad. Press, New York.
Racker, E., and Horstman, L. L., 1967, J.Biol.Chem., 242:2547-2551.
Racker, E., and Kandrach, A., 1971, J.Biol.Chem., 246:7069-7071.
Racker, E., and Kandrach, A., 1973, J.Biol.Chem., 248:5841-5847.
Racker, E., and Krimsky, I., 1958, Fed.Proc., 17:1135-1141.
Racker, E., Tyler, D. D., Estabrook, R. W., Conover, T. E., Parsons, D. F., and Chance, B., 1965, in: "Oxidases and Related Redox Systems", T. E. King, H. S. Mason, and M. Morrison, eds., Vol.2, pp.1077-1094, John Wiley & Sons, Inc., New York - London - Sydney.
Racker, E., Horstman, L. L., King, D., Fessenden-Raden, J. M., 1969, J.Biol.Chem., 244:6668-6674.
Racker, E., Loyter, A., and Christiansen, R. O., 1971, in: "Probes of Structure and Function of Macromolecules and Membranes", Vol. I: Probes and Membrane Function, B. Chance, C. P. Lee, and J. K. Blasie, eds., pp.407-410, Acad. Press, New York - London.
Ragan, C. I., 1976a, Biochem.J., 154:295-305.
Ragan, C. I., 1976b, Biochem.J., 156:367-374.
Ragan, C. I., and Heron, C., 1978, Biochem.J., 174:783-790.
Ragan, C. I., and Hinkle, P. C., 1975, J.Biol.Chem., 250:8472-8476.
Ragan, C. I., and Racker, E., 1973, J.Biol.Chem., 248:2563-2569.
Ragan, C. I., and Widger, W. R., 1975, Biochem.Biophys.Res.Commun., 62:744-749.
Ragan, C. I., Galante, Y. M., Hatefi, Y., and Ohnishi, T., 1982a, Biochemistry, 21:590-594.
Ragan, C. I., Galante, Y. M., and Hatefi, Y., 1982b, Biochemistry, 21:2518-2524.
Rajwade, M. S., Katyare, S. S., Fatterpaker, P., and Sreenivasan, A., 1975, Biochem.J., 152:379-387.
Rao, N. A., Felton, S. P., Huennekens, F. M., and Mackler, B., 1963, J.Biol.Chem., 238:449-455.
Rapoport, S. M., Schewe, T., Wiesner, R., Halangk, W., Ludwig, P., Janicke-Höhne, M., Tannert, C., Hiebsch, C., and Klatt, D., 1979, Eur.J.Biochem., 96:545-561.
Rapoport, S., Dubiel, W., and Müller, M., 1982, FEBS Lett., 147: 93-96.
Rascati, R. J., and Parsons, P., 1979, J.Biol.Chem., 254:1594-1599.
Raymond, Y., and Shore, G. C., 1979, J.Biol.Chem., 254:9335-9338.
Raymond, Y., and Shore, G. C., 1981, J.Biol.Chem., 256:2087-2090.
Rhodin, T. R., and Racker, E., 1974, Biochem.Biophys.Res.Commun., 61:1207-1212.
Riccio, P., Aquila, H., and Klingenberg, M., 1975a, FEBS Lett., 56: 129-132.
Riccio, P., Aquila, H., and Klingenberg, M., 1975b, FEBS Lett., 56: 133-138.
Riccio, P., Schägger, H., Engel, W. D., and von Jagow, G., 1977, Biochim.Biophys.Acta, 459:250-262.
Richter, D., 1971, Biochemistry, 10:4422-4425.

Rieske, J. S., 1976, Biochim.Biophys.Acta, 456:195-247.
Rieske, J. S., Hansen, R. E., and Zaugg, W. S., 1964a, J.Biol.Chem., 239:3017-3022.
Rieske, J. S., MacLennan, D. H., and Coleman, R., 1964b, Biochem. Biophys.Res.Commun., 15:338-344.
Rieske, J. S., Zaugg, W. S., and Hansen, R. E., 1964c, J.Biol.Chem., 239:3023-3030.
Rieske, J. S., Baum, H., Stoner, C. D., and Lipton, S. H., 1967, J.Biol.Chem., 242:4854-4866.
Righetti, P., and Cerletti, P., 1971, FEBS Lett., 13:181-183.
Ringler, R. L., Minakami, S., and Singer, T. P., 1963, J.Biol.Chem., 238:801-810.
Rogers, P., and Küntzel, H., 1977, Molec.Cell Biochem., 14:87-89.
Romani, R., and Monadjem, A., 1970, Proc.Natl.Acad.Sci.U.S.A., 66: 869-873.
Romani, R. J., and Ozelkök, S., 1973, Plant Physiol., 51:702-707.
Romani, R. J., Tuskes, S. E., and Ozelkök, S., 1974, Arch.Biochem. Biophys., 164:743-751.
Roodyn, D. B., and Wilkie, D., 1968, The Biogenesis of Mitochondria, Methuen, London.
Roodyn, D. B., Suttie, J. W., and Work, T. S., 1962, Biochem.J., 83: 29-40.
Ross, E., and Schatz, G., 1976a, J.Biol.Chem., 251:1991-1996.
Ross, E., and Schatz, G., 1976b, J.Biol.Chem., 251:1997-2004.
Rossi, C. R., Sartorelli, L., Tato, L., and Siliprandi, N., 1964, Arch.Biochem.Biophys., 107:170-175.
Rossi, C. R., Sartorelli, L., Tato, L., Baretta, L., and Siliprandi, N., 1965, Biochim.Biophys.Acta, 98:207-209.
Rossi, C. R., Alexandre, A., Carignani, G., and Rossi, C. S., 1974, in: "Membrane Proteins in Transport and Phosphorylation", G. F. Azzone, M. E. Klingenberg, E. Quagliariello, and N. Siliprandi, eds., pp.171-176, North-Holland/American Elsevier, Amsterdam - London - New York.
Rothman, J. E., and Fine, R. E., 1980, Proc.Natl.Acad.Sci.U.S.A., 77: 780-784.
Rouslin, W., 1977, Arch.Biochem.Biophys., 178:408-418.
Rouslin, W., and Schatz, G., 1969, Biochem.Biophys.Res.Commun., 37: 1002-1007.
Rowe, M. J., Lensman, R. A., and Woodward, D. O., 1974, Eur.J. Biochem., 41:25-30.
Ruben, G. C., Telford, J. N., and Carroll, R. C., 1976, J.Cell Biol., 68:724-739.
Rubin, M. S., and Tzagoloff, A., 1973a, J.Biol.Chem., 248:4269-4274.
Rubin, M. S., and Tzagoloff, A., 1973b, J.Biol.Chem., 248:4275-4279.
Rubin, M. S., and Tzagoloff, A., 1978, Methods Enzymol., 53:73-79.
Rubio, V., and Grisolia, S., 1977, FEBS Lett., 75:281-284.
Russell, L. K., Kirkley, S. A., Kleyman, T. R., and Chan, S. H. P., 1976, Biochem.Biophys.Res.Commun., 73:434-443.
Rydström, J., 1975, Fed.Proc., 34:577.
Rydström, J., 1976, Biochim.Biophys.Acta, 455:24-35.

Rydström, J., 1979, J.Biol.Chem., 254:8611-8619.
Rydström, J., Kanner, N., and Racker, E., 1975, Biochem.Biophys.Res. Commun., 67:831-839.
Ryrie, I. J., 1975, Arch.Biochem.Biophys., 168:712-719.
Ryrie, I. J., and Gallagher, A., 1979, Biochim.Biophys.Acta, 545: 1-14.
Sabatini, D. D., Kreibich, G., Morimito, T., and Adesnik, M., 1982, J.Cell Biol., 92:1-22.
Saccone, C., and Quagliariello, E., 1975, Intern.Rev.Cytol., 43: 125-165.
Sacher, R., Steffens, G. J., and Buse, G., 1979, Hoppe-Seyler's Z.Physiol.Chem., 360:1385-1392.
Sadler, M. H., Hunter, D. R., and Haworth, R. A., 1974, Biochem. Biophys.Res.Commun., 59:804-812.
Saheki, T., and Holzer, H., 1975, Biochim.Biophys.Acta, 384:203-214.
Saito, K., and Hanahn, D. J., 1962, Biochemistry, 1:521-532.
Sakakibara, R., Huynh, Q. K., Nishida, Y., Watanabe, T., and Wada, H., 1980, Biochem.Biophys.Res.Commun., 95:1781-1788.
Salach, J., Tisdale, H. D., Singer, T. P., and Bader, P., 1967, Biochim.Biophys.Acta, 143:281-284.
Salach, J. I., Seng, R., Tisdale, H., Singer, T. P., 1971, J.Biol. Chem., 246:340-347.
Saltzgaber-Müller, J., and Schatz, G., 1978, J.Biol.Chem., 253:305-310.
Sandri, G., Panfili, E., and Sottocasa, G. L., 1974, Acta Vitamin Enzymol., 28:317-322.
Saraste, M., Penttilä, T., Coggins, J. R., and Wikstrom, M., 1980, FEBS Lett., 114:35-38.
Saraste, M., Penttilä, T., and Wikstrom, M., 1981, Eur.J.Biochem., 115:261-268.
Sarzala, M. G., Van Golde, L. M. G., De Kruyff, B., and Van Deenen, L. L. M., Biochim.Biophys.Acta, 202:106-119.
Satav, J. G., Katyare, S. S., Fatterpaker, P., and Sreenivasan, A., 1977, Biochem.J., 73:287-296.
Sato, N., Wilson, D. F., and Chance, B., 1971, Biochim.Biophys.Acta, 253:88-97.
Sato, N., Chance, B., Kato, K., and Kletman, W., 1973, FEBS Lett., 29:222-226.
Satre, M., De Jerphanion, M. -B., Huet, J., and Vignais, P. V., 1975, Biochim.Biophys.Acta, 387:241-255.
Scarpa, A., and Azzi, A., 1968, Biochim.Biophys.Acta, 150:473-481.
Scarpa, A., and Lindsay, J. G., 1972, Eur.J.Biochem., 27:401-407.
Schatz, G., 1963, Biochem.Biophys.Res.Commun., 12:448-451.
Schatz, G., 1968, J.Biol.Chem., 243:2192-2199.
Schatz, G., and Mason, T. L., 1974, Annu.Rev.Biochem., 43:51-87.
Schatz, G., and Saltzgaber, J., 1969, Biochem.Biophys.Res.Commun., 37:996-1001.
Schatz, G., Tuppy, H., and Klima, J., 1963, Z.Naturforsch., 18b: 145-153.
Schatz, G., Penefsky, H. S., and Racker, E., 1967, J.Biol.Chem., 242: 2552-2560.

Schatz, G., Groot, G. S. P., Mason, T., Rouslin, W., Wharton, D. C., and Saltzgaber, J., 1972, Fed.Proc., 31:21-29.
Scherphof, G. L., and Van Deenen, L. L. M., 1965, Biochim.Biophys. Acta, 98:204-206.
Scherphof, G. L., Waite, M., and Van Deenen, L. L. M., 1966, Biochim. Biophys.Acta, 125:406-409.
Scherphof, G. L., Scarpa, A., and van Toorenenbergen, A., 1972, Biochim.Biophys.Acta, 270:226-240.
Schewe, T., Halangk, W., Hiebsch, Ch., Rapoport, S., 1977, Acta Biol. Med.Germ., 36:363-372.
Schewe, T., Albracht, S. P. J., and Ludwig, P., 1981, Biochim. Biophys.Acta, 636:210-217.
Schimke, R. T., Sweeney, E. W., and Berlin, C. M., 1965, J.Biol. Chem., 240:4609-4620.
Schleyer, M., Schmidt, B., and Neupert, W., 1982, Eur.J.Biochem., 125:109-116.
Schmelzer, E., and Heinrich, P. C., 1980, J.Biol.Chem., 255:7503-7506.
Schmitt, H., 1972, FEBS Lett., 26:215-220.
Schnaitman, C. A., 1969, Proc.Natl.Acad.Sci.U.S.A., 63:412-419.
Schnaitman, C. A., and Greenawalt, J. W., 1968, J.Cell Biol., 38: 158-175.
Schneider, D. L., and Racker, E., 1971, Fed.Proc., 30:1190.
Schneider, D. L., and Racker, E., 1973, in: "Oxidases and Related Redox Systems", T. E. King, H. S. Mason, and M. Morrison, eds., Vol.2, University Park Press, Baltimore - London - Tokyo.
Schneider, D. L., Kagawa, Y., and Racker, E., 1972, J.Biol.Chem., 247:4074-4079.
Schneider, H., Lemasters, J. J., Höchli, M., and Hackenbrock, C. R., 1980a, Proc.Natl.Acad.Sci.U.S.A., 77:442-446.
Schneider, H., Lemasters, J. J., Höchli, M., and Hackenbrock, C. R., 1980b, J.Biol.Chem., 255:3748-3756.
Schneider, W. C., Striebich, M. J., and Hogeboom, G. H., 1956, J.Biol.Chem., 222:969-977.
Scholte, H. R., 1973, Biochim.Biophys.Acta, 330:283-293.
Scholte, H. R., Weijers, P. J., and Wit-Peeters, E. M., 1973, Biochim.Biophys.Acta, 291:764-773.
Schwab, A. J., Sebald, W., and Weiss, H., 1972, Eur.J.Biochem., 30: 511-516.
Schweyen, R., and Kaudewitz, F., 1970, Biochem.Biophys.Res.Commun., 38:728-735.
Scragg, A. H., 1971, FEBS Lett., 17:111-114.
Sebald, W., 1977, Biochim.Biophys.Acta, 463:1-27.
Sebald, W., and Jackl. G., 1975, in: "Electron Transfer Chains and Oxidative Phosphorylation", E. Quagliariello, S. Papa, F. Palmieri, E. C. Slater, and N. Siliprandi, eds., pp.193-198, North-Holland/American Elsevier, Amsterdam - Oxford - New York.
Sebald, W., Bücher, Th., Olbrich, B., and Kaudewitz, F., 1968, FEBS Lett., 1:235-240.

Sebald, W., Hofstöffer, Th., Hacker, D., and Bücher, Th., 1969a, FEBS Lett., 2:177-180.
Sebald, W., Schwab, A. J., and Bücher, Th., 1969b, FEBS Lett., 4: 243-246.
Sebald, W., Weiss, H., and Jackl, G., 1972, Eur.J.Biochem., 30:413-417.
Sebald, W., Machleidt, W., and Otto, J., 1973, Eur.J.Biochem., 38: 311-324.
Sebald, W., Graf, T., and Lukins, H. B., 1979a, Eur.J.Biochem., 93: 587-599.
Sebald, W., Hoppe, J., and Wachter, E., 1979b, in: "Function and Molecular Aspects of Biomembrane Transport", E. Quagliariello, F. Palmieri, S. Papa, and M. Klingenberg, eds., pp.63-74, Elsevier/North-Holland.
Senior, A. E., 1973, Biochim.Biophys.Acta, 301:249-277.
Senior, A. E., and Brooks, J. C., 1970, Arch.Biochem.Biophys., 140: 257-266.
Senior, A. E., and Brooks, J. C., 1971, FEBS Lett., 17:327-329.
Serrano, R., Kanner, B. I., and Racker, E., 1976, J.Biol.Chem., 251: 2453-2461.
Sevarino, K., and Poyton, R. O., 1980, Proc.Natl.Acad.Sci.U.S.A., 77: 142-146.
Severina, E. P., and Evtodienko, Yu. V., 1981, Biokhimiya, 46:1199-1201.
Shakhov, Yu. A., Mansurova, S. E., and Kulaev, I. S., 1974, Dokl. Akad.Nauk SSSR, 219:1017-1019.
Shaw, A. B., Anderson, M. M., and McCarty, R. E., 1976, Plant Physiol., 57:724-729.
Sherman, F., Stewart, J. W., Margoliash, E., Parker, J., and Campbell, W., 1966, Proc.Natl.Acad.Sci.U.S.A., 55:1498-1504.
Shertzer, H. G., and Racker, E., 1976, J.Biol.Chem., 251:2446-2452.
Shertzer, H. G., Kanner, B. I., Banerjee, R. K., and Racker, E., 1977, Biochem.Biophys.Res.Commun., 75:779-784.
Shore, G. C., and Tata, J. R., 1977, J.Cell Biol., 72:714-725.
Shore, G. C., Carignan, P., and Raymond, Y., 1979, J.Biol.Chem., 254:3141-3144.
Shore, G., Power, F., and Bendayan, M., 1981, J.Biol.Chem., 256: 8761-8766.
Sidhu, A., and Beattie, D. S., 1982, J.Biol.Chem., 257:7879-7886.
Siedow, J. N., Power, S., de la Rosa, F. F., and Palmer, G., 1978, J.Biol.Chem., 253:2392-2399.
Siegel, M. R., and Sisler, H. D., 1965, Biochim.Biophys.Acta, 103: 558-567.
Siegenthaler, P. -A., 1970, Experientia, 26:1308-1310.
Siegenthaler, P. -A., 1972, Biochim.Biophys.Acta, 275:182-191.
Siegenthaler, P. -A., 1974, FEBS Lett., 39:337-340.
Siekevitz, P., and Potter, R., 1955, J.Biol.Chem., 215:221-235.
Sierra, M., and Tzagoloff, A., 1973, Proc.Natl.Acad.Sci.U.S.A., 70:3155-3159.
Sies, H., and Moss, K. M., 1978, Eur.J.Biochem., 84:377-383.

Silman, H. I., Rieske, J. S., Lipton, S. H., and Baum, H., 1967, J.Biol.Chem., 242:4867-4875.
Singer, T. P., 1966, in: "Comprehensive Biochemistry", M. Florkin and E. H. Stotz, eds., Vol.14, pp.127-198, Elsevier, Amsterdam - London - New York.
Singer, T. P., and Kearney, E. B., 1963, in: "The Enzymes", Vol.7, pp.383-445, Academic Press, New York - London.
Sjöstrand, F. S., 1963, J.Ultrastruct.Res., 9:340-361.
Skrede, S., and Bremer, J., 1970, Eur.J.Biochem., 14:465-472.
Skulachev, V. P., 1972, Energy Transformation in Biomembranes, (in Russian), Izd.Nauka, Moscow.
Skulachev, V. P., and Kozlov, I. A., 1977, Proton Adenosine Triphosphatases, (in Russian), Izd.Nauka, Moscow.
Skulachev, V. P., Chistyakov, V. V., Jasaitis, A. A., and Smirnova, E. G., 1967, Biochem.Biophys.Res.Commun., 26:1-6.
Slater, E. C., 1971, Quart.Rev.Biophys., 4:35-71.
Slater, E. C., 1972, in: "The Molecular Basis of Electron Transport", J.Schultz and B. F. Cameron, eds., pp.95-114, Academic Press, New York - London.
Slater, E. C., 1973, in: "Mechanisms in Bioenergetics", G. F. Azzone, L. Ernster, S. Papa, E. Quagliariello, and N. Siliprandi, eds., pp.405-431, Academic Press, New York - London.
Slater, E. C., and Cleland, K. W., 1962, Nature, (London), 170:118-119.
Slater, E. C., and Cleland, K. W., 1953, Biochem.J., 55:566-580.
Slater, E. C., and Lee, I. -Y., 1973, in: "Oxidases and Related Redox Systems", Vol.2, pp.823-841, University Park Press, Baltimore - London - Tokyo.
Slater, E. C., Berden, J. A., and Lee, I. -Y., 1972, in: "Biochemistry and Biophysics of Mitochondrial Membranes", G. F. Azzone, E. Carafoli, A. Lehninger, E. Quagliariello, and N. Siliprandi, eds., pp.123-145, Academic Press, New York - London.
Slonimski, P. P., 1956, Proc.3me Congr.Intern.Biochim., Brussels, pp.242-252, Academic Press, New York.
Slonimski, P. P., Claisse, M. L., Foucher, M., Jacq, C., Kochko, A., Lamouroux, A., Pajot, P., Perrodin, G., Spyridakis, A., and Wambier-Kluppel, M. L., 1978, in: "Biochemistry and Genetics of Yeast", M. Bacila, B. L. Horeker, and A. O. M. Stoppani, eds., pp.391-401, Academic Press, New York - San Francisco - London.
Smith, D. S., 1963, J.Cell Biol., 19:115-133.
Smith, R. J., and Capaldi, R. A., 1977, Biochemistry, 16:2629-2632.
So, A. G., and Davie, E. W., 1963, Biochemistry, 2:132-141.
Sogawa, K., and Takahashi, K., 1978, J.Biochem., (Tokyo), 84:763-770.
Sogawa, K., and Takahashi, K., 1979, J.Biochem., (Tokyo), 86:1313-1322.
Sogawa, K., Ichihara, Y., and Takahashi, K., 1981, J.Biochem., (Tokyo), 90:1243-1248.
Somasundaram, T., and Jayaraman, J., 1981a, Biochemistry, 20:5369-5373.

Somasundaram, T., and Jayaraman, J., 1981b, Biochemistry, 20:5373-5380.
Somlo, M., Cosson, J., Clavilier, J., Krupa, M., and Laporte, I., 1982, Eur.J.Biochem., 122:369-374.
Sonderegger, P., Jaussi, R., Christen, P., and Gehring, H., 1982, J.Biol.Chem., 257:3339-3345.
Sone, N., Furuya, E., and Hagihara, B., 1969, J.Biochem., (Tokyo), 65:935-943.
Soper, J. W., Decker, G. L., and Pedersen, P. L., 1979, J.Biol.Chem., 254:11170-11176.
Soslau, G., and Nass, M. M. K., 1971, J.Cell Biol., 51:514-524.
Sottocasa, G. L., Kuylenstierna, B., Ernster, L., and Bergstrand, A., 1967a, J.Cell Biol., 32:415-438.
Sottocasa, G. L., Kuylenstierna, B., Ermster, L., and Bergstrand, A., 1967b, in: "Methods in Enzymology", Vol.X, R. W. Estabrook and M. E. Pullman, eds., p.448, Academic Press, New York - London.
Sottocasa, G., Sandri, G., Panfili, E., de Bernard, B., Gazzotti, P., Vasington, F. D., and Carafoli, E., 1972, Biochem.Biophys.Res. Commun., 47:808-813.
Spithill, T. W., Trembath, M. K., Lukins, H. B., and Linnane, A. W., 1978, Mol.Gen.Genet., 164:155-162.
Stade, S., and Brambl, R., 1981, J.Bacteriol., 147:757-767.
Stasny, J. T., and Crane, F. L., 1964, J.Cell Biol., 22:49-62.
Steffens, G., and Buse, G., 1976, Hoppe-Seyler's Z.Physiol.Chem., 357:1125-1137.
Steffens, G. J., and Buse, G., 1979, Hoppe-Seyler's Z.Physiol.Chem., 360:613-619.
Steffens, G. C. M., Steffens, G. J., and Buse, G., 1979, Hoppe-Seyler's Z.Physiol.Chem., 360:1641-1650.
Stein, Y., and Stein, O., 1969, Israel J.Med.Sci., 5:985-990.
Stekhoven, F. S., Waitkus, R. F., and Moerkerk, H. T. B., 1972, Biochemistry, 11:1144-1150.
Stephenson, G., Marzuki, S., and Linnane, A. W., 1981, Biochim. Biophys.Acta, 636:104-112.
Stevens, B., 1977, Biol.Cellulaire, 28:37-56.
Stiggall, D. L., Galante, Y. M., and Hatefi, Y., 1978, J.Biol.Chem., 253:956-964.
Stiles, J. W., and Crane, F. L., 1966, Biochim.Biophys.Acta, 126: 179-181.
Stoeckenius, W., 1963, J.Cell Biol., 17:443-454.
Stoffel, W., and Schiefer, H. -G., 1968, Hoppe-Seyler's Z.Physiol. Chem., 349:1017-1026.
Stone, A. B., and Wilkie, D., 1975, J.Gen.Microbiol., 91:150-156.
Stratman, F. W., Zahlten, R. N., Hochberg, A. A., and Lardy, H. A., 1972, Biochemistry, 11:3154-3162.
Stuchell, R. N., Weinstein, B. I., and Beattie, D. S., 1975, J.Biol. Chem., 250:570-576.
Stutterheim, E., Henneke, M. A. C., and Berden, J. A., 1981, 634: 271-278.
Subramanian, M., Katyare, S. S., Fatterpaker, P., and Sreenivasan, A., 1975, Indian J.Biochem.Biophys., 12:307-310.

Sun, F. F., Prezbindowski, K. S., Crane, F. L., and Jacobs, E. E., 1968, Biochim.Biophys.Acta, 153:804-818.
Suyama, Y., 1967, Biochemistry, 6:2829-2839.
Svardal, A. M., and Pryme, I. F., 1980, Subcell.Biochem., 7:117-170.
Swanljung, P., Frigeri, L., Ohlson, K., and Ernster, L., 1973, Biochim.Biophys.Acta, 305:519-533.
Sweetman, A. J., Green, A. P., and Hooper, M., 1974, Biochem.Biophys. Res.Commun., 58:337-343.
Swick, R. W., Rexroth, A. K., and Stange, J. L., 1968, J.Biol.Chem., 243:3581-3587.
Swift, H., and Hruban, Z., 1964, Fed.Proc., 23:1026-1037.
Swift, H., Rabinowitz, M., and Getz, G., 1967, J.Cell Biol., 35: 131A-132A.
Szabolczi, G., and Biszku, E., 1961, Biochim.Biophys.Acta, 48:335-341.
Tada, M., Suzuki, H., Tanaka, M., and Ozawa, T., 1981, Biochem. Internat., 2:495-502.
Tait, A., 1968, Nature, (London), 219:941.
Tait, A., 1970, Nature, (London), 225:181-182.
Takahama, U., and Nishimura, M., 1975, Plant Cell Physiol., 16:737-748.
Takayanagi, R., Takeshige, K., and Minakami, S., 1980, Biochem.J., 192:853-860.
Tanaka, M., Hanui, M., Zeitlin, S., Yasunoby, K. T., Yu, C., Yu, L., and King, T. E., 1975, Biochem.Biophys.Res.Commun., 66:357-367.
Tanaka, M., Haniu, M., Yasunoby, K. T., Yu, C., Yu, L., Wei, Y. -H, and King, T. E., 1977, Biochem.Biophys.Res.Commun., 76:1014-1019.
Tanaka, M., Haniu, M., Yasunoby, K., Yu, C., Yu, L., Wei, Y. -H., and King, T. E., 1979, J.Biol.Chem., 254:3879-3885.
Tapley, D. F., 1956, J.Biol.Chem., 222:325-339.
Tappell, A. L., 1973, Fed.Proc., 32:1870-1874.
Taylor, C. B., Bailey, E., and Bartley, W., 1967, Biochem.J., 105: 605-609.
Tedeschi, H., 1981, Biochim.Biophys.Acta, 639:157-196.
Teintze, M., Slaughter, M., Weiss, H., and Neupert, W., 1982, J.Biol. Chem., 257:10364-10371.
Teleford, J. M., and Racker, E., 1973, J.Cell Biol., 57:580-586.
Terpstra, P., and Butow, R. A., 1979, J.Biol.Chem., 254:12662-12669.
Terpstra, P., Zanders, E., and Butow, R. A., 1979, J.Biol.Chem., 254:12653-12661.
Thalenfeld, B. E., and Tzagoloff, A., 1980, J.Biol.Chem., 255:6173-6180.
Tisdale, H. D., Wharton, D. C., and Green, D. E., 1963, Arch.Biochem. Biophys., 102:114-119.
Todd, R. D., and Douglas, M. G., 1981, J.Biol.Chem., 256:6984-6989.
Todd, R. D., Griesenbeck, T. A., and Douglas, M. G., 1980, J.Biol. Chem., 255:5461-5467.
Tokunaga, M., Loranger, J. M., Wolfe, P. B., and Wu, H. C., 1982, J.Biol.Chem., 257:9922-9925.

Topping, T. M., and Travis, D. F., 1974, J.Ultrastruct.Res., 46:1-22.
Touraille, S., Briand, Y., Durand, R., Bonnafous, J. -C., and Mani, J. -C., 1981, FEBS Lett., 128:142-144.
Travis, D. F., and Travis, A., 1972, J.Ultrastruct.Res., 39:124-148.
Trumpower, B. L., 1981, Biochim.Biophys.Acta, 639:129-155.
Trumpower, B. L., and Edwards, C. A., 1979, J.Biol.Chem., 254:8697-8706.
Trumpower, B. L., and Katki, A., 1975, Biochemistry, 14:3635-3642.
Trumpower, B. L., Edwards, C. A., and Ohnishi, T., 1980, J.Biol.Chem. 255:7487-7493.
Turner, G., and Lloyd, D., 1971, J.Gen.Microbiol., 67:175-188.
Turovetskii, V. B., Sitkovskii, M. V., Afanas'eva, N. N., Danilov, V. S., and Kozlov, Yu. P., 1972, Izv.Akad.Nauk SSSR, Ser.Biol. No 4:599-602.
Tustanoff, E. R., and Bartley, W., 1964, Can.J.Biochem., 42:651-665.
Tyler, D. D., 1975, Biochim.Biophys.Acta, 396:335-346.
Tzagoloff, A., 1969, J.Biol.Chem., 244:5027-5037.
Tzagoloff, A., 1970, J.Biol.Chem., 245:1545-1551.
Tzagoloff, A., 1972, J.Bioenergetics, 3:39-45.
Tzagoloff, A., 1982, The Mitochondrion, Plenum Press, New York.
Tzagoloff, A., and Akai, A., 1972, J.Biol.Chem., 247:6517-6523.
Tzagoloff, A., Meagher, P., 1971, J.Biol.Chem., 246:7328-7336.
Tzagoloff, A., Meagher, P., 1972, J.Biol.Chem., 247:594-603.
Tzagoloff, A., Yang, P. C., Wharton, D. C., and Rieske, J. S., 1965, Biochim.Biophys.Acta, 96:1-8.
Tzagoloff, A., MacLennan, D. H., McConnell, D. G., and Green, D. E., 1967, J.Biol.Chem., 242:2051-2061.
Tzagoloff, A., Byington, K. H., and MacLennan, D. H., 1968a, J.Biol. Chem., 243:2405-2412.
Tzagoloff, A., MacLennan, D. H., and Byington, K. H., 1968b, Biochemistry, 7:1596-1602.
Tzagoloff, A., Akai, A., and Sierra, M. F., 1972, J.Biol.Chem., 247: 6511-6516.
Tzagoloff, A., Rubin, M. S., and Sierra, M. F., 1973, Biochim. Biophys.Acta, 301:71-104.
Tzagoloff, A., Akai, A., and Needleman, R. B., 1975, J.Biol.Chem., 250:8228-8235.
Tzagoloff, A., Foury, F., and Akai, A., 1976, Mol.Gen.Genet., 149: 33-42.
Tzagoloff, A., Macino G., and Sebald, W., 1979, Annu.Rev.Biochem., 48:419-441.
Tzagoloff, A., Bonitz, S. G., Coruzzi, G., Thalenfend, B. E., and Macino, G., 1980, in: "The Organization and Expression of the Mitochondrial Genome", A. M. Kroon, and C. Saccone, eds., pp.181-190, Elsevier/North-Holland Biomedical Press, Amsterdam - New York - Oxford.
Ulane, R. E., and Cabib, E., 1876, J.Biol.Chem., 251:3367-3374.
Umezawa, H., 1973, Pure and App.Chem., 33:129-144.
Utter, M. F., Duell, E. A., and Bernofsky, C., 1967, in: "Aspects of Yeast Metabolism", A. K. Mills, ed., pp.197-212, Blackwell Scientific Publications, Oxford -Edinburgh.

Vaganova, T. I., Lastovetskaya, L. V., Strongin, A. Ya., Lyublinskaya L. A., and Stepanov, V. M., 1976, Biokhimiya, 41:2229-2238.
Vail, W. J., and Riley, R. K., 1974, FEBS Lett., 40:269-273.
Vail, W. J., Riley, R. K., and Williams, C. H., 1972, J.Bioenergetics 3:467-479.
Vail, W. J., Riley, R. K., and Rieske, J. S., 1975, FEBS Lett., 58: 33-38.
Van Deenen, L. L. M., and De Haas, G. H., 1964, Advances in Lipid Res., 2:167-234.
Van den Bergh, S. G., 1967, in: "The Round Table Discussion on Mitochondrial Structure and Compartmentation", E. Quagliariello, S. Papa, E. C. Ernster, and J. M. Tager, eds., pp.400-410, Adriatica Editrice, Bari.
Van den Bergh, S. G., and Slater, E. C., 1962, Biochem.J., 82:362-371
Vanderkooi, G., 1974, Biochim.Biophys.Acta, 344:307-345.
Vanderkooi, G., and Green, D. E., 1970, Proc.Natl.Acad.Sci.U.S.A., 66:615-621.
Van de Stadt, R. J., Kraaipoel, R. J., and Van Dam, K., 1972, Biochim.Biophys.Acta, 267:25-36.
Van de Stadt, R. J., de Boer, B. L., and Van Dam, K., 1973, Biochim. Biophys.Acta, 292:338-349.
Van't Sant, P., Mak, J. F. C., and Kroon, A. M., 1981, Eur.J.Biochem. 121:21-26.
Van Tuyle, G. C., and Kalf, G. F., 1972, Arch.Biochem.Biophys., 149: 425-434.
Vartapetian, B. B., Andreeva, I. N., and Maslova, I. P., 1971, Dokl. Akad.Nauk SSSR, 196:1231-1233.
Vartapetian, B. B., Maslova, I. P., and Andreeva, I. N., 1972, Fiziol.Rast., 19:106-112.
Vartapetian, B. B., Maslov, A. I., and Andreeva, I. N., 1975, Plant Sci.Lett., 4:1-8.
Vartapetian, B. B., Andereeva, I. N., and Kozlova, G. I., 1976, Protoplasma, 88:215-224.
Vartapetian, B. B., Andereeva, I. N., Kozlova, G. I., and Agapova, L. P., 1977, Protoplasma, 91:243-256.
Vary, M. J., Edwards, C. L., and Stewart, P. R., 1969, Arch.Biochem. Biophys., 130:235-243.
Vary, M. J., Stewart, P. R., and Linnane, A. W., 1970, Arch.Biochem. Biophys., 141:430-439.
Vasdev, S. C., Biro, G. P., Narbaitz, R., and Kako, K. J., 1980, Can.J.Biochem., 58:1112-1119.
Vazquez, D., 1964, Biochem.Biophys.Res.Commun., 15:464-468.
Velours, J., Guerin, M., and Guerin, B., 1980, Arch.Biochem.Biophys., 201:615-628.
Verheul, F. E. A. M., Draijer, J. W., Dentener, I. K., and Muijsers, A. O., 1981, Eur.J.Biochem., 119:401-408.
Vignais, P. V., 1976, Biochim.Biophys.Acta, 456:1-38.
Vignazia, E., Pagani, S., and Galante, Y. M., 1981, Biochem.Inter., 3:281-289.
Vinogradov, A. D., and King, T. E., 1979, Methods Enzymol., 55: 118-127.

Vinogradov, A. D., Ackrell, B. A. C., and Singer, T. P., 1975, Biochem.Biophys.Res.Commun., 67:803-809.
Vinogradov, A. D., Gavrikova, E. V., and Goloveshkina, V. G., 1976, Biokhimiya, 41:1155-1168.
Vinogradov, A. D., Gavrikov, V. G., and Gavrikova, E. V., 1980, Biochim.Biophys.Acta, 592:13-27.
Vladimirov, Yu. A., and Archakov, A. I., 1972, Lipid Peroxidation in Biological Membranes (in Russian), Izd.Nauka, Moscow.
Volodina, T. V., Kozel'tsev, V. L., Rebrov, L. B., and Debov, S. S., 1977, Vopr.Med.Khim., 23:338-342.
Von Jagow, G., and Klingenberg, M., 1970, Eur.J.Biochem., 12:583-592.
Von Jagow, G., and Klingenberg, M., 1972, FEBS Lett., 24:278-282.
Von Jagow, G., Schägger, H., Engel, W. D., Machleidt, W., Machleidt, I., and Kolb, H. J., 1978, FEBS Lett., 91:121-125.
von Rücker, A. A., 1976, Uber die Koordination der Mitochondrialen und Cytoplasmischen Proteinsynthese bei der Bildung von Mitochondrialen Membranproteinen im Besonderen der Cytochromoxidase, Ph.D.Thesis, Ludwig-Maximilians-University, München.
von Rücker, A. A., Michel, R., and Neupert, W., 1973, in: "Intracellular Protein Catabolism", pp.165-170, Johann Ambrosius Barth Verlag, Leipzig.
Waite, M., 1969, Biochemistry, 8:2536-2542.
Waite, M., and Sisson, P., 1971, Biochemistry, 10:2377-2383.
Waite, M., and Sisson, P., 1972, Biochemistry, 11:3098-3105.
Waite, M., and Van Deenen, L. L. M., 1967, Biochim.Biophys.Acta, 137:498-517.
Waite, M., Scherphof, G. L., Boshouwerst, F. M. G., and Van Deenen, L. L. M., 1969a, J.Lipid Res., 10:411-420.
Waite, M., Van Deenen, L. L. M., Ruigrock, T. J. C., and Elbers, P. F., 1969b, J.Lipid Res., 10:599-608.
Wakiyama, S., and Ogura, Y., 1972, J.Biochem., (Tokyo), 71:295-300.
Waksman, A., Hubert, P., Crémel, G., Rendon, A., and Burgun, C., 1980, Biochim.Biophys.Acta, 604:249-296.
Wallace, P. G., Huang, M., and Linnane, A. W., 1968, J.Cell Biol., 37:207-220.
Walter, P., and Blobel, G., 1980, Proc.Natl.Acad.Sci.U.S.A., 77:7112-7116.
Walter, P., and Blobel, G., 1981a, J.Cell Biol., 91:551-556.
Walter, P., and Blobel, G., 1981b, J.Cell Biol., 91:557-561.
Walter, P., and Blobel, G., 1982, Nature, (London), 299:691-698.
Walter, P., and Lardy, H. A., 1964, Biochemistry, 3:812-816.
Walter, P., Ibrahimi, I., and Blobel, G., J.Cell Biol., 91:545-550.
Wang, H. -P, and Kimura, T., 1976, Biochim.Biophys.Acta, 423:374-381.
Warshaw, J. B., Lam, K. W., Nagg, B., and Sanadi, D. R., 1968, Arch. Biochem.Biophys., 123:385-396.
Watanabe, K., and Kubo, S., 1982, Eur.J.Biochem., 123:587-592.
Watari, H., Kearney, E. B., Singer, T. P., Basinski, T. P., Hauber, J., and Lusty, C. J., 1962, J.Biol.Chem., 237:1731-1733.
Watson, M. L., and Siekevitz, P., 1956, J.Biophys.Biochem.Cytol., 2:639-652.

Watson, K., Haslam, J. M., and Linnane, A. W., 1970, J.Cell Biol., 46:88-96.
Weber, R., 1964, J.Cell Biol., 22:481-487.
Weinbach, E. C., 1959, J.Biol.Chem., 234:1580-1586.
Weinbach, E., and Garbus, J., 1968, Biochem.J., 106:711-717.
Weisiger, R. A., and Fridowich, I., 1973a, J.Biol.Chem., 248:3582-3592.
Weisiger, R. A., and Fridowich, I., 1973b, J.Biol.Chem., 248:4793-4796.
Weinstein, J. N., Blumenthal, R., van Renswoude, J., Kempf, C., and Klausner, R. D., 1982, J.Membrane Biol., 66:203-212.
Weiss, H., 1976, Biochim.Biophys.Acta, 456:291-313.
Weiss, H., and Sebald, W., 1978, Methods Enzymol., 53:66-73.
Weiss, H., and Ziganke, B., 1974, Eur.J.Biochem., 41:63-71.
Weiss, H., Sebald, W., and Bücher, Th., 1971, Eur.J.Biochem., 22:19-26.
Wenzler, H., and Brambl, R., 1981, J.Biol.Chem., 256:7166-7172.
Werner, S., 1977, Eur.J.Biochem., 79:103-110.
Werner, S., and Bertrand, 1979, Eur.J.Biochem., 99:463-470.
Werner, S., Schwab, A. J., and Neupert, W., 1974, Eur.J.Biochem., 49:607-617.
Wheeldon, L. W., Dianoux, A. -C., Bof, M., and Vignais, P. V., 1974, Eur.J.Biochem., 46:189-199.
Whereat, A. F., Orishimo, M. W., and Nelson, J. B., 1968, Fed.Proc., 27:362.
Whereat, A. F., Orishimo, M. W., and Nelson, J. B., 1969, J.Biol. Chem., 244:6498-6506.
Whiting, M. J., 1976, Biochm.J., 158:391-400.
Whittaker, P. A., 1979, Subcell.Biochem., 6:175-232.
Whittaker, V. P., 1966, in: "Regulation of Metabolic Processes in Mitochondria, B.B.A. Library, Vol.7, pp.1-24, Elsevier, Amsterdam - London - New York.
Wickner, W., 1979, Annu.Rev.Biochem., 48:23-45.
Wielburski, A., Kuzela, and Nelson, B. D., 1982, Biochem.J., 204:239-245.
Wikström, M. K. F., 1971, Biochim.Biophys.Acta, 253:332-345.
Wikström, M. K. F., 1973, Biochim.Biophys.Acta, 301:155-193.
Wikström, M. K. F., Harmon, H. J., Ingledew, W. J., and Chance, B., 1976, FEBS Lett., 65:259-277.
Williams, J. N., 1968, Biochim.Biophys.Acta, 162:175-181.
Wilson, M. A., and Cascarano, J., 1970, Biochim.Biophys.Acta, 216:54-62.
Wilson, D. F., Dutton, P. L., and Wagner, M., 1973, in: "Current Topics in Bioenergetics", Vol.5, D. R. Sanadi and L. Packer, eds., pp.233-265, Academic Press, New York - London.
Winter, D. B., Bruyninckx, W. J., Foulke, F. G., Grinich, N. P., and Mason, H. S., 1980, J.Biol.Chem., 255:11408-11414.
Wintersberger, E., 1965, Biochem.Z., 341:409-419.
Wintersberger, E., 1967, Hoppe-Seyler's Z.Physiol.Chem., 348:1701-1704.

Wit-Peeters, E. M., Biochim.Biophys.Acta, 176:453-462.
Wlodawer, P., Parsons, D. F., Williams, G. R., and Wojtczak, L., 1966, Biochim.Biophys.Acta, 128:34-47.
Wohlrab, H., 1980, J.Biol.Chem., 255:8170-8173.
Wojtczak, L., and Lehninger, A. L., 1961, Biochim.Biophys.Acta, 51:442-456.
Wojtczak, L., and Wojtczak, A. B., 1960, Biochim.Biophys.Acta, 39: 277-286.
Wojtczak, L., Wlodawer, P., and Zborowski, J., 1963, Biochim.Biophys. Acta, 70:290-305.
Wolf, D. H., and Fink, G. R., 1975, J.Bacteriol., 123:1150-1156.
Wood, D. D., and Luck, D. J. L., 1969, J.Mol.Biol., 41:211-224.
Woodrow, G., and Schatz, G., 1979, J.Biol.Chem., 254:6088-6093.
Woods, R. A., Sanders, H. K., Briquet, M., Foury, F., Drysdale, B. -E., and Mattoon, J. R., 1975, J.Biol.Chem., 250:9090-9098.
Wrigglesworth, J. M., Packer, L., and Branton, D., 1970, Biochim. Biophys.Acta, 205:125-135.
Wu, T. -W., and Tinker, D. O., 1969, Biochemistry, 8:1558-1568.
Yamamoto, T., and Orii, Y., 1974, J.Biochem., (Tokyo), 75:1081-1089.
Yamanaka, T., Fujii, K., and Kamita, Y., 1979, J.Biochem., (Tokyo), 86:821-824.
Yamashita, S., and Racker, E., 1969, J.Biol.Chem., 244:1220-1227.
Yates, D. W., and Garland, P. B., 1967, Biochem.J., 102:40P.
Yatscoff, R. W., and Freeman, K. B., 1977, Can.J.Biochem., 55:1065-1074.
Yotsuyanagi, Y, 1962, J.Ultrastruct.Res., 7:141-158.
Yu, C. -A., and Yu, L., 1977, Biochim.Biophys.Acta, 495:248-259.
Yu, C., and Yu, L., 1980a, Biochemistry, 19:3579-3585.
Yu, C., and Yu, L., 1980b, Biochim.Biophys.Acta, 591:409-420.
Yu, L., and Yu, C. A., 1980c, Biochim.Biophys.Acta, 593:24-38.
Yu, C., and Yu, L., 1981, Biochim.Biophys.Acta, 639:99-128.
Yu, C. -A., Yu, L., and King, T. E., 1972, J.Biol.Chem., 247:1012-1019.
Yu, C. -A., Yu, L., and King, T. E., 1975, Biochem.Biophys.Res. Commun., 66:1194-1200.
Yu, C. -A., Yu, L., and King, T. E., 1977a, Biochem.Biophys.Res. Commun., 78:259-265.
Yu, C. -A., Yu, L., and King, T. E., 1977b, Biochem.Biophys.Res. Commun., 74:670-676.
Zakowski, J. J., and Tappel, A. L., 1978, Biochim.Biophys.Acta, 526:65-76.
Zalman. L. S., Nikaido, H., and Kagawa, Y., 1980, J.Biol.Chem., 255:1771-1774.
Ziegler, D. M., and Doeg, K. A., 1962, Arch.Biochem.Biophys., 97: 41-50.
Ziegler, D. M., Linnane, A. W., and Green, D. E., 1958, Biochim. Biophys.Acta, 28:524-538.
Zimmerman, R., and Neupert, W., 1980, Eur.J.Biochem., 109:217-229.
Zimmerman, R., Paluch, U., and Neupert, W., 1979a, FEBS Lett., 108: 141-151.

Zimmerman, R., Paluch, U., Sprinzl, M., and Neupert, W., 1979b, Eur. J.Biochem., 99:247-252.

Zimmerman, R., Hennig, B., and Neupert, W., 1981, Eur.J.Biochem., 116:455-460.

Zubatov, A. S., 1972, Studies on the Degradation and Restoration of Mitochondria of Facultatively Anaerobic Yeast Saccharomyces cerevisiae in vivo, Ph.D.Thesis, Moscow State University, Moscow.

Zubatov, A. S., Rainina, E. I., Bakeeva, L. E., and Luzikov, V. N., 1971, Biokhimiya, 36:1156-1160.

Zubatov, A. S., Mikhailova, A. E., and Luzikov, V. N., 1983, submitted for publication.

Zurini, M., Hugentobler, G., and Gazzotty, P., 1981, Eur.J.Biochem., 119:517-521.

Zvyagil'skaya, R. A., Korosteleva, N. L., and Kotel'nikova, A. V., 1977, Biokhimiya, 42:1888-1895.

Index